Energy and Transport Processes

Presentations at the 2009 AIChE Annual Meeting

Nashville, Tennessee, USA
8-13 November 2009

ISBN: 978-1-61567-931-7

Printed from e-media with permission by:

Curran Associates, Inc.
57 Morehouse Lane
Red Hook, NY 12571

Some format issues inherent in the e-media version may also appear in this print version.

Printed by Curran Associates, Inc. (2010)

For permission requests, please contact AIChE
at the address below.

AIChE
3 Park Avenue
New York, NY 10016-5991

Phone: (203) 702-7660
Fax: (203) 775-5177

www.aiche.org

Additional copies of this publication are available from:

Curran Associates, Inc.
57 Morehouse Lane
Red Hook, NY 12571 USA
Phone: 845-758-0400
Fax: 845-758-2634
Email: curran@proceedings.com
Web: www.proceedings.com

TABLE OF CONTENTS

Kinetic Study of Biomass Gasification at High Pressures

Pradeep K. Agrawal, School of Chemical & Biomolecular Engineering, Georgia Institute of Technology, Atlanta, GA

Kristiina Iisa, National Renewable Energy Laboratory, Golden, CO

Scott A. Sinquefield, Ipst, Georgia Institute of Technology, Atlanta, GA

Steven J. Lien, Ipst, Georgia Institute of Technology, Atlanta, GA

A major advantage of biomass gasification in thermochemical platform for biomass utilization is that all parts of any type of biomass can be converted to syngas, making it attractive for unmerchantable wood, harvest residue, and waste from forest and paper industry. During gasification (using steam and/or CO2), biomass is converted to syngas (desired components CO and H2). There are two major constraints in the development of gasification protocols: (1) besides the desired CO and H2, a number of contaminants including hydrocarbons and tars are also formed during biomass gasification which adds significantly to the costs of downstream processing and gas-conditioning, and (2) carbon gasification reaction is inhibited by CO and H2. The mechanism of biomass gasification is complicated and depends on the gasification agent (steam, air, or CO2). Langmuir-Hinshelwood type kinetic models are widely used to describe these gasification processes. Reducing the amount of these contaminants in the raw syngas will decrease overall costs.

This study has two goals: (i) to study biomass gasification rates at high pressures as well as measure the rates of formation of hydrocarbons and tar contaminants, and (ii) to develop kinetic models that quantitatively predict the rates of biomass gasification and the rates of contaminant formation.

Two different reactor types are being utilized in this study: pressurized entrained flow reactor (PEFR) and pressurized thermogravimetric analyzer (PTGA). These two reactors complement each other in terms of the biomass heating rates and residence times. In addition, experiments conducted at atmospheric pressure will provide a broader range of model applicability and enable a better mechanistic understanding of the gasification process. The three biomass candidates (loblolly pine, switch grass, and corn stover) would differ in their elemental composition and ash content which will likely affect the gasification behavior. The biomass gasification variables include pressure, temperature, gas-phase composition, and particle size.

See more of this Session: Advances in Gasification Research - I
See more of this Group/Topical: Energy and Transport Processes

Modeling and Simulation of Downdraft Biomass Gasifier

Pratik N Sheth[a] and B V Babu[b1]
Birla Institute of Technology and Science (BITS), PILANI-333 031
Rajasthan, India
[a]Lecturer, Chemical Engineering Group, Email: pratik@bits-pilani.ac.in
[b]Dean - Educational Hardware Division & Professor of Chemical Engineering
Phone: +91-1596-245073 Ext. 259; Fax: +91-1596-244183
Email: bvbabu@bits-pilani.ac.in; Homepage: http://discovery.bits-pilani.ac.in/~bvbabu

Abstract

Kinetics-free equilibrium models can predict the exit gas composition, given the solid composition and the equilibrium temperature, but they cannot be used for reactor design Hence, there is a need to develop a combined transport and kinetic model which takes into account of the kinetics of homogeneous and heterogeneous chemical reactions, transport of volatiles produced, heat and mass transfer between solid and gaseous phase and pyrolysis reactions. Taking into account of the importance of downdraft biomass gasifier and its commercial applications, it is essential to have a complete model for such a configuration. In the present study, a transient one-dimensional model is developed for the throated close-top downdraft biomass gasifier. The model takes into account of the pyrolysis, secondary tar reactions, homogeneous gas reactions and heterogeneous combustion/gasification reactions. The experimental data obtained in our earlier study are used to validate the simulation results of the complete combined transport and kinetic model. The fraction of initial moisture content, air flow rate, temperature of the pyrolysis zone, and chemical composition of the biomass are required as input data for the model to predict the composition of producer gas. The variation of molar fraction of producer gas components with time is predicted and compared with the experimental data. It is concluded from the present study that the developed model can predict the performance of the biomass gasifier, *a priori*. The results of this study are also useful in the design of a downdraft biomass gasifier.

1. Introduction

Biomass is the oldest form of energy used by human beings. Biomass is a term used to describe all biologically produced matter and it is the name given to all earth's living matter. The chemical energy contained in the biomass is derived from solar energy using the process of photosynthesis. The thermo chemical conversion of biomass (pyrolysis, gasification, combustion) is one of the most promising non-nuclear forms of future energy. It is a renewable source of energy and has many advantages from an ecological point of view (Babu and Chaurasia, 2003a). The pyrolysis process consists of the thermal degradation of the biomass feedstock, in the absence of oxygen/air, leading to the formation of solid, liquid and gaseous products. Gasification enables conversion of this material into combustible gas, mechanical and electrical power and synthetic fuels and chemicals (Babu and Sheth, 2006). Gasification is a process of conversion of solid carbonaceous fuel into combustible gas by partial combustion. Downdraft gasification is comparatively cheap method of gasification that can produce a

[1] Corresponding Author

product gas with very low tar content (Giltrap et al., 2003). The understanding of the interaction between chemical and physical mechanisms during gasification is of fundamental importance, for the optimal design of biomass gasifier. Modeling of biomass gasification implies the representation of chemical and physical phenomena constituting all the four zones of the gasifier (pyrolysis, combustion, reduction, and drying) in the mathematical form. The models of downdraft biomass gasification can be categorized into two groups: (1) Equilibrium models and (2) Combined transport and kinetic models. Equilibrium models are important in order to predict the thermodynamic limits of chemical reactions describing the gasification process. Kinetics-free equilibrium models can predict the exit gas composition, given the solid composition and the equilibrium temperature, but they cannot be used for reactor design (Di Blasi, 2000). An equilibrium model can not predict the concentration or temperature profiles across the axis of gasifier and hence results generated using an equilibrium model would give the same final composition for different lengths of reduction zone of biomass gasifier. Hence, there is a need to develop a combined transport and kinetic model which takes into account of the kinetics of homogeneous and heterogeneous chemical reactions, transport of volatiles produced, heat and mass transfer between solid and gaseous phase and pyrolysis reactions. Various models that have been reported for different gasifier configurations include: (1) unsteady one-dimensional model for stratified downdraft gasification (Di Blasi, 2000), (2) transient single particle and fuel bed model for crosscurrent moving bed furnace (Wurzenberger et al., 2002), (3) steady-state reduction zone model for downdraft gasification (Giltrap et al., 2003) and (4) steady state fluid flow and heat transfer model for open top throat-less downdraft gasification (Sharma, 2007). However, for throated close-top downdraft biomass gasifier, commonly known as an Imbert downdraft gasifier, a complete model including pyrolysis, combustion and reduction zones has not been reported in the literature. In a survey of gasifier manufacturers, it is reported that 75% of gasifiers offered commercially were downdraft, 20% were fluid beds (including circulation fluid beds), 2.5% were updraft, and 2.5% were of other types (Bridgwater, 2002). Taking into account of the importance of downdraft biomass gasifier and its commercial applications, it is essential to have a complete model for such a configuration.

In the present study, a transient one-dimensional model is developed for the throated close-top downdraft biomass gasifier. The model takes into account of the pyrolysis, secondary tar reactions, homogeneous gas reactions and heterogeneous combustion/gasification reactions. The developed model is divided into three parts according to three prevailing zones in the gasifier: (1) pyrolysis, (2) oxidation, and (3) reduction. The drying zone is indirectly incorporated in the developed model. The experimental data obtained in our earlier study (Sheth and Babu, 2009) are used to validate the simulation results of the combined transport and kinetic model.

2. Mathematical Model

A transient one dimensional model is developed for the throated close-top downdraft biomass gasifier. The model takes into account of the pyrolysis, secondary tar reactions, homogeneous gas reactions and heterogeneous combustion/gasification reactions. Eight gaseous species namely O_2, N_2, CO_2, CO, H_2O, H_2, CH_4 and tar are considered in the gas phase. In the pyrolysis and combustion zone, the solid phase is a biomass, whereas in the reduction zone it is a charcoal. The developed model is divided into three parts according to three different zones formed: (1) pyrolysis, (2) oxidation, and (3) reduction.

2.1 Pyrolysis

Pyrolysis is a process by which a biomass feedstock is thermally degraded in the absence of oxygen/air to produce solid (charcoal), liquid (tar and other organics) and gaseous (H_2, CO_2, CO, etc.) products. Released volatiles from each biomass particle flow downward in packed pyrolysis bed. Rate of volatiles release depends on the particle size and temperature within a single particle. Pyrolysis bed is modeled as a stack of particles in one dimension. The model presented here considers temperature gradient both in the bed and also inside the single particles. Thus the entire bed is divided into two subsystems, i.e., gas phase inside the bed and the individual particles. The assumptions made in this model are: (1) the volatiles released from biomass particles flow downward, (2) unsteady state operation, (3) released volatiles is a mixture of CO, CH_4, CO_2, H_2, H_2O and tar, (4) velocity varies along the bed, (5) solid biomass does not move downward during the operation, and (6) porosity of the bed remains constant.

2.1.1 Transport Model for Gas Phase in Pyrolysis Zone

The continuity equation for the gas phase in a packed bed is given by Eq. (1).

$$\frac{\partial \varepsilon_{g,bed}\rho_{g,bed}}{\partial t} = -\frac{\partial}{\partial z}\left[\varepsilon_{g,bed}\rho_{g,bed}v_{g,bed}\right] + a_p \dot{m}_p \qquad (1)$$

where, $\varepsilon_{g,bed}$ is the porosity of the bed, $v_{g,bed}$ is the gas velocity (m/s), $\rho_{g,bed}$ is the density of the gas phase (kg/m^3), $\dot{m}_p$ is the mass transfer rate from particle to the gas phase (kg/m^2 s), and a_p is the specific surface area of particle (m^2/m^3).

The species conservation equation is given by Eq. (2).

$$\varepsilon_{g,bed}\frac{\partial \rho_{ig,bed}}{\partial t} = -\varepsilon_{g,bed}\left[v_{g,bed}\frac{\partial \rho_{ig,bed}}{\partial z} + \rho_{ig,bed}\frac{\partial v_{g,bed}}{\partial z}\right] + \frac{\partial}{\partial z}\left[\varepsilon_{g,bed}D_{i,eff,bed}\frac{\partial \rho_{ig,bed}}{\partial z}\right] + a_p \dot{m}_{i,p} \qquad (2)$$

where, $\rho_{ig,\,bed}$ is the density of the species *i* in packed bed gas phase, $D_{i,eff,bed}$ is the effective diffusivity of species *i* in packed bed gas phase, and $\dot{m}_{i,p}$ is the mass transfer rate of species *i* from particle to the gas phase (kg/m^2 s).

The initial and boundary conditions are given below by Eq. (3) and Eq. (4) respectively.

Initial Conditions:

at $t = 0$; $\rho_{ig,\,bed} = 0$; $v_{g,\,bed} = 0$ (3)

Boundary Conditions:

at $z = 0$; $\rho_{ig,\,bed} = 0$; $v_{g,\,bed} = 0$ for all $t > 0$ (4a)

at $z = L$; $\frac{\partial \rho_{ig,bed}}{\partial z} = 0$ for all $t > 0$ (4b)

Pyrolysed gas generated from the biomass particle flows downward and its velocity is computed using Eq. (5).

$$v_{g,bed} = \left(\frac{\text{Flow rate}}{\text{Area}}\right) = \left(\frac{\rho_{g,bed}A\Delta z}{\text{Density of gaseous mixture}}\right)\left(\frac{1}{\Delta t}\right)\left(\frac{1}{\text{Area}}\right) \qquad (5)$$

2.1.2 Single Particle

To describe the chemical process of pyrolysis in a single solid particle, an unsteady state one dimensional variable property model of transport phenomena is required. It should include heat (conductive, convective and radiative modes), mass (diffusive and convective modes) and momentum transport of the products formed within the solid (volatiles and gases). Babu and Chaurasia (2003 a-b; 2004 a-d) carried out exhaustive modeling and simulation studies on the pyrolysis of single solid particle. The model developed and modified by Babu and Chaurasia (2003 a-b; 2004 a-d) uses physically measurable parameters and practically explainable kinetic scheme, incorporating the convective and diffusion effects. Babu and Chaurasia (2004c) proposed three models for the pyrolysis of biomass under two categories namely, (1) Generalized reference model (Model-I), and (2) Simplified models (Model-II & Model-III). In most practical situations of industrial pyrolysis reactions, the contributions of the bulk motion of gases inside the pores of the particle are insignificant. Because the resistance offered by the pores in the solid particle is so high, the transport of these gases would take place essentially by a diffusion mechanism but not by bulk motion (i.e. convection). Taking this situation into consideration, Model –II is used in the present study, which is based on the assumption that there is no bulk motion contribution (i.e. convective transport is neglected) to the temperature profile and the product yield predictions.

2.2 Combustion

The biomass present in the oxidation zone reacts with oxygen to form carbon dioxide, which provides heat for the subsequent gasification reactions. In complete combustion, carbon present in biomass is completely converted to carbon dioxide while hydrogen is converted to water vapor. It is an exothermic reaction and yields temperatures in the range of 1000 °C to 1500 °C. In the present model, complete combustion of biomass is assumed which can be ensured by supplying excess air (usually around 20%) than stoichiometrical requirement. The combustion reaction therefore is represented as given by Eq. (6)

$$C_6H_{HB}O_{OB} + \left(6 + \frac{HB}{4} - \frac{OB}{2}\right) O_2 \longrightarrow 6CO_2 + \frac{HB}{2} H_2O \qquad (6)$$

The volatile products generated in the pyrolysis zone flow downwards and enter into the oxidation zone where a part of volatiles gets oxidized. It is assumed that the tar present in the pyrolysed gas mixture completely gets decomposed due to very high temperature present in the oxidation zone. The tar decomposition is represented by Eq. (7).

$$\mathrm{C_6H_{6.2}O_{0.2} + 7.45O_2 \longrightarrow 6CO_2 + 3.1H_2O} \qquad (7)$$

2.3 Reduction

The main components of the gaseous mixture leaving the combustion zone are carbon dioxide, water vapor, inert nitrogen, carbon monoxide, hydrogen and some amount of low molecular weight hydrocarbons such as methane, ethane, ethylene etc. In the reduction zone, the gaseous mixture passes through the hot porous charcoal bed resting above the grate. The reduction zone is often referred as gasification zone. Giltrap et al. (2003) developed a model of reduction zone of downdraft biomass gasifier to predict the composition of producer gas under steady state operation. The accuracy of the model is limited by the availability of data on the initial conditions at the top of the reduction zone. Moreover they assumed that the char reactivity factor (*CRF*) which represents the reactivity of char and the key variable in simulation is constant throughout the reduction zone. Giltrap's model (2003) is modified by incorporating the variation of *CRF* along the reduction zone of downdraft biomass gasifier in our earlier study (Babu and Sheth, 2006). It is assumed that *CRF* is exponentially increasing along the bed length of the reduction zone. Solid carbon in the form of char is assumed to be present throughout the reduction zone. Modeling equations of single particle model (Babu and

Chaurasia, 2004c) and reduction model proposed by Babu and Sheth (2006) are presented in the present article to reduce the length of the article.

3. Experimental Studies

The main equipment of the experimental set up is an Imbert downdraft biomass gasifier. An Imbert downdraft biomass gasifier is the one, which has throated combustion zone and different diameter for pyrolysis and reduction zones unlike stratified downdraft biomass gasifier in which diameter of the gasifier is uniform through out (Reed and Das, 1988). Biomass is fed from top of the gasifier and air is introduced through nozzle in the combustion zone. The downdraft gasifier converts the solid biomass into a combustible gas, generally known as a producer gas. Biomass undergoes pyrolysis and gets oxidized in the combustion zone near air inlet. The pyrolysed gas mixture and the gases produced due to combustion passes over the charcoal bed resting above the grate and generate producer gas. The details of the experimental set up, procedure and operating variables are reported in our earlier study (Sheth and Babu, 2009).

4. Results and Discussion

The experimental data reported in our earlier study (Sheth and Babu, 2009) are used to validate the simulation results of the complete combined transport and kinetic model. The initial moisture content fraction, air flow rate, temperature of the pyrolysis zone, and chemical composition of the biomass are required as an input data for the model to predict the composition of producer gas. The combined transport and kinetic model is validated using the parametric values as given in Table-1.

Table 1 Details of the biomass gasification experiments

Experimental No	Air flow rate (m^3/h)	Initial moisture content (wt fraction, wet basis)	Biomass consumption Rate (kg/h)	Equivalence ratio(Φ)
1	2.7765	0.1145	2.10	0.2533
2	3.3935	0.0437	3.63	0.1791
3	1.8510	0.0437	2.12	0.1673
4	2.7765	0.0437	2.67	0.1992
5	2.7765	0.073	2.59	0.2054
6	1.8510	0.10	1.00	0.3546
7	2.7765	0.1518	2.20	0.2418
8	2.1595	0.07	1.488	0.278
9	2.1595	0.044	2.12	0.1951
10	2.1595	0.1167	1.1626	0.3558
11	2.1595	0.164	1.0424	0.3968

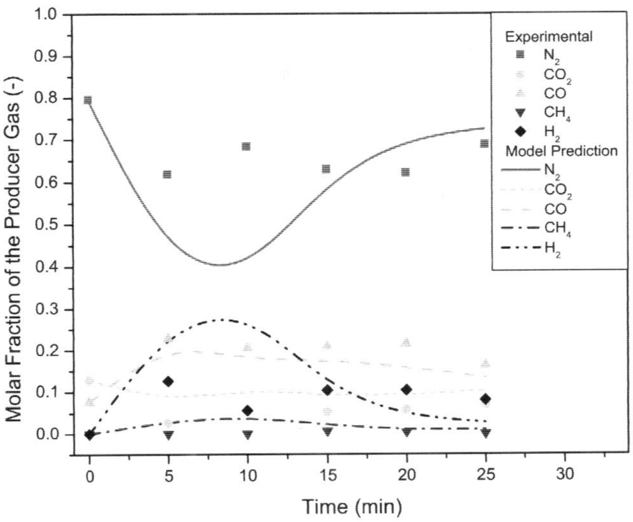

Fig. 1 Comparison of model predicted producer gas composition with experimental values of the present study (Φ = 0.2533)

The variation of molar fraction of producer gas components with time is predicted and compared with the experimental data. Fig. 1 show the comparison of model predicted molar fractions of producer gas components with those found experimentally for an equilibrium ratio of 0.2533. It is found that the model predicted molar fraction of nitrogen decreases first during the few initial minutes (5-10 min) of gasification. After that it increases and attains a steady value (after 10-15 min). The molar amount of nitrogen is constant for a particular flow rate of air as nitrogen acts as an inert but its composition varies due to the changes in molar amount of other components of gaseous mixture. It is found that the model predicted molar fractions of carbon monoxide and hydrogen increase first with time and after that it decreases and attains a steady value. It is observed that the model predicted molar fraction of methane is very less and almost remains constant. It is also found that the model predicted molar fraction of carbon dioxide decreases a little and attains a steady value.

The simulation results of the molar composition of various components of the producer gas match well with the experimental data of 10 minutes or higher from the start of run. For the experimental data of 5 min and 10 min, the simulation results differ more. This is because of the assumption taken in the model that all the gas generated in pyrolysis or reduction zone travel downwards in the gasifier. However, it is observed while carrying out the experiments that the gas produced in the pyrolysis zone first travels upwards and occupies the empty space above the biomass. After 5 - 10 minutes from the start of the run, the accumulated gas builds up a pressure and the producer gas gets started flowing downwards. Because of this, the model predicts higher concentration of hydrogen and carbon monoxide and lower concentration of nitrogen in comparison to the experimental data for initial 5 -10 minutes from the start of a particular experimental run. Fig. 2 shows the model predicted composition profile of gaseous phase across the gasifier at 5 min from the start of the experiment for the first run of the present study. Figs. 3 and 4 show the similar profiles at different times from the start of the experiment (run 1). The pyrolysis zone in the gasifier is represented by first 100 mm, oxidation zone by 100 to 140 mm and reduction zone by 140 to 270 mm in Figs. 2 to 4.

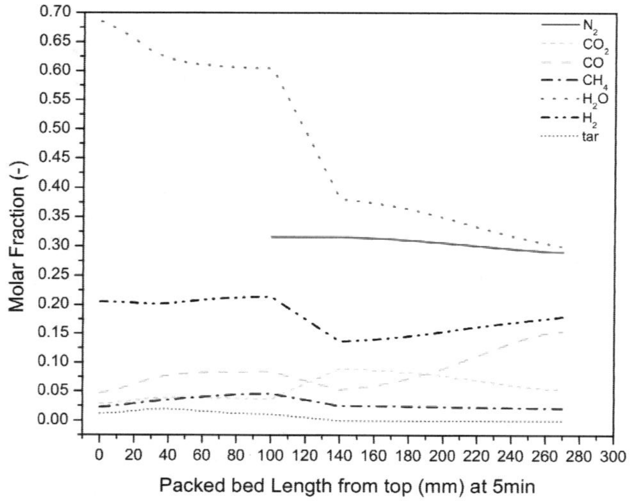

Fig. 2 Simulated composition profile across the gasifier at 5 min from the start of the experiment for the present study (Φ = 0.2533)

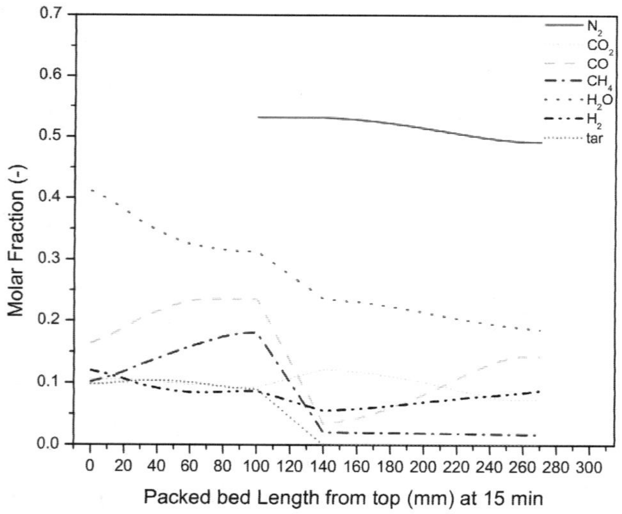

Fig. 3 Simulated composition profile across the gasifier at 15 min from the start of the experiment for the present study (Φ = 0.2533)

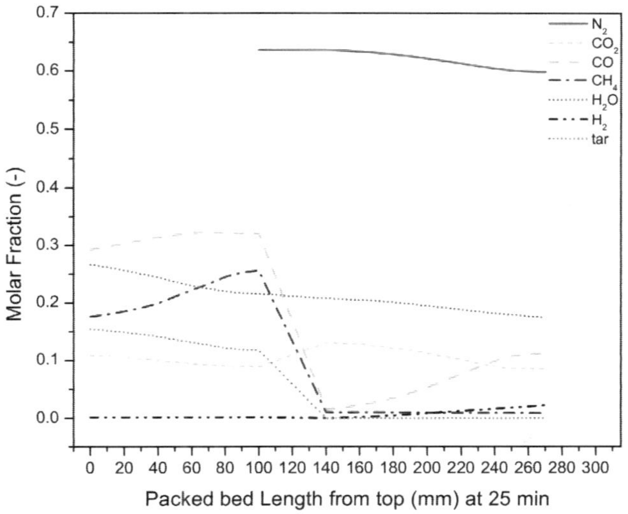

Fig. 4 Simulated composition profile across the gasifier at 25 min from the start of the experiment for the present study (Φ = 0.2533)

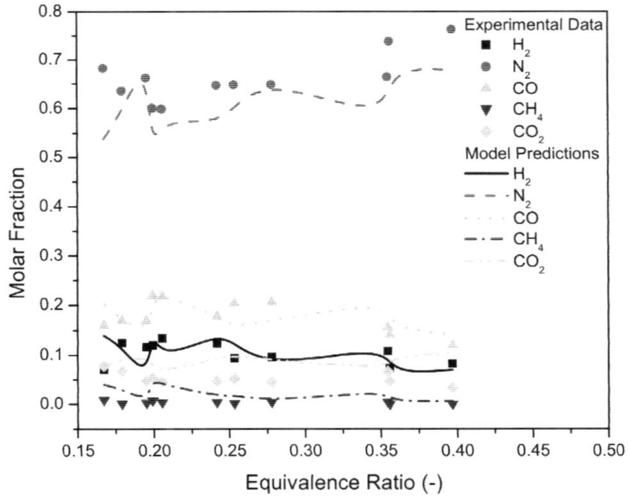

Fig. 5 Comparison of experimental data of present study with model predicted composition of producer gas

It shows the variation of molar fraction of the main components of the producer gas, i.e. nitrogen, carbon monoxide, carbon dioxide, methane, water vapor, hydrogen, and tar, with time in the gasifier. It is found that the composition of water vapor in the gaseous phase is the highest in pyrolysis zone (0-100 mm) and decreases in oxidation zone (100-140 mm) and reduction zone (140-270 mm) of the gasifier, when it travels through these zones (Figs. 2 to 4). It is also observed that the molar fraction of water vapor decreases with time in all zones of the gasifier. The molar fractions of CO and CH_4 in the gaseous phase increase in the pyrolysis zone as gas travels through it.

It is found that the molar fractions of CO, H_2, and CH_4 decrease in oxidation zone due to combustion. The molar fraction of CO_2 increases and that of tar decreases due to the high temperature oxidation. The molar fraction of nitrogen almost remains constant in the oxidation zone due to the constant supply of air. In the reduction zone of gasifier, the molar fractions of CO and H_2 increase and those of CO_2 and N_2 decrease. The molar flow rate of nitrogen does not change but due to the water gas reaction, methane formation reaction, and Boudouard reaction, formation of hydrogen and carbon monoxide occurs and the composition of nitrogen decreases in the reduction zone.

It is also found that the composition of carbon monoxide, methane, hydrogen and tar increases and that of water vapor decreases with time in the pyrolysis zone of the gasifier due to successive increase in the temperature of the pyrolysis zone. The simulated composition profiles shown in Figs. 2 to 4 at different times from the start of the experiment give proper insight of the gasifier. These simulated variations of the molar fraction of gaseous components with time match with the theory of downdraft biomass gasification. To study the variation of the simulated composition of the combined transport and kinetic model with equivalence ratio, the composition of producer gas is averaged over 25 minutes. Fig. 5.shows the comparison of the average composition of producer gas with those found experimentally as a function of equivalence ratio.

5. Conclusions

Based on the results obtained in the present study, the following conclusions are drawn:

1. The proposed combined transport and kinetic model is successfully validated with the experimental data reported in the literature.
2. The model predicted composition of producer gas matches very well with the experimental data reported in the literature.
3. The molar fractions of CO, CH_4 and tar increase and that of water vapor decreases with time in the pyrolysis zone of the gasifier due to a subsequent increase in the temperature of pyrolysis zone.
4. The molar fraction of water vapor in the gaseous phase is the highest in the pyrolysis zone and decreases as it travels through the oxidation and reduction zones of the gasifier.
5. The molar fractions of CO and CH_4 in the gaseous phase increase in the pyrolysis zone as gas travels through it. However, the molar fractions of CO, H_2, tar, and CH_4 decrease in the oxidation zone due to high temperature oxidation and hence the molar fraction of CO_2 increases.

6. References

Babu, B.V., Chaurasia A.S., 2003a. Modeling, Simulation, and Estimation of Optimum Parameters in Pyrolysis of Biomass. Energy Conversion and Management, 44, 2135-2158.

Babu, B.V., Chaurasia A.S., 2003b. Modeling for Pyrolysis of Solid Particle: Kinetics and Heat Transfer Effects. Energy Conversion and Management, 44, 2251-2275.

Babu, B.V., Chaurasia A.S., 2004a. Parametric Study of Thermal and Thermodynamic Properties on Pyrolysis of Biomass in Thermally Thick Regime. Energy Conversion and Management, 45, 53-72.

Babu, B.V., Chaurasia A.S., 2004b. Dominant Design Variables in Pyrolysis of Biomass Particles of Different Geometries in Thermally Thick Regime. Chemical Engineering Science, 59, 611-622.

Babu, B.V., Chaurasia A.S., 2004c. Pyrolysis of Biomass: Improved Models for Simultaneous Kinetics & Transport of Heat, Mass, and Momentum. Energy Conversion and Management, 45, 1297-1327.

Babu, B.V., Chaurasia A.S., 2004d. Heat Transfer and Kinetics in the Pyrolysis of Shrinking Biomass Particle, Chemical Engineering Science, 59, 1999-2012.

Babu, B.V., Sheth, P.N., 2006. Modeling and Simulation of Reduction Zone of Downdraft Biomass Gasifier: Effect of Char Reactivity Factor, Energy Conversion and Management, 47, 2602-2611.

Bridgwater, A.V., 2002. Bio-Energy Research Group, Aston University, Birmingham B47ET, UK,(2002). http://www.icheme.org/.../conferences/gasi/Gasification% 20Conf%20Papers/ Session%202%20presentation-Bridgewater.

Di Blasi, C., 2000. Dynamic Behaviour of the Stratified Downdraft Gasifiers. Chemical Engineering Science, 55, 2931-2944.

Giltrap, D.L., McKibbin, R.; Barnes, G.R.G., 2003. A Steady State Model of Gas-Char Reactions in a Down Draft Biomass Gasifier. Solar Energy, 74, 85-91.

Sharma, A.K., 2007. Modeling Fluid and Heat Transport in the Reactive, Porous Bed of Downdraft (Biomass) Gasifier. International Journal of Heat and Fluid Flow, 28, 1518–1530.

Sheth, P.N., Babu, B.V. 2009. Experimental studies on producer gas generation from wood waste in a downdraft biomass gasifier. Bioresource Technology, 100, 3127-3133

Wurzenberger, J.C., Wallner, S., Raupenstrauch, H., Khinst, J.G., 2002. Thermal Conversion of Biomass: Comprehensive Reactor and Particle Modeling, American Institute of Chemical Engineers Journal, 48, 2398-2411.

Syngas Compositions From Coal/Biomass Co-Gasification

Maruthi Pavani, Center for Energy Technology, Research Triangle Institute, Research Triangle Park, NC

Pradeepkumar Sharma, Center for Energy Technology, Research Triangle Institute, Research Triangle Park, NC

David Dayton, Center for Energy Technology, Research Triangle Institute, Research Triangle Park, NC

Raghubir Gupta, Center for Energy Technology, Research Triangle Institute, Research Triangle Park, NC

Co-gasification of coal and biomass helps to mitigate some of the non-technical barriers impeding the development of independent coal or biomass gasification technologies. Co-gasification of coal and biomass can be considered as a potential fuel base for future energy needs. By mixing coal and biomass together, a renewable component is added to coal gasification to offset fossil CO2 emissions and biomass can be used at a much larger scale to improve process economics. When integrated with appropriate downstream process, syngas obtained from coal/biomass co-gasification could surface as the technology of choice for economically producing a variety of fuels and chemicals (methanol, mixed alcohols, gasoline etc.). The difference in the chemical properties of coal and biomass provides an opportunity to tailor the syngas composition by changing the operating parameters (temperature, pressure and coal/biomass ratio) as per the requirement of the end product.

The chemical and physical property differences between coal and biomass contribute to the technical challenges associated with co-gasification. Higher hydrogen to carbon ratio and oxygen content in biomass compared to coal makes it more reactive leading to less severe (lower temperature and pressure) conditions for gasification. Unfortunately, the lower severity for biomass gasification also leads to tar formation that is a major gas cleanup challenge in these processes. The trace elements in coal and biomass are quite different leading to different gas cleanup requirements for each solid feedstock. Sulfur and other heavy metals like As, Cd, Hg and Pb are major concern for coal gasification but char and tars more important for biomass gasification.

RTI's Center for Energy Technology has been conducting R&D associated with coal gasification for over 15 years. In the present research, a detailed parametric coal/biomass co-gasification study was conducted varying temperature, pressure and coal/biomass ratio. Effect of above process parameters on syngas composition was determined at atmospheric and high pressure in entrained flow gasification systems. Data collected from the parametric gasification testing were used to develop empirical correlations to interpret the effect of feedstock, coal/biomass blending ratio, and process conditions on measured product gas, liquid, and solid concentrations and composition. These results will be presented and discussed.

See more of this Session: Advances in Gasification Research - I
See more of this Group/Topical: Energy and Transport Processes

Performance Evaluation of Biomass Co-gasification with Coal

Xiaotao Zhang[1,2], D. Salerno[2], H. Arellano-Garcia[2], G. Wozny[2]
[1]North China University of Water Conservancy and Electric Power, Road Beihuan 36, 450011 Zhengzhou, China
[2]Technische University Berlin, Sekr. KWT-9, Str. des 17. Juni 135, D-10623 Berlin, Germany

Abstract

Biomass co-gasification with coal in existing entrained flow coal-gasifiers can benefit to sustainable energy development and improve local environment by reducing CO2 and sulfur emissions of fossil fuels. Based on an equilibrium Gibbs reactor model, the performances of two types of entrained flow gasification technologies have been evaluated, namely dry feed gasifier and slurry feed gasifier with oxygen as agent. For this purpose, biomass replaces 5% to 20% of coal in weight as the gasifier's feedstock while the biomass moisture is varied from 5% to 30%. This work represents a feasibility study to find changes of product gas's compositions when coal is replaced by biomass according to a certain ratio. The obtained results have an important significance for the development of biomass energy.

Keywords: Biomass; Oxygen-co-gasification; Performance

1 Introduction

Biomass has great potential as a clean, renewable feedstock for generating electricity or producing chemicals. Biomass energy is expected to become increasingly important in the future because of the improvement of local environment and the possibility to obtain security of energy supply.

In OECD Countries, a target for bioelectricity is to achieve 15% of Electricity from Biomass by 2020[1]. In China, the nation's total biomass resource that can potentially be converted into energy was about 500 million ton in 2007. Following a social and economic development, the biomass resource that could potentially be converted into energy is expected to increase to 1 billion ton in the future [2].

Recently, gasification technology with a high performance and biomass feedstock flexibility has been the research focus in the field of renewable energy. There are many types of gasifiers: fixed-bed, fluidized bed, and entrained flow gasifiers [3].

Entrained flow gasification processes have already been developed and demonstrated on large-scale applications for coal. Furthermore, biomass co-gasification with coal in entrained flow gasifiers allows the reduction of fossil fuel consumption and the accompanying CO_2-emissions.

Different entrained flow gasifiers are operated worldwide, but only few have experience with biomass. Antonio Valero [4], validated an entrained flow gasifier's model with nearby 3000 actual steady state operation data (4800 h), which was used to study co-gasification of coal, petroleum coke and up to 10% of several types of biomass. Currently, 30 wt% of co-gasification biomass operations is done at the Buggenum IGCC power plant [5].

To meet the OECD's bioelectricity or China's biomass energy targets, biomass co-gasification with coal will be a good choice to enhance the utilization of biomass. In China, biomass resources include mainly straw and other agricultural wastes, forestry wastes, and etc. Several entrained flow gasifiers, such as GE, Shell, and Siemens gasification technologies, have been imported to China and used to produce chemicals now [6].

The main task of this work is to evaluate the performance of two types of entrained flow gasification technologies, dry feed gasifier and slurry feed gasifier with oxygen as agent: Here, biomass replaces 5% to 20% of coal as the gasifier's feedstock while the biomass moisture content changes from 5% to 30%.

The composition of product gas can be obtained in every case and the performance of two gasification technologies can be compared as well. The results are based on the assumption that biomass as the gasifier's feedstock has been pretreated. This work represents a feasibility study in order to elucidate what will ensue, when coal is replaced by biomass up to a certain ratio. Consequently, it has a significant implication for the development of biomass energy, and can supply suitable study results for biomass co-gasification operation.

2 Co-gasification process design

2.1 Entrained flow equilibrium gasifier model

A chemical equilibrium Gibbs reactor model is applied to predict the product gas composition in gasifiers using Aspen Plus software. Equilibrium models are valuable because they predict the thermodynamic limits of the gasification reaction system [7]. The gasifier is regarded as a perfectly insulated apparatus, i.e. heat losses are neglected. The residence time of the solids in the large-scale entrained flow gasifiers is much shorter. The model assumes that gasification reaction rates are fast enough and residence time is sufficiently long to reach the equilibrium state.

The entrained flow gasifiers typically use oxygen as the oxidant and operate at high temperatures well above ash slagging conditions in order to assure reasonable carbon conversion and to provide a mechanism for slag removal. Moreover, entrained-flow gasification has the advantage in that it can gasify almost all types of coals and biomass regardless of fuel rank, characteristics, or the amount of fuel fines.

Two cases are considered as following:

Case1: coal dry feed gasifier. The coal with a stream of oxygen is fed into the gasifier.

Case2: slurry feed gasifier. The coal/water slurry containing about 66.5 weight percent solids is fed into the gasifier, along with a feed stream of oxygen.

In both cases, the performance of gasifier has been evaluated with replacing 5% to 20% coal by several types of biomass. Entrained flow gasifier model is described in Fig.1.

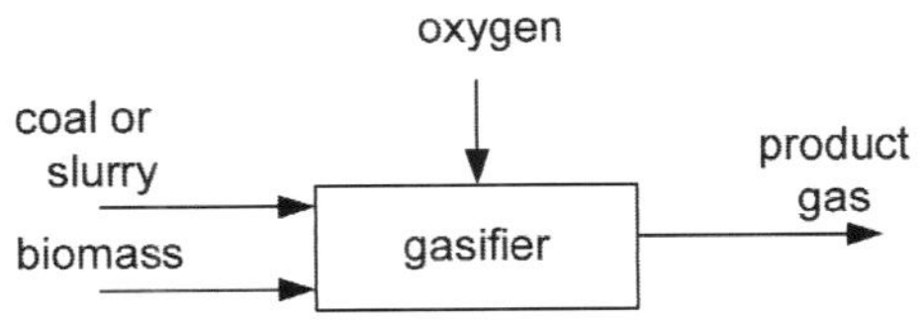

Fig.1 Schematic diagram of gasifier

The pressure of the gasifier is designed as 3.0 Mpa and the temperature modified near at 1350°C by adjusting oxygen mass flow. Gasification entails partial oxidation of the feedstock, so chemical energy of biomass is converted into chemical and thermal energy of synthesis gas. The input variables are biomass/fuel mass flow ratio, biomass moisture content, O_2/fuel mass flow ratio; and the results are the product gases composition and the gasifier's temperature.

2.2 Characteristics of feedstock

The coal as feedstock has the following characteristics in table1 [8]:

Table 1 Typical analyses of coal

Proximate analysis (wt%, as received)				Ultimate analysis (wt%, dry)					
Moisture	Volatile matter	Fixed carbon	Ash	C	H	O	N	S	Cl
11,9	28.1	50.9	9.1	75.68	3.94	8.5	0.81	0.72	0.02

A wide range of biomass sources, such as traditional agricultural crops, dedicated energy crops, residues from agriculture and foresting as well as organic wastes can be gasified. These biomass types differ in chemical composition, heating value, ash and moisture content. The selected biomass resources are straw, wood and sludge with moisture changing from 5% to 30%. High moisture content biomass fuel could influence system efficiency because the vaporization of water to steam requires heat and higher moisture content in the fuel means a lower net heating value of the fuel [9]. Table2 shows the proximate and ultimate analyses for the biomasses that are considered [3,10].

Table 2 Typical analyses of biomass

Fuel	Proximate analysis (wt%, dry)			Ultimate analysis (wt%, dry)					
	Volatile matter	Fixed carbon	Ash	C	H	O	N	S	CL
Straw	80.86	14.08	5.06	46.60	5.71	41.62	0.65	0.10	0.26
Wood	85.07	13.71	1.22	49.42	6.00	43.07	0.23	0.05	0.01
Sludge	51.03	2.47	46.5	25.73	5.03	18.19	3.3	1.14	0.11

Biomass as feedstock to power plant is cleaner than traditional coal because most biomass resources, such as straw and wood, contains minimal sulfur (<0.1 percent by weight). And biomass has higher amounts of volatile matter and lower fixed carbon than coal. Biomass reactivities are very different from coal under gasification conditions. Wood and straw are desirable because they are abundant and clean. Sludge is considered due to its high ash content, which can show different performances from wood and straw when co-gasification with coal in gasifiers.

3 Results and discussion

Table 3 shows the product gas results when coal or slurry without biomass mixture as feedstock is gasified in two types of entrained flow gasifiers. They are the basic data to be compared with biomass replacing coal in a certain ratio. The main product gas composition, such as H_2O, H_2, CO, and CO_2, consists of more than 97% of whole product gases. The other gases compositions are negligible due to their lower value.

Table 3 The basic results of gasifier

Project	Case1	Case2
Gasifier Temperature(°C)	1350	1350
Gasifier Pressure(MPa)	3.0	3.0
Product gas (Mole,%)		
H_2O	0.6	22.6
H_2	29.9	24.6
CO	66.7	39.3
CO_2	0.5	11.7
others	2.3	1.7

3.1 Influence of biomass ratio and biomass moisture content to gasifier's performance

The product gas compositions are important results that can show the performance of the gasification process. More composition of CO and H_2 accompanies with higher efficiency.

Fig. 2 and Fig. 3 show that the mole compositions of CO, H_2, CO_2 and H_2O change with biomass(wood)/fuel mass flow ratio and the biomass (wood) moisture content in Case1.

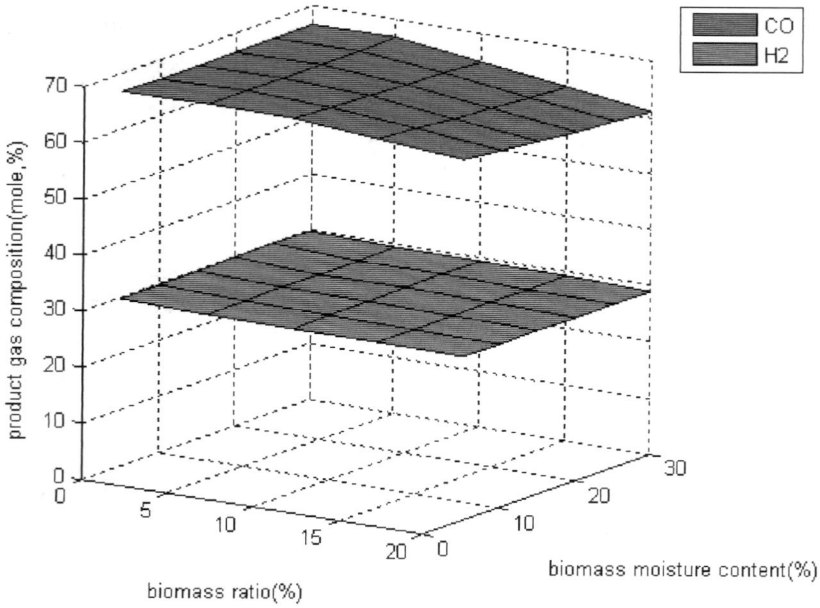

Fig. 2. CO and H_2 mole composition versus wood/fuel mass flow ratio and wood moisture content (Case1)

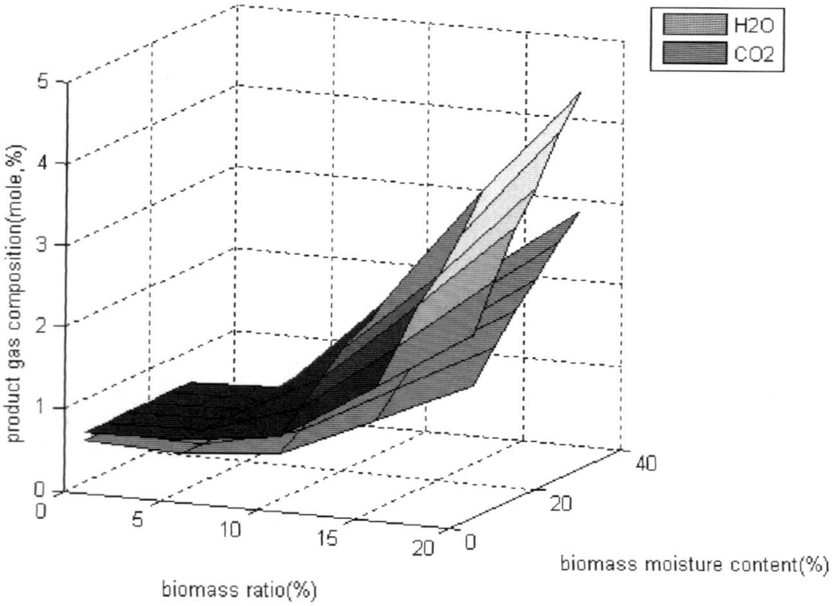

Fig. 3. CO_2 and H_2O mole composition versus wood/fuel mass flow ratio and wood moisture content (Case1)

CO mole composition decreases with growth of wood/fuel mass flow ratio. When there is only 5% moisture in biomass, CO mole composition changes from 66.8% to 64.5% (reduction of 2.3%) as the wood/fuel mass flow ratio increases from 5% to 20%. However, CO mole composition decreases to 61.0% as the biomass moisture content reach to 30% and wood/fuel mass flow ratio reach to 20% simultaneously. When there is only 5% and 10% biomass replacing coal as feedstock, CO mole composition changes a little as the biomass moisture content increases from 5% to 30%. H_2 mole composition changes not so apparently when wood/fuel mass flow ratio and wood moisture content increase. H_2 mole composition is nearly 30% as only coal as feedstock, the value reduces less than 0.5% as wood/fuel mass flow ratio and biomass moisture content reach to 20% and 30%.

CO_2 and H_2O mole compositions increase from less than 1% to 2.8% and 4.3% when wood/fuel mass flow ratio changes from 5% to 20% and wood moisture content from 5% to 30%. The value of CO_2 and H_2O mole compositions change not so apparently when less than 10% wood instead of coal in weight and the wood contains less than 10% moisture.

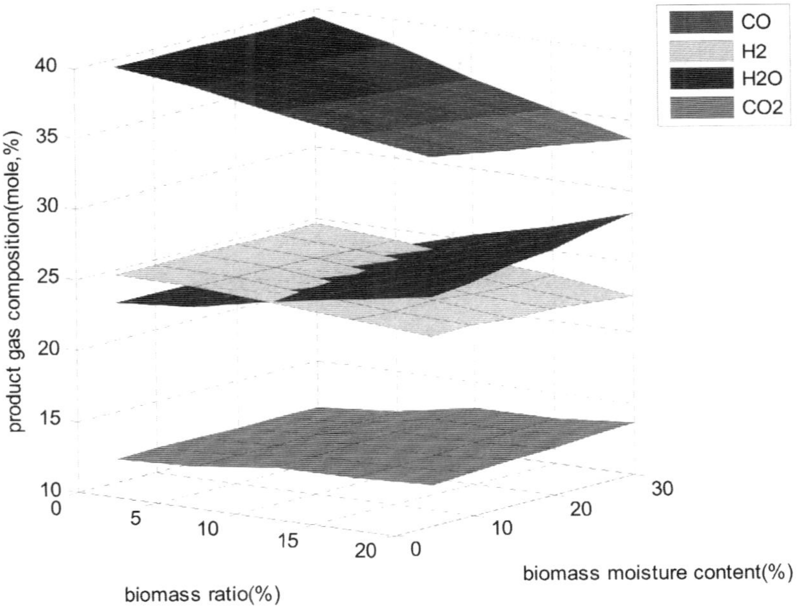

Fig. 4. Product gas mole composition versus wood/fuel mass flow ratio and wood moisture content (Case2)

In Case 2, the slurry feed is water/fuel mass flow ratio (33.4/66.6). Simulation results are shown in Fig. 4. CO and H_2 mole compositions decrease with the growth of wood/fuel mass flow ratio and wood moisture content. CO mole composition changes from 38.8% to 36.0% when wood/fuel mass flow ratio increases from 5% to 20% as well as there is only 5% moisture content in wood. However, CO mole composition reduces to 33.7% when there is 20% wood in weight instead of coal and the wood moisture content reach to 30%. H_2 mole composition decreases from 24.5% to 22.6% when both wood/fuel mass flow ratio and wood moisture content increase to 20% and 30%.

CO2 and H_2O mole compositions are all higher compared with them in Case1. CO_2 mole composition changes from 11.9% to 13.6% when wood/fuel mass flow ratio increases from 5% to 20% and wood moisture content from 5% to 30% and H_2O mole composition changes from 23.1% to 28.4% at the same time.

When there is only 5% biomass replacing coal in weight in Case2, product gas compositions change not so apparently.

3.2 Influence of biomass types to gasifier's performance

There are different product gas compositions when different biomass types instead of coal in certain ratio as feedstock. Fig.5 and Fig.6 show CO and H_2 mole composition of three kinds of biomass, such as wood, straw and sludge, replacing part of coal to be gasified. CO mole compositions of wood and straw are very close due to their nearly characteristics. There is a little difference of CO mole composition between wood and sludge when they replace coal in the same ratios. CO mole composition difference of different kinds of biomass replacing coal changes little when the biomass/fuel mass flow and biomass moisture content increases. CO mole composition reduces about 6% in three cases.

H_2 mole compositions fluctuate between 29% to 30%, and change less compared with CO mole compositions. H_2 mole composition as sludge replacing coal is a little higher.

The CO and H_2 mole composition differences in Case2 show in Fig.7 and Fig.8 when three kinds of biomass replacing part of coal are gasified. CO mole compositions of wood and straw co-gasification coal are very close too. But the CO mole composition differences between different feedstock become more and more when the biomass/fuel mass flow and biomass moisture content increases. The H_2 mole compositions show almost no difference using three kinds of feedstock.

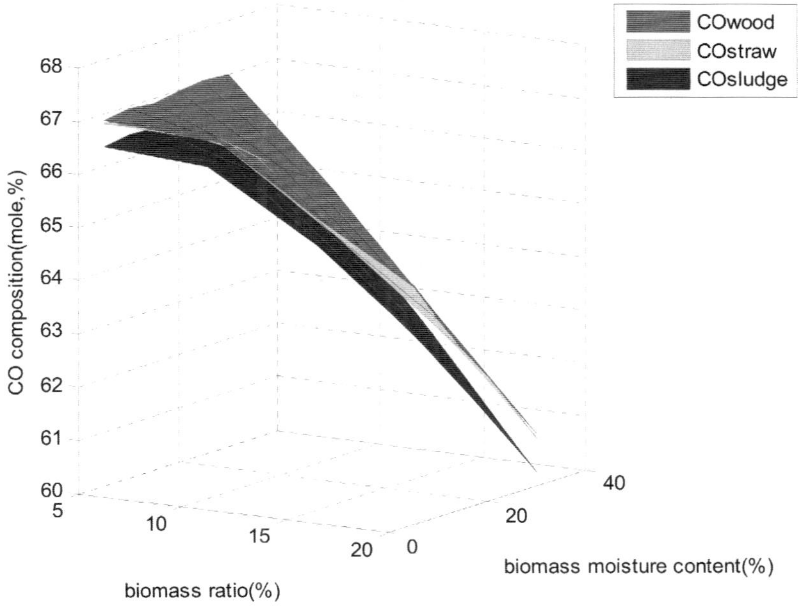

Fig. 5. CO mole composition versus different biomass (Case1)

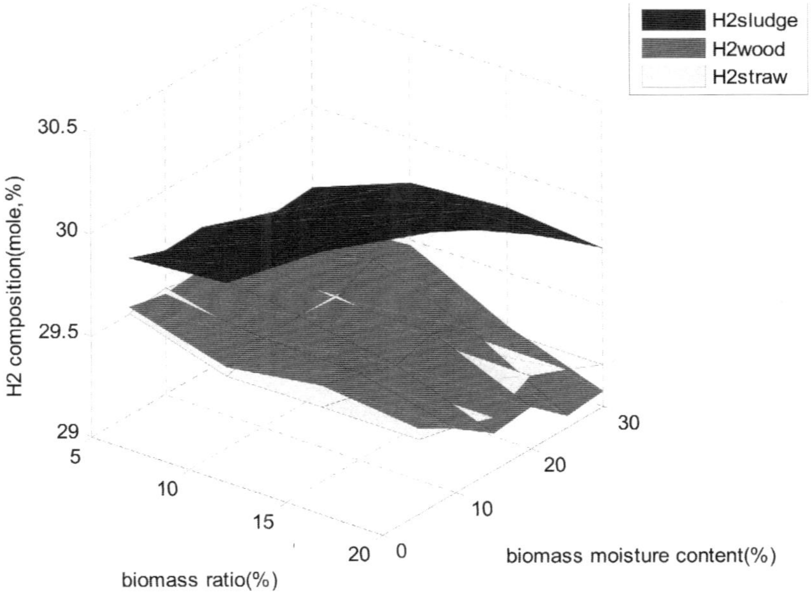

Fig. 6. H_2 mole composition versus different biomass (Case1)

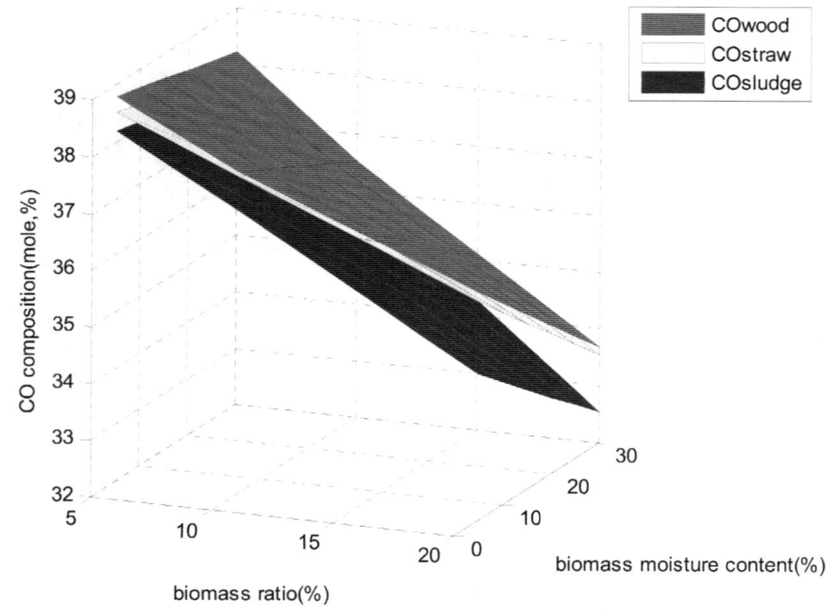

Fig. 7. CO mole composition versus different biomass (Case2)

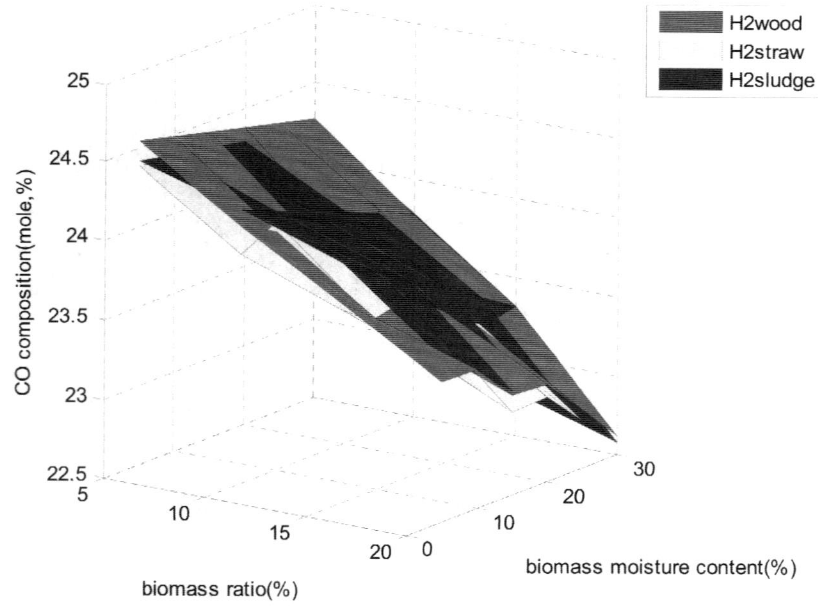

Fig. 8. H_2 mole composition versus different biomass (Case2)

4 Conclusion

Biomass co-gasification with coal in existing entrained flow gasifier may dig and utilize the potential for sustainable energy development and can benefit to the CO_2 reduction and sulfur emission of fossil fuel plant.

CO and H_2 mole compositions change with biomass/fuel mass flow ratio and the biomass moisture content. When more biomass replacing coal feeds in gasifier, the mole compositions of CO and H_2 in product gas will decrease; when ash content is higher in biomass, CO mole composition in product gas will reduce accordingly.

When there is only 5% biomass instead of coal, these changes in feedstock influence little to performance of gasification process. However, when both the biomass/fuel mass flow ratio and the biomass moisture content increase, the results change more.

Reference:

[1] Ausilio Bauen, Jeremy Woods and Rebecca Hailes. Bioelectricity Vision: Achieving 15% of Electricity from Biomass in OECD Countries by 2020, Imperial College London, Centre for Energy Policy and Technology, 2004

[2] National Development and Reform Commission. Medium and Long-Term Development Plan for Renewable Energy in China, People's Republic of China, 2007

[3] Brian Ricketts, Richard Hotchkiss. Technology Status Review of Waste/Biomass Co-Gasification with Coal, IChemE Fifth European Gasification Conference, Netherlands, 2002

[4] Antonio Valero, Sergio Uso´n. Oxy-co-gasification of coal and biomass in an integrated gasification combined cycle (IGCC) power plant. Energy 31: 1643–1655, 2006

[5] Ad van Dongen and Marco Kanaar. Co-gasification at the Buggenum IGCC power plant . Energy utilitzation from biomass, DGMK-meeting, 2006

[6] Natural Sciences Sector. Report on applying coal gasification technology in China's coal chemical industry. UNESCO Office Beijing, 2007

[7] Mark Jan Prins. Thermodynamic analysis of biomass gasification and torrefaction. Eindhoven, 2005

[8] Higman, C.; van der Burgt, M. Gasification; Elsevier: New York,2003.

[9] Clean Energy Decision Support Centre. Biomass heating project analysis. Minister of Natural Resources Canada, 2001 - 2005.

[10] R.Stahl, E. Henrich, H.J. Gehrmann, S. Vodegel, M. Koch. Definition of a standard biomass. Report prepared under the framework of project: Renewable fuels for advanced power trains. 2003

Techno-Economic Analysis of Effects of Alternative Gasification Configurations Assumptions On Lignocellulosic Biomass to Ethanol Systems

Yunhua Zhu, Energy and Environment Directorate, Pacific Northwest National Laboratory, Richland, WA

Mark A. Gerber, Energy and Environment Directorate, Pacific Northwest National Laboratory, Richland, WA

Susanne B. Jones, Energy and Environment Directorate, Pacific Northwest National Laboratory, Richland, WA

Don J. Stevens, Energy and Environment Directorate, Pacific Northwest National Laboratory, Richland, WA

Abstract The conversion of biomass to liquid fuels plays an important role, as part of a sustainable energy portfolio, for managing oil shortage and environmental problems being confronted today. Currently, lignocellulosic biomass gasification technologies have been developed to the point of large-scale demonstrations. However, widespread commercialization of large-scale biomass gasifiers and their integration with syngas conversion to ethanol technologies have not been realized because of technical and economic challenges. With these technologies still in the early development phase, system modeling and economic analysis is a powerful tool to provide insights for decision making about the research and development priorities of these technologies. Techno-economic assessments have been conducted for ethanol production from biomass systems. As an important process in a gasifier based biomass-to-fuel system, gasifiers with different configurations or operating conditions affect not only the gasifier equipment cost and size, but also impact downstream processes. The purpose of this study is to examine alternative gasification configurations and determine their effects on the overall biomass-to-ethanol performance and economics, and thus to identify potential opportunities for system improvement and optimization.

In this study, an indirectly-heated gasifier based wood chips to ethanol system was investigated. The system consists of five major processes: (1) gasification, (2) syngas purification, (3) syngas adjustment, (4) mixed alcohol synthesis and product purification, and (5) steam cycle. In this system, wood chips are dried and gasified to produce syngas. The syngas is cleaned and the H2/CO ratio is adjusted. The syngas is then converted to mixed alcohols and separated into products. Unconverted syngas is used for processes fuel and electricity production. Gasifier capacity was assumed to be 2000 dry tonne/day wood chips. Process models of systems with alternative gasification configurations were developed in CHEMCAD. Excel spreadsheet models were developed to estimate the capital and operating costs of different systems. In this study, the major configuration assumptions for the gasification process are investigated in this study, including operating temperatures, pressures, and different raw syngas compositions. Two gasification temperatures of 1600 oF and 1800 oF were considered. Gasification pressures were assumed to be 23 psi to 147 psi. Two different indirectly-heated gasifiers, generating syngas with H2:CO ratios of 0.6 and 1.7, respectively, were considered to study the impact of raw syngas composition.

The process design basis and modeling approach are described in the paper. Major performance and cost results of case studies are listed and analyzed. The effects of cases with alternative gasification configuration assumptions are compared and their results are discussed, including carbon conversion efficiencies, power consumption and output, and final product yields. The cost analysis results include estimated capital and operating costs, and ethanol selling prices. By comparing the performance and cost results of alternative configurations, potential opportunities for improving the performance and costs of gasification based biomass-to-ethanol systems are identified. As an example, results given in this work indicate that high pressure gasifier operation results in an ethanol selling price that is decreased by approximately 20% due to the cost savings associated with syngas compression. The impacts of alternative gasification temperatures and different raw syngas compositions are also discussed. By identifying those changes that have the greatest impacts, insights for decision making on research and development priorities of these technologies are obtained. Research can be more effectively directed toward achieving economic viability by focusing research in the key areas and setting processing milestones to measure technical progress.

Keywords: Biomass, Gasification, Syngas, Mixed Alcohol

See more of this Session: Advances in Gasification Research - I
See more of this Group/Topical: Energy and Transport Processes

Biowastes-to-SNG Via Gasification. An Integrated Economic, Environmental and Process Efficiency Evaluation

A. Sues, Laboratory for Environmental Technology. Department of Chemical Engineering & Chemistry, Eindhoven University of Technology, Eindhoven, Netherlands

H.J. Veringa, Eindhoven University of Technology, Eindhoven, Netherlands

K.J. Ptasinski, Eindhoven University of Technology, Eindhoven, Netherlands

Biomass is a promising renewable energy source for electricity, heat and transport biofuels production. Electricity and heat can also be produced from other renewable alternatives (such as solar or wind energy), but biomass is the only available renewable option to produce transport biofuels that can be stored in large quantities. This is of a great importance since the transport sector represents a significant share of the world's energy consumption, which is almost exclusively fuelled by fossil oils. In 2005, transport sector accounted for 27% of the global CO2 emissions.

However, biomass availability is also rather limited and stochastically distributed, especially in densely populated regions where biomass production could compete with other land uses, notably agriculture. In fact, this is the case of the so-called first-generation biofuels (e.g. bioethanol and biodiesel). However, when taking into account emissions from transport and conversion treatments, life-cycle analyses reveal that first-generation biofuels frequently exceed the emissions threshold of fossil fuels. Second generation biofuels are now being developed as a possible better alternative to the first generation, as they can use non-food crops or biowastes from different origins in order to avoid land competition for food production. Second generation biofuels are also expected to achieve higher efficiencies in larger conversion plants. Biowastes-to-biofuels conversion involves several challenges. Firstly, the existing technology must be re-designed and optimized to become cost and efficiency competitive with fossil fuels. Moreover, due to the wide diversity of biowastes and biofuels that can be obtained, the most sustainable conversion routes must be properly selected. Hence, an inherent challenge is to develop a reliable model to evaluate the sustainability of any process.

In this paper we present the design and evaluation Synthetic Natural Gas (SNG) production from woody wastes via gasification and later catalytic conversion of syngas.

Evaluation is made in terms of process efficiency, economic and environmental impact, and these 3 key parameters are later integrated and compared with other biofuels and fossil fuels.

In the biowaste-to-SNG process, pre-treatment steps are required in order to enhance the low energy density of the feedstock prior to gasification. However, since pre-treatment is directly affected by local conditions, the Dutch Friesland province is taken as a case study. The production chain is modeled in Aspen Plus and Aspen Icarus for efficiency (exergy) and economic evaluation, whereas a LCA is conducted for the environmental impact analysis. Moreover, all analyses are conducted under two different scenarios, a "full-renewable" and a "semi-renewable" production plant. In the first case, energy demand of the plant is covered by heat integration but extra biomass is also needed to fulfill heat and electricity demand. In the second case, the extra amount of biomass is substituted by a fossil fuel, i.e., natural gas from the grid. Both analyses are extended until the final use of SNG in vehicles. Results will show whether a fully or semi-renewable production plant is preferred and if SNG can become a primary biofuel for the transport sector. The methodology used will be also used to create an own sustainability model that could be applied to evaluate other biomass-to-biofuels conversion routes which will be discussed in this paper as well.

See more of this Session: Advances in Gasification Research - I

See more of this Group/Topical: Energy and Transport Processes

Ni-Co/La-γAl_2O_3 Oxygen Carrier for Fluidized Bed Chemical-Looping Combustion

Mohammad R. Quddus, Chemical and Biochemical Engineering, The University of Western Ontario, London, ON, Canada

Mohammad M. Hossain, Chemical and Biochemical Engineering, The University of Western Ontario, London, ON, Canada

Hugo de Lasa, Chemical and Biochemical Engineering, The University of Western Ontario, London, ON, Canada

Chemical-looping combustion (CLC) has been considered as one of the most promising technologies in capturing CO_2 from fossil fuel based power generation. In CLC, fuel combustion is carried out in two interconnected circulating fluidized bed reactors: a fuel and an air reactor. An oxygen carrier (OC) circulates between these two reactors supplies oxygen for fuel combustion, thereby preventing dilution of the CO_2 in the flue gas with the nitrogen from the air. Consequently, CO_2 is delivered without any further energy penalty for separation. CLC also helps minimizing NO_x emissions, since the fuel is burned in the absence of nitrogen without a flame. Even though, the technology is feasible enough for commercialization, large scale operation is still contingent upon the availability of suitable OC. In accordance with the prompt ability to react with fuel gas, a good OC should exhibit quick re-oxidation in contact with air, substantial oxygen storage, fluidizability and mechanical strength. This study was aimed at developing a material with the above characteristics. To this end a Co promoted Co-Ni/La-gAl_2O_3 carrier material was synthesized and evaluated.

The oxygen carriers of this study were prepared by impregnation method under vacuum conditions. Prior to nickel/cobalt loading, g$Al_2O_3$was first stabilized using 5 % La. The resultant support material La-gAl_2O_3 was then modified with 1 % Co. Finally, 20 % Ni was loaded using successive impregnation, 2.5 % Ni in each cycle.

As indicated above, the most important characteristic of oxygen carrier is its reactivity and stability at the high temperature of the cyclic redox process. To investigate these matters successive TPR and TPO experiments were carrier out at 950°C. In cyclic TPR/TPO experiments, the reduction and oxidation profiles of Ni-Co/La-gAl_2O_3 remains unchanged and the amount of Ni reduction also shows a stable behavior. Approximately 80 % nickel conversion was achieved using the Co modified sample, while this value was 71 % range for the unmodified carrier. Therefore, the addition of Co aids the formation of easily reducible nickel oxides species minimizing nickel support interaction and formation of non-reactive nickel aluminates. The pulse chemisorption results further confirms the stable behavior of the sample in consecutive redox cycles. A well dispersed state of Ni crystal, observed in pulse chemisorption, also signifies the absence of agglomeration of Ni crystals. This also indicates that Co helps to increase the dispersion of nickel on La-gAl_2O_3 support. Thus, the addition of Co modifies the metal surface changing the degree of interaction between Ni and the alumina support, maintaining a consistent metal dispersion during the repeated redox process. It is also apparent that the metal crystal size of the Ni/La-gAl_2O_3 sample remains unchanged over repeated cycles, a further indication of the absence of agglomeration of nickel crystals.

See more of this Session: Chemical Looping Processes - I
See more of this Group/Topical: Energy and Transport Processes

Attrition and Reactivity Analysis of $NiO/NiAl_2O_4$ Catalyst Particles Used in Chemical Looping Combustion

Saber Naserifar, Naoko Ellis and C. Jim Lim
Chemical and Biological Engineering, University of British Columbia, Vancouver, BC, Canada

Introduction

Chemical Looping Combustion (CLC) is one of the promising methods for inherent CO_2 separation with minimal energy penalty in combustion systems. This process can be done in two interconnected reactors where metal oxide transfers required oxygen from the air to fuel reactors. Selection of a metal oxide is crucial for the continuous operation of CLC and must ensure high attrition resistivity and reactivity. In this study, the standard attrition test method (ASTM 5757) has been modified to operate the air jet attrition unit at elevated temperatures (> 600°C) with steam injection, to assess the attrition behaviour of NiO catalysts at the typical operating conditions of the CLC. In addition, thermogravemetric analyzer (TGA) experiments are conducted at different temperatures (600-1000°C) to investigate the optimum reactivity temperature for the CLC catalyst. The optimum temperature is chosen based on the temperature at which the catalyst support (NiAl2O4) does not react during the redox reactions. The results indicate that increasing the operating temperature increases the attrition. Furthermore, the injection of steam is found to be effective in suppressing the attrition.

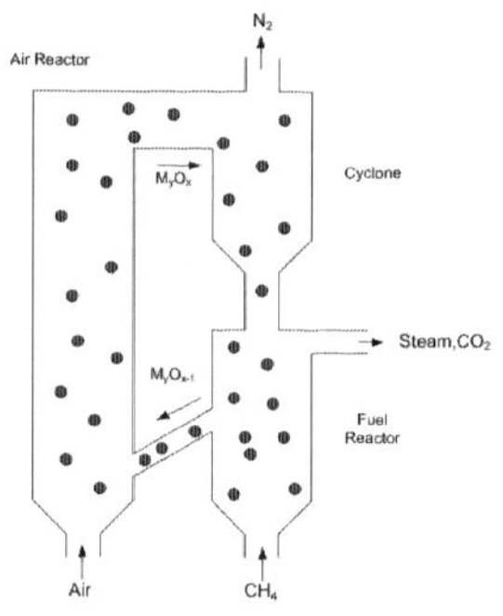

Figure 1. Schematic of a Chemical Looping Combustion unit.

As shown in Figure 1, two reactors, consisting of a fuel reactor and an air reactor are interconnected in a CLC. A catalyst (solid oxygen carrier) circulates between the two reactors and transfers required oxygen for the hydrocarbon combustion; hence, the air and fuel are not in direct contact. The combustion exhaust mainly consists of CO_2 and steam, allowing the separation of CO_2 with much less energy than that required for a conventional combustion unit.

Selection of a metal oxide is crucial for the continuous operation of CLC. A suitable oxygen carrier must ensure high attrition resistance, high reactivity, low carbon deposition and good durability in cyclic reactions. In addition being environmentally benign and economically viable can increase the tendency towards the selection as the oxygen carrier, as depicted in Figure 2.

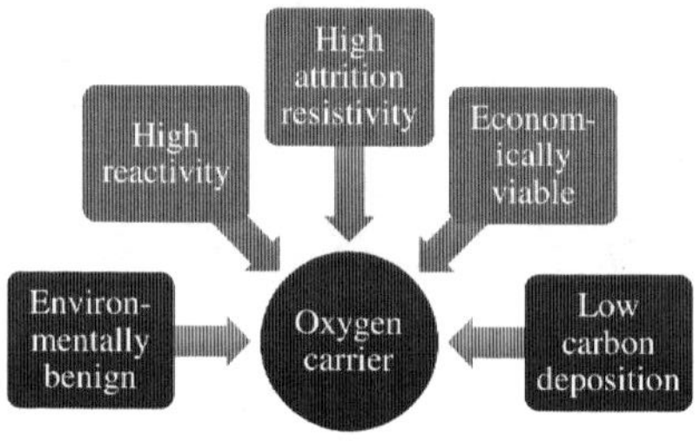

Figure 2. Requirements for a desirable oxygen carrier in CLC.

Attrition of particles changes the particle size distribution affecting the hydrodynamics of the system. Low reactivity of particles decreases the conversion of fuel and the regeneration of oxides. Carbon deposition on the surface of oxygen carrier prevents complete reduction of catalyst in the fuel reactor.

Experimental Section

In order to study the attrition at high temperatures, the following set-up, as shown in Figure 3, has been constructed. The test section consists of an attrition tube, settling chamber and a fine collector assembly. Attrition tube is a 710 mm long 35 mm ID stainless-steel tube with two 750 W external heaters. High velocity and efficient collision of particles are achieved by the distributor plate which consists of three drilled square edge nozzles each with 4.37 mm long and 0.4 ± 0.25 mm ID. The plate is attached to the bottom of the attrition tube and the top surface of the wind box. Couplers and seals are appropriately sized to ensure tight and leek free connections in the system. Settling chamber is a 300 mm long, 110 mm ID stainless-steel cylinder with heaters of 750 W (Zesta semi-cylindrical ceramic fiber heater) and 1500 W (Omega round cross section) attached to the top section of the attrition tube.

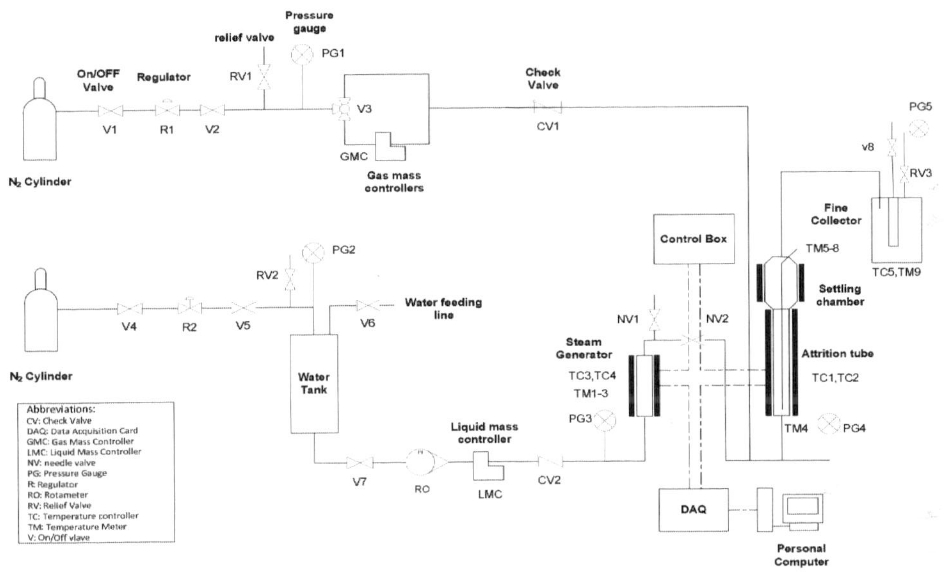

Figure 3. Process flow diagram of the attrition test apparatus.

Results and Discussions

Catalyst Attrition Study

$NiO/NiAl_2O_4$ catalysts used in this study were prepared by VITO, Flemish Institute for Technological Research, Belgium, using the freeze granulation technique. A water-based slurry of NiO and α-Al_2O_3 was prepared and poly acrylic acid and polyethylene were added as dispersant and binder, respectively. The slurry was pumped to a spray nozzle and quench in liquid nitrogen to form spherical particles. The drying step removed the residual water, followed by sintering the particles at 1400°C. Supporting $NiAl_2O_4$ was produced by reaction between α-Al_2O_3 and part of the NiO. Extra NiO was used to reach the mass ratio of about 4:6 between NiO and $NiAl_2O_4$. Finally particles were cooled and sieved to obtain particle size distribution of 106-180 μm.

As shown in Figure 4, the extent of attrition is expressed as the Air Jet Index (*AJI %*) defined as:

$$AJI, \% = \frac{f(5)}{S} \times 100$$

where *f(5)* is the weight of $NiO/NiAl_2O_4$ particles collected from the fine collector assembly after 5 hrs of continuous operation and *S* is the initial amount of sample (50 g). It is observed that total weight of attrition is increased with increasing temperature from 150 to 600°C. The increment of particle attrition at higher temperatures could be explained by the effect of thermal stress which is induced within the particles when contraction or expansion is limited. At high temperatures, uneven expansion and contraction may result due to the temperature gradients within particles which results in decrepitation. Figure 4 also compares the attrition of catalyst at different temperatures with and without steam injection. It is observed that the injection of steam is an important factor for suppressing the amount of attrition. In fact, for experiments that steam is not used the attrition of catalyst increases by 61.3%, 45.7%, and 22.8% for temperatures of 150, 400, and 600°C, respectively.

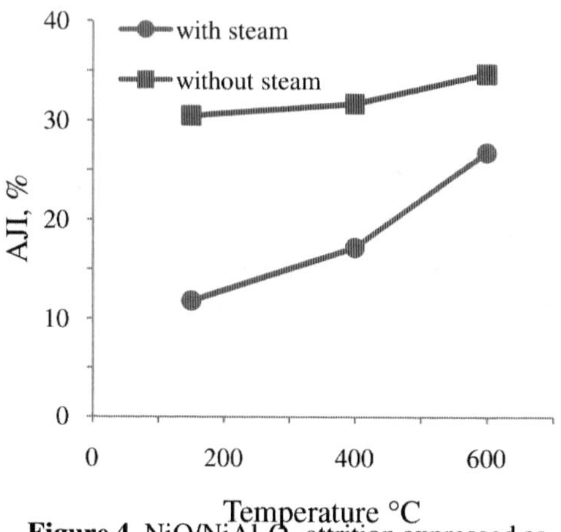

Figure 4. $NiO/NiAl_2O_4$ attrition expressed as Air Jet Index (AJI %) with initial particle size of 106-180 μm at different operating temperatures (150, 400 and 600°C) with and without steam.

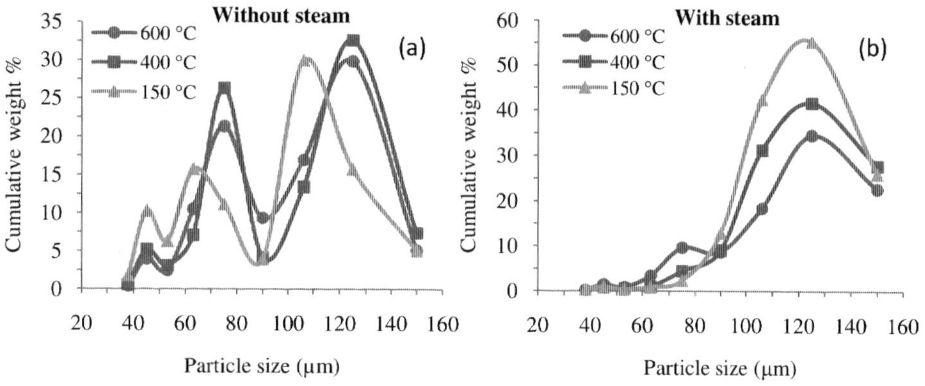

Figure 5. Particle size distribution of $NiO/NiAl_2O_4$ catalyst particles after attrition at temperatures of 150, 400, and 600 °C (a) with and (b) without steam in N_2 atmosphere.

Figure 5 shows the particle size distribution of $NiO/NiAl_2O_4$ catalysts collected from the attrition tube and the settling chamber measured by sieve analysis for experiments with and without steam. The broad ranges of catalyst particles are formed after the attrition (>20-150μm). The particle size distribution of particles in the fine collector assembly is between 1-20 μm as the fluidizing gas flow rate was set to transfer only particles less than 20 μm to the collector. For all temperatures, it is observed that from the initial particle size distribution of 106 - 180μm, the presence of steam suppresses the production of particles sizes <90 μm, compared to without the steam injection. The increase in the operating temperature increases the generation of smaller size particles for both with and without the steam.

Thermogravimetric Analyzer (TGA) Study

The reactivity experiments has been performed in a Thermogravimetric analyzer (TA instrument SDT Q600). For each experiment, the temperature is increased from ambient temperature to the desired temperature in a N_2 atmosphere at 100°C/min heating rate. Purge gas flow of N_2 at the flow rate of 25-90 ml/min is used to avoid diffusion of reactant gas to the balance chamber. After the temperature becomes equilibrated, three-port switching valve changes the N_2 stream to reactant gas (CH_4 or O_2) with concentration of 10% and N_2 as carrier gas with the total flow rate of 100 ml/min.

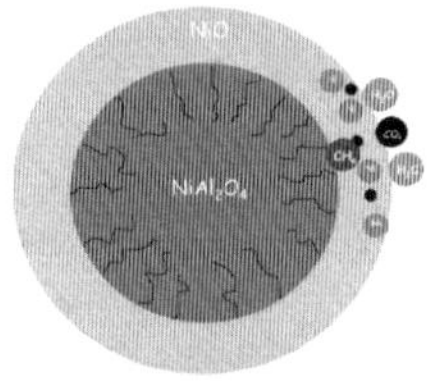

Figure 6. Representation of NiO/NiAl2O4 catalyst reduction by CH_4.

In this set of experiments, CH_4 is being passed through the TGA while the NiO/$NiAl_2O_4$ catalyst reduces, resulting in reduction of mass. Further reduction of particles results in the catalyst support material, $NiAl_2O_4$, participating in the reaction, as depicted in Figure 6. The theoretical oxygen of NiO/$NiAl_2O_4$ (assuming NiO is the only active oxygen source) was found to be around 9 wt.% of the total catalyst, calculated from the elemental analysis conducted by EDX (Energy dispersive X-ray spectroscopy). The reduction reaction involves the surface reaction, pore diffusion and eventually support material reaction and coking:

$$CH_4\ (g) + 4NiO(s) \rightarrow CO_2\ (g) + 2H_2O\ (g) + 4Ni(s) + coke$$

Based on the elemental analysis, complete oxidation of NiO will result in 91 wt.% compared to the initial fully oxidized state.

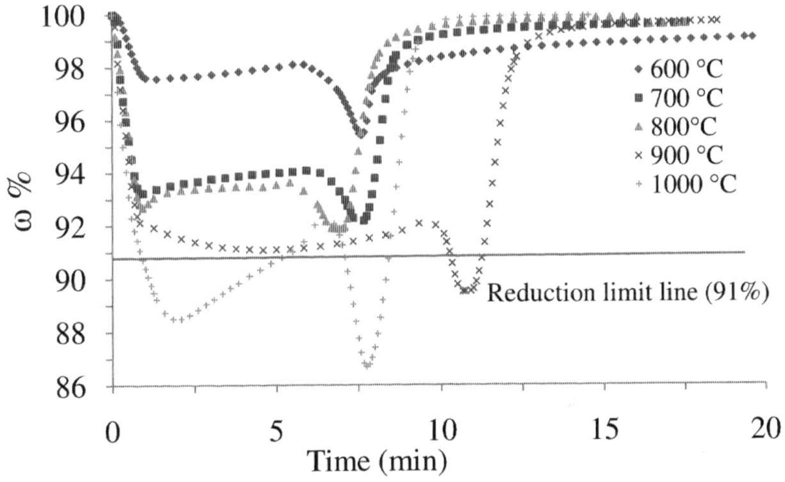

Figure 7. TGA results from seven reduction-oxidation reaction of NiO/$NiAl_2O_4$ using 10% of reactant gases (CH_4 and O_2) in N_2 as purge gas. Measurements were performed at temperature range of 600-1000 °C at atmospheric pressure.

The mass conversions of NiO/$NiAl_2O_4$ particles for temperatures in the range of 600-1000°C using 10% of reactant gases (CH_4 and O_2) are compared in Figure 7. The mass conversion percentage, ω%, is plotted against the reaction time. It is shown that reduction of catalyst is a strong function of temperature which not only influences the rate of reaction, but also the other parameters such as carbon deposition. The redox curves cross the theoretical reduction limit line for temperatures of 900 and 1000°C indicating that the support ($NiAl_2O_4$) participates in the reactions. Therefore, the optimum temperature selection is more favourable to 600, 700 and 800°C.

Figure 8 compares the reduction of catalyst for temperature range of 600-800°C with 10% CH_4 as a reactant gas. Owing to the lower conversion and reduction rate at 600°C, this temperature is not considered as ideal. Comparing the reduction curves, the reduction at 800°C results in higher conversion and reduction rate, especially at the initial stage. Although, slightly less carbon deposition at 700°C is considered to be an advantage, water vapour which is very effective in preventing carbon deposition is produced as one of the byproduct of reduction reaction, or injected in the fuel reactor during the operation of the CLC system. Therefore, 800°C is considered to be the optimum temperature for reduction and oxidation of $NiO/NiAl_2O_4$.

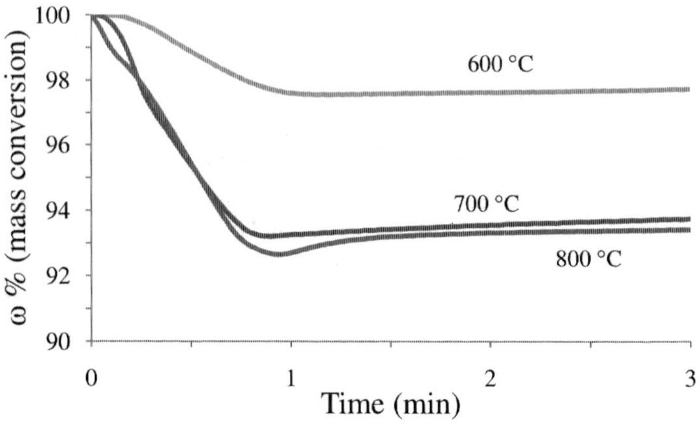

Figure 8. TGA results from three reduction-oxidation reaction of $NiO/NiAl_2O_4$ using 10% of reactant gases CH_4 in N_2 as purge gas. Measurements were performed at temperature range of 600-800 °C at atmospheric pressure.

Conclusions

The attrition of $NiO/NiAl_2O_4$ catalyst particles were determined by improved ASTM D5757 standard test method to investigate the attrition of particles at elevated temperatures and in the presence of steam. The attrition was increased by increasing temperature which could be explained by higher thermal stress at higher temperatures. Steam was found to be very effective in decreasing the attrition compared to the experiments without steam conducted at same temperatures. Thermogravimetric analysis (TGA) was used to investigate the reactivity of $NiO/NiAl_2O_4$ catalyst (106-125 μm) as one of the most common used catalyst in CLC process. Several reduction and oxidation reactions were performed with methane and oxygen as reducing and oxidizing agents, respectively. The reduction shows two distinct steps; first, fast reduction of catalyst where NiO sites convert to Ni rapidly and then a step that rate retards, where the slow reduction probably is caused by severe limitations of reactant gas to reach the layers underneath the catalyst surface. The optimum temperature was found to be 800°C for operation of CLC system with this type of catalyst.

Acknowledgement

The authors gratefully acknowledge financial support from the Total S.A. of France and Natural Sciences and Engineering Research Council of Canada (NSERC).

An Iron-Based Chemical Looping Coal Combustion Island

Felipe Vargas-Villamil, National Energy Technology Laboratory, West Virginia University Research Corporation, Morgantown, WV

Stephen Zitney, Collaboratory for Process and Dynamic Systems Research, Office of Research & Development, National Energy Technology Laboratory, U.S. Department of Energy, Morgantown, WV

Coal is a carbon intensive energy source. Almost a third of the anthropogenic CO2 emissions are produced by coal-fired power plants, which produce more than 50% of the electricity generated in the US and about 40% of the electricity generated worldwide [1]. Among them, the pulverized coal-based power plants account for more than 90% of the electricity generated from coal [2]. Nowadays, there is both a growing concern about the effect of the greenhouse gases into the global climate and, a need to improve the efficiency of power plants. Coal Chemical Looping Combustion (CLC) is a promising technology to address these issues.

CLC is a two step process that provides inherent separation of carbon dioxide by avoiding direct contact between air and fuel. It integrates the combustion, the CO2 separation, and the water gas shift reaction into the chemical loop. An oxygen carrier, which transports oxygen and energy, circulates between an air reactor and a fuel reactor. In the air reactor, the oxygen carrier is oxidized with air while it provides the oxygen and energy for combustion in the fuel reactor. Generally, the oxidation of oxygen carrier is exothermic while its reduction is endothermic. In the fuel reactor, the fuel is never mixed with air so that almost pure CO2 is produced which may be compressed for sequestration.

At NETL we are working to model and simulate a CLC island that may be used for existing pulverized coal-based power plants. The aim is to develop plants with zero CO2 emissions. In this work we provide an optimized design of an iron-based CLC island. An Illinois No. 6 coal was used.

References: [1] U.S. Department of Energy, Annual Energy Outlook 2008, revised to Include the Impact of H.R. 6, "Energy Independence and Security Act of 2007" Enacted in December 2007. DOE Energy Information Administration, Washington, DC, Report #: DOE/EIA-0383, 2008. [2] B. G. Miller, Coal Energy Systems, Elsevier, Amsterdam, Boston, 2005. [3] J. N. Armor, Catal. Lett., Vol. 114, 115-121, 2007. [4] F. Li and L. Fan, Clean coal conversion processes - progress and challenges, Energy Environ. Sci., Vol. 1, 248-267, 2008.

See more of this Session: Chemical Looping Processes - I
See more of this Group/Topical: Energy and Transport Processes

The Effect of Ash on Oxygen Carriers for Chemical Looping Combustion of Solid Fuels

Aurora Rubel*, Kunlei Liu, and Jim Neathery
University of Kentucky
Center for Applied Energy Research
2540 Research Park Drive
Lexington, KY 40511
rubel@caer.uky.edu

Introduction

Chemical looping combustion (CLC) is a promising technology [1-4] being investigated at the University of Kentucky Center for Applied Energy Research (CAER) for the combustion of solid fuels. CLC's potential is two-fold in that it can provide a high-purity, sequestration-ready stream of CO_2 and theoretically provide a substantial improvement in thermal conversion efficiency relative to conventional energy conversion technologies with other CO_2 control processes such as amine scrubbing and oxy-combustion. CLC uses a solid oxygen carrier (OC), usually a metal oxide to transfer the oxygen from the air to the fuel in a reducing reactor without the direct contact between the fuel and air. The reduced OC is recycled by using air in a second reactor to oxidize the carrier to its original state. Air is not mixed with the fuel, and so the product CO_2 is not diluted by the nitrogen in the flue gas. Nearly pure CO_2 is obtained in the exit gas stream from the reducer following condensation of H_2O. The pure CO_2 is ready for subsequent sequestration without costly purification [5]. This technology was first proposed primarily for the combustion of gaseous fuels and only recently been considered for solid fuels such as coal without gasification of the coal first to syngas [6-9].

Previous work at CAER indicated that iron oxide powder and catalysts were the best candidates for use as OC for CLC of solid materials from the OCs tested. They did not agglomerate; are relatively inexpensive; retained activity; and remained durable through multiple oxidation – reduction cycles. The results suggested that access of oxidizing – reducing gases to the active sites of the OC through adequate porosity was important. It was found that increased H_2, CO concentrations resulted in increased reduction rate for the iron oxides which may prove useful in a CLC process for solid materials as it provides a method to control the reduction rate during processing. Fe_2O_3 powder was used to directly combust a beneficiated gasification char which contained 60% carbon. A mixture of Fe_2O_3: char mixture (90:10) heated to 950°C in Argon resulted in 88% removal of the carbon with the OC as the only source of oxygen. This was encouraging for the application of CLC to the combustion of solid fuels [8, 9].

This work was to determine the effect of fly ash on the reactivity of two of the iron oxide oxygen carriers previously studied at the CAER since this is a very important consideration for their potential use for CLC of solid fuels.

Experimental

Two iron oxide catalysts, Wustite and fused iron, were studied. The two oxygen carriers and the fly ash were characterized by elemental analysis and XRD. A Netzsch Jupiter 449C thermal analyzer-differential scanning calorimeter-mass spectrometer (TG-DSC-MS) was used for oxidation/reductions cycles of the OCs. The TG-DSC-MS was fitted with a multiple gas mixing/controlling system comprised of 5 mass flow controllers (valves and controllers), a Lab View control system (software and interfaces), and NResearch switching valves. This system could be operated unattended continuously through multiple oxidation – reduction cycles. The OCs were mixed with fly ash obtained from a KY power plant in the following ratios of OC to Ash: 75:25, 50:50, and 25:75. Oxidation/reductions were all carried out to completion at 950°C for a minimum of 5 cycles. All samples were oxidized first followed by reduction. The gas flow rate through the TGA was 200 ml/min metered at standard conditions. The oxidizing gas was 20% O_2 in a balance of Ar. The reducing gas composition was 10% H_2, 15% CO, 20% CO_2, balance Ar. A 5 minute purge of Ar was used between oxidation and reduction. Weight gain/loss and maximum rates during oxidation/reduction were determined.

Results and Discussion

XRD analysis of the ash indicated it contained mostly quartz and aluminum alumosilicate with a small amount of Fe_2O_3. The XRD analysis was qualitative so quantitative amounts were not known. The ash also contained 6.22% C, 0.05%H, 0.25%S and 1.3%O. The elemental analysis suggests that S is present as SO_4.

First step in oxidation/reduction cycling was always oxidation. All carbon in the ash should be removed during this first oxidation so the first oxidation result was discounted. The weight gain/loss on oxidation/reduction in response to the ash was similar for both catalysts. Corrected for dilution of the OC by the ash, the amount of weight gain and loss increased as the amount of ash increased (Figure 1, results for Wustite). This increase was most likely the result of the presence of Fe_2O_3 and SO_4 in the ash contributing as OCs. Corrected for weight loss related to the ash (Figure 2, results for fused iron), ash had no detrimental effect on weight gain/loss on oxidation/reduction of the OCs. Again both OCs behaved the same. Additionally, there was an increase in the porosity of the pellet formed by the oxidation/reduction cycling of both Wustite and fused iron. With increasing ash, the pellet was more porous and more fragile. Dependant on the surface area measurement theory used, BJH or BET, there was a 12 and 3 fold increase respectively in pore surface area between the spent 100% fused iron and spent 75:25 blend of ash with fused iron (Table 1). BJH pore volume increased 7 times with addition of ash. This would result in increased access (penetration into the pellet) of reactive gases to the OC particles and more complete oxidation and reduction of the OC. This was also supported by the fact that the rates of oxidation and reduction increased 4 and 3 fold respectively as the ash increased from 0 to 75% of the OCs. Again both OCs tested behaved very similarly. Figure 3 give the results for Wustite.

Summary and conclusions

This work was very encouraging for the use of iron oxide OCs for the CLC of solid fuels. The ash was not detrimental to the reactivity of the two OCs studied and in fact it appeared that some components, Fe_2O_3 and SO_4, in the ash may have functioned as OCs contributing to additional weight gain/loss on oxidation/reduction cycling. The increase in the rates of oxidation/reduction with increased concentration of ash again suggests the importance of access to active sites of the reactive gases.

References

[1] Croiset, E Thambimuthu, K. Coal Combustion with Flue Gas Recirculation for CO_2 Recovery. In: Riemer, P, Eliasson, BWokauun, A, Editors. Greenhouse Gas Control Technologies, New York: Elsevier; 1999, p. 581-6.
[2] Marion, J, Naskala, Y N, Griffin, T. Controlling Power Plant CO_2 Emissions: a Long Range Approach. In: The First National Conference on Carbon Sequestration, Washington, DC,2001; DOE/NETL-2001/1114
[3] Herzog, H, Eliasson, B, Kaarstad, O. Capturing Greenhouse Gases. Scientific American 2000; 282(2):72-9.
[4] Hatanaka, T, Matsuda, S, Hatano, H. A New-concept Gas–solid Combustion System 'MERIT' for High Combustion Efficiency and Low Emissions. In: Intersociety Energy Conversion Engineering Conference 1997:944-8.
[5] Freund, P. Abatement and Mitigation of Carbon Dioxide Emissions from Power Generation. Powergen 98 Conference. 1998 [cited; Available from: http://www.ieagreen.org.uk/pge98.htm.
[6] Richter, H Knoche, K. Reversibility of Combustion Processes. In: ACS,1983:71-86.
[7] Anthony, E J. Solid looping cycles: A new technology for coal conversion. Industrial and Engineering Chemistry Research 2008; 47(6):1747-1754.
[8] Rubel, A, Lui, K, Neathery, J, Taulbee, D. Evaluation of oxygen carriers for chemical looping combustion of coal and solid fuels. In: 24th Pittsburgh Coal Conference, Johannesburg, SA,2007:paper 21-1.
[9] Rubel, A, Liu, K, Neathery, J, Taulbee, D. Oxygen carriers for chemical looping combustion of solid fuels. Fuel 2009; In Press, Corrected Proof.

Table 1. Surface area and pore volume measurements for fused Fe and fused Fe:ash (25:75).

Sample	BJH SA*, m^2/g	BJH pore vol, m^3/g	BET SA**, m^2/g
100% Fused Fe	0.0036	0.00014	0.0212
Fused Fe:ash (25:75)	0.0427	0.00104	0.0699

*Barrett, Joyner, Halenda method for surface area measurements
**Brunauer, Emmett, Teller method for surface area measurements

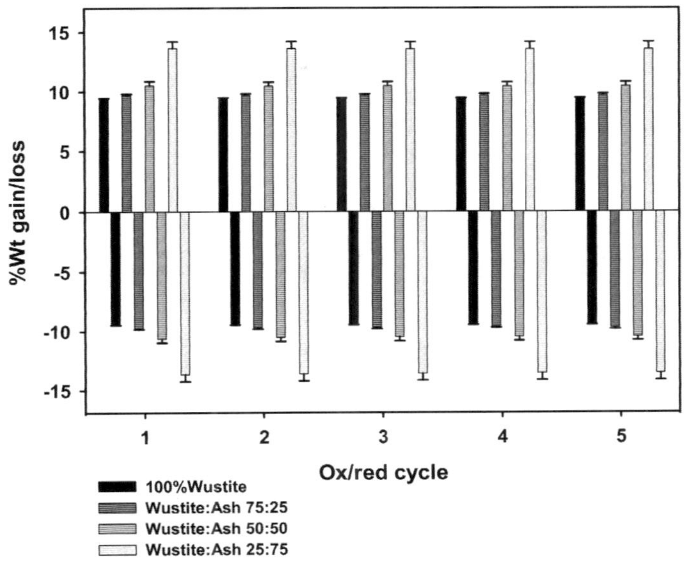

Figure 1. Weight gain/loss on oxidation/reduction of Wustite with increasing ash concentration corrected for dilution of the Wustite.

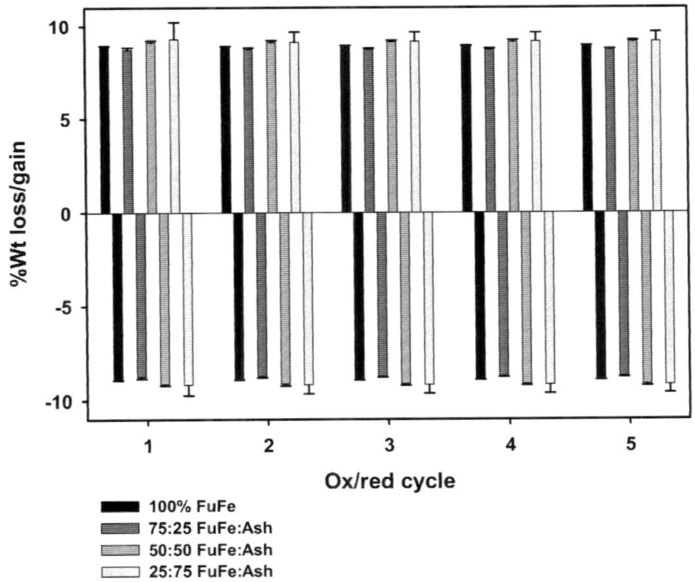

Figure 2. Weight gain/loss on oxidation/reduction of fused iron with increasing ash concentration corrected for the weight gain/loss by the ash.

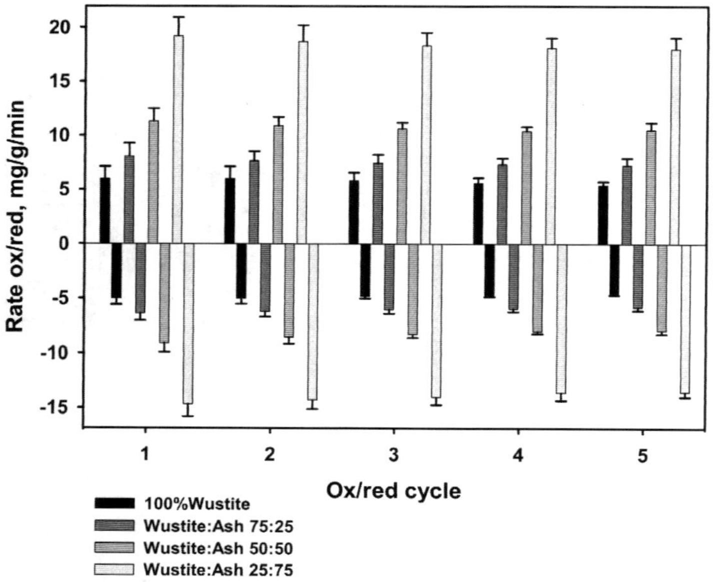

Figure 3. Effect of ash on the rate of oxidation/reduction of Wustite.

Design, Construction and Preliminary Tests of the Sub-Pilot Scale Syngas Chemical Looping System

Deepak Sridhar, William. G. Lowrie Dept. of Chemical and Biomolecular Engineering, The Ohio State University, Columbus, OH

Andrew Tong, William. G. Lowrie Dept. of Chemical and Biomolecular Engineering, The Ohio State University, Columbus, OH

Fanxing Li, William. G. Lowrie Dept. of Chemical and Biomolecular Engineering, The Ohio State University, Columbus, OH

Fei Wang, William. G. Lowrie Dept. of Chemical and Biomolecular Engineering, The Ohio State University, Columbus, OH

Hyung Rae Kim, William. G. Lowrie Dept. of Chemical and Biomolecular Engineering, The Ohio State University, Columbus, OH

Liang Zeng, William. G. Lowrie Dept. of Chemical and Biomolecular Engineering, The Ohio State University, Columbus, OH

L. -S. Fan, William. G. Lowrie Dept. of Chemical and Biomolecular Engineering, The Ohio State University, Columbus, OH

The national energy strategy focus is on the reduction of greenhouse gas emissions and energy security. Therefore, clean energy technologies that are feasible for industrial applications are extensively pursued. Fossil fuels are projected to remain a dominant energy source for the next several decades. Therefore, a process capable of reducing greenhouse gas emissions from fossil fuel as well as utilizing renewable energy sources is highly attractive. A promising candidate is the Syngas Chemical Looping (SCL) process. The SCL system developed at the Ohio State University demonstrates flexibilty in fuel source and products. A unique iron based composite particle synthesized at OSU makes this process possible. The particles are used as a looping medium which undergo redox cycles passing through three reactors namely; the reducer, oxidizer and combustor. In the reducer, the gaseous fuel is fully combusted while reducing the iron oxide particles. Steam reacts with the reduced particles in the oxidizer to yield high purity H_2. Finally, the partially oxidized particles enter the combustor to complete the re-oxidation step of the looping medium accompanied by heat generation. Prior work on the individual reactor studies indicated the possibility of producing Hydrogen and/or heat based on the reaction scheme. With encouraging results, initiative was taken to further scale up the process to demonstrate the continuous operation of all three reactors. In this article, we will focus on the design, construction and preliminary tests of a 25 kW_{th} sub-pilot SCL system. Details on the safety features and startup operations are also discussed. Particle attrition and reactivity results are highlighted. The ability of SCL to simultaneously produce high purity hydrogen and a concentrated CO_2 stream ready for sequestration is validated.

See more of this Session: Chemical Looping Processes - I
See more of this Group/Topical: Energy and Transport Processes

Investigation of Mechanism of Chemical Looping Combustion of Coal Utilizing

Ranjani V. Siriwardane, National Energy Technology Laboratory, U.S. Department of Energy, Morgantown, WV

Hanjing Tian, Parsons Infrastructure & Technology Group, Inc, Morgantown, WV

George Richards, National Energy Technology Laboratory, U.S. Department of Energy, Morgantown, WV

James Poston, National Energy Technology Laboratory, U.S. Department of Energy, Morgantown, WV

Thomas Simonyi, RDS, Parsons Infrastructure & Technology Group, Inc, Morgantown, WV

Chemical looping combustion (CLC) is an emerging technology for clean energy production from fossil and renewable fuels. In CLC, an oxygen carrier (typically a metal oxide) is first oxidized with air. The hot metal oxide is then reduced in contact with a fuel in a second reactor, thus combusting the fuel. Finally, the reduced oxygen carrier is transferred back to the oxidizer. CLC thus allows for flame-less, NOx-free combustion without requiring expensive air separation. Most importantly, CLC produces sequestration-ready CO2-streams without significant energy penalty. A solid fuel such as coal is rarely used in CLC since the process with solid fuels faces many challenges. Few studies have been reported on the combustion of coal with oxygen carriers. Several mechanisms for the reaction of coal have been presented but the mechanism is not clearly understood. Research work focused in understanding the reaction mechanism of coal with oxygen carriers will be presented in this paper.

Thermo gravimetric studies and flow reactor studies were conducted with coal, coal char and carbon mixed with various metal oxide oxygen carriers in the presence of both nitrogen and carbon dioxide. X-ray diffraction studies and X-Ray photo electron studies were also conducted to understand the solid-solid interaction at various temperatures. Identification of reaction mechanisms and rate information obtained from these studies will be presented.

See more of this Session: Chemical Looping Processes - I
See more of this Group/Topical: Energy and Transport Processes

Modeling pyrite behavior in gasification environments

Esam Jassim, Steven A. Benson, Wayne S. Seames, and Michael D. Mann, Department of Chemical Engineering, 241 Centennial Drive Stop 7101, Grand Forks, ND 58202-7101

Abstract

The behavior of pyrite and other sulfide minerals in gasification processes are of fundamental importance to efficient and clean coal utilization technologies. Previous work has established that the physical and chemical transformations of pyrite and other impurities depend on mineral composition, mineral physical properties, the mineral association with coal particles, and the presence of organically associated inorganic elements. Further, the studies pyrite transformations found that under gasification conditions the existence of sticky reduced forms iron phases is prolonged and sulfur is released producing hydrogen sulfide. The properties of the iron rich pyrite derived phases impact the release of other elements and their ability to interact with other particles.

This paper is aimed at developing a more quantitative model of pyrite particle transformations during combustion and gasification processes. Modeling studies are being conducted to determine the impact of residence time and temperature on particle fragmentation, coalescence and release of volatile elements. The detailed particle model provides an accurate characterization of particle temperature, sulfur release rates, sulfur reactions, and pyrite conversion rates to intermediate species over wide ranges of gas temperatures, pressures, and compositions. This information can be used more accurately permit the accurately predict of pyrite mass loss rates, pollutant formation, and iron rich intermediate ash particle physical characteristics.

Preliminary results show that sulfur release is greatest early in the oxidation process as a result of the conversion of pyrite to pyrrhotite and pyrrhotite decomposition. In addition, the process of pyrite particle fragmentation was found to enhance the release of sulfur during the devolatilization process and to increase particle temperature as a result of sulfur oxidization. However, when the amount of oxygen in the bulk environment is reduced, the formation of hydrogen sulfide increases and pyrite particle fragmentation decreased.

1. Introduction

Coal will continue to play a major role in meeting energy demands for the next several decades in United States, China and India (Mitchell et al., 2007). Future coal-fired power plants must be highly efficient and likely equipped with carbon dioxide separation and capture systems to reduce greenhouse gas emission. Such aims can only be achieved by improving the coal conversion technologies through a clear understanding of the chemical and physical processes of coal gasification and combustion combined with accurate computational modeling of key processes.

Raw coal contains various types of impurities such as minerals included in the carbon matrix, organically associated inorganic elements, and excluded minerals not associated with the coal matrix (Benson et al., 1993 and Seames, 2003). The included mineral particles and organically associated impurities tend to interact and coalesce inside a coal particle during combustion (Yan et al., 2002). The excluded minerals particles tend to fragment during combustion (Lawrence et al., 1995).

The most common mineral forms in coal include: quartz, alumino-silicate clays, kaolinite, illite, calcite, and pyrite (Srinivasachar S. et al, 1989). The pyrite mineral grains have long been recognized as a host site for trace elements (Finkelman, 1994) and a key contributor to wall slagging in combustion systems (Lawrence et al., 1995).

Experiments have been performed to describe the thermal history of pyrite, using drop tube furnace (Groves S.J. et al, 1987; Asaki et al, 1985). These studies have developed a better understanding of the progress of decomposition of individual pyrite particles and showed that pyrite initially decomposes to pyrrhotite, and then melts to form liquid iron sulfides at temperatures > 1373K (Srinivasachar and Boni, 1989).

However since recent studies suggest that pulverized coal fly ash particle distribution is best described as a tri-model particle size distribution including: a submicron fume region, a fine fragmentation region centered at approximately 2.0 μm diameter, and a bulk fragmentation region (Seames, 2003), it becomes essential to include the fragmentation process in the excluded minerals combustion progress, which is the main theme of this study.

Fragmentation of excluded minerals into a number of smaller particles is important as it influences the particle size distribution of fly ash, particularly for a coal with a significant fraction of excluded minerals. For instant, Helble et al (1990) reported North Dakota lignite with a significant fraction of excluded pyrite exhibited a much finer PSD of combustion ash compared to that of minerals in the parent coal.

Yan et al (2001) have examined the fragmentation behavior of pure pyrite and calcite during coal combustion at 1300 °C using Poisson distribution to simulate the particle size variation resulting from the fragmentation of minerals grains. Their study concluded that pyrite likely generated four pieces of fragments, whereas calcite expected to produce three segments during fragmentation.

2. Model of Excluded Mineral Transformation

The fragmentation mechanisms associated with excluded mineral grain transformations has not been examined in detail for coal conversion systems. Although numerous studies have been performed to address the thermal history of excluded minerals, particularly pyrite, there is no complete model for predicting the progress of oxidation of exclusion element in coal gasification and combustion environments. In most studies, a limited number of steps in the transformation process have been included and sometimes the model is not consistent with experimental observations. For example, the model developed by Asaki et al. (1985) that assumed the formation of sulfur dioxide instead of sulfur monoxide, as the gaseous oxidation during pyrrhotite oxidation (Srinivasachar and Boni, 1989)

Figure 1 illustrates pathway for the excluded pyrite particle showing the particle transformations during combustion. The process can be summarized as follows:

i. Decomposition of pyrite to pyrrhotite

ii. Fragmentation

iii. Oxidation to molten oxysulphide, then magnetite and finally hematite.

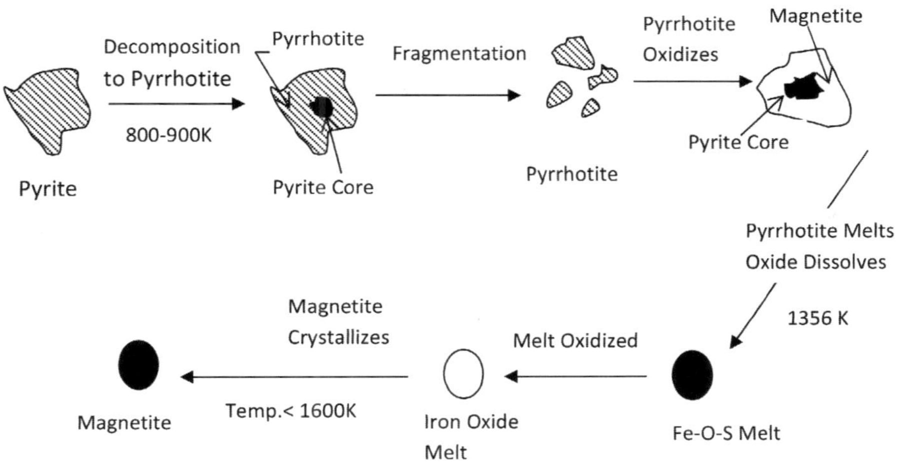

Figure 1. Reaction Pathway history for Excluded Pyrite (Sirnivasachar and Boni, 1989)

3. Mathematical formulas

The general transport equation used to track the particle temperature variation in terms of energy transferred to the particle by conduction and radiation, and the rate at which it is generated or consumed due to chemical reactions or phase changes is written as:

$$m_p \overline{C}_p \frac{dT_p}{dt} = \pi d_p^2 \left[h(T_g - T_p) + \sigma\varepsilon(T_w^4 - T_p^4)\right] + (-\Delta H)\dot{R} \tag{1}$$

where $\dot{R}$ is the reaction rate obtained by combining the intrinsic reaction and pore diffusion steps:

$$\dot{R} = \frac{(\pi d_p^2)(p_{O_2} / RT_g)(\phi)}{\frac{1}{k_i} + \frac{1}{k_{ext}} + \frac{1}{k_{int}}} \text{ where: } k_{ext} = \frac{ShD_m}{d_p} \text{ ;: } k_{int} = \frac{2D_{s2-pore}}{d_p} \tag{2}$$

and for small particle Reynolds number, Sh=2 and $Nu = \frac{hd_p}{k_m} = 2$ where Nu is the Nusselt number and k_m is the thermal conductivity of the gas.

The stages of the pyrite transformation process starting form ambient conditions are:

Stage-1 Pyrite preheat

Solid pyrite is heated from ambient temperature to decomposition temperature (~870K as suggested by Srinivasachar, 1989). No reactions occur during this stage. Hence the particle mass and size is assumed to remain constant.

Stage-2 Pyrite conversion to pyrrhotite.

Once the pyrite particle has reached the decomposition temperature, it decomposes to form pyrrhotite.

$$FeS_{2(s)} \rightarrow 1.14Fe_{0.877}S_{(s)} + 0.43S_{2(g)} \tag{3}$$

Sulfur is released until complete pyrite conversion to pyrrhotite (Fe1-xS) by 970K (Mitchell, 2002). Gaseous sulfur resulting from the dissociation of pyrite reacts with oxygen in the boundary layer. The heat derived from the exothermic reaction (ΔH=-109 kJ/mol O_2) is conducted back to the particle surface. Assuming the rate of sulfur leaving the particle equals the rate of diffusion of oxygen from bulk gas to the reaction front, the sulfur release rate is given by:

$$\dot{m}_s = \frac{\dot{Q}_{in}}{\Delta H_1} \text{ where} \tag{4}$$

$$\dot{Q}_{in} = \frac{4\pi k_{f1}(T_f - T_p)}{\frac{1}{r_p} - \frac{1}{r_f}} + 4\pi r_p^2 \sigma\varepsilon(T_f^4 - T_p^4) \tag{5}$$

$$\text{with } r_f = \frac{\dot{m}_s}{4\pi D_m C_{O_2}} \tag{6}$$

The reaction front temperature (T_f) can be obtained by solving the equations for the sulfur release rate and the quasi-steady state heat balance at the reaction front simultaneously, using iteration scheme of Newton-Raphson. The heat balance is based on the fact that the energy released by the sulfur reaction is transferred to particle and bulk environment:

$$\dot{Q}_{in} + \dot{Q}_{out} = 4\pi . D_m . C_{O_2} . r_f . (-\Delta H_2) \quad (7)$$

$$\dot{Q}_{out} = 4\pi r_f k_{f2} (T_f - T_g) + 4\pi r_f^2 \sigma (T_f^4 - T_w^4) \quad (8)$$

Stage-3 Fragmentation

Fragmentation of pyrite grains was observed during in combustion experiments. The fragmentation of a mineral grain during combustion is stochastic since the number of offspring particles formed from parent mineral grain is random (Yan L., 2000).

Physically, fragmentation is based on Poisson distribution. Assuming that particles produced during fragmentation are in the same size and composition, the probability that the original mineral grain fragments into $b + 1$ segment is given by (Yan L., 2000):

$$P_b = \frac{e^{-\lambda} \lambda^b}{b!} \quad \text{(b= 0, 1, 2, ...)} \quad (9)$$

where the Poisson parameter λ is the average net growth number of particles per parent particle after fragmentation, which was found from fitting experiment data. Yan et al (2001) suggested that pyrite generated four fragments as an average, while calcite produced three fragments.

Stage-4 Pyrrhotite melting

When the particle temperature reaches the melting point of pyrrhotite (~1356K), the pyrrhotite portion of the particle starts to melt (ΔH=341.15 kJ/kg) whereas magnetite on the surface that formed during oxidation, dissolves in the pyrrhotite melt.

$$1.14 Fe_{0.877} S_{(l)} + 1.2367 O_2 \rightarrow 0.33 Fe_3 O_{4(l)} + 1.14 SO_{(g)} \quad (10)$$

and ΔH= -190.1 kJ/mol O_2

Stage-5 Molten sulfide oxidation

During combustion process, the remaining sulfur in the Fe-S-O droplet is oxidized. Oxidization and particle temperature continue to increase in this stage until all the sulfide is consumed, leaving behind an oxide melt. Under gasification process, the remaining of the sulfur would be reacted with the hydrogen producing hydrogen sulfide.

4. Assumptions and Numerical Approach

For small pyrite particles, it is reasonable that the temperature distribution on the particle surface is uniform. Although appreciable loss of sulfur has been observed to begin at 700K and end by 970K (Srinivasachar and Boni, 1989), the decomposition processes are rapid enough to assume that it occurs at constant temperature (Lawrence et al, 1995).

The numerical solution involves solving the unsteady state heat balance for single particle using the Runge-Kutta-Fehlberg method in which the error in the solution can be estimated and controlled and an appropriate step size of time can be determined automatically.

After sulfur devolatilization, the pyrrhotite is exposed to the exothermic oxidation reaction that results in increasing the particle temperature to pyrrhotite melting point. As described model of fragmentation, the exothemic transformation stage causes the particle to breakup producing fragments.

4.1 CFD simulation

The pyrite particle is fixed in space in a freely convective air flow of high temperature and 1 atmosphere pressure environment. It is intended to study the process in both combustion and gasification perspectives. Accordingly, the mole concentration of O_2 in the study is chosen to be 21%, 5%, and 0.5% so that covers both combustion and gasification processes.

The CFD simulation is illustrated in Figure1a where a spherical particle placed in an open atmosphere, defined as length and width with about 10 particle diameters. The particle is exposed to standard atmospheric pressure (101.3 kPa) and temperature of 1400 K. The computational domain has to be selected in such a way so that it has minimal influence on the result. It was found from our studies that the far field effect becomes insignificant on the particle film temperature at a domain outer radius of $1.10r_p$ from the particle center. A Cartesian mixed grid of 7x4 in the boundary layer was employed to describe the temperature distribution in the film layer and compare it with the value estimated form the mathematical model. The thickness of the grid in the boundary layer was chosen to be $0.05r_p$ as concluded from Figure 1b.

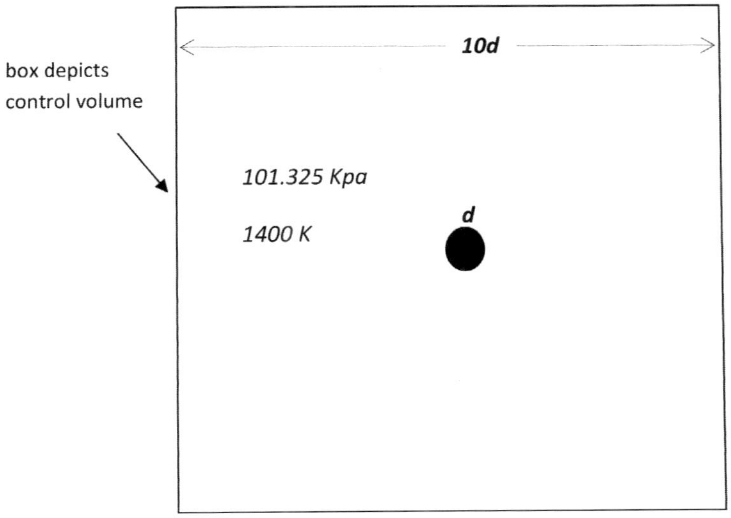

Figure 1a. Control volume shows the pyrite particle as injected into a hot gas environment

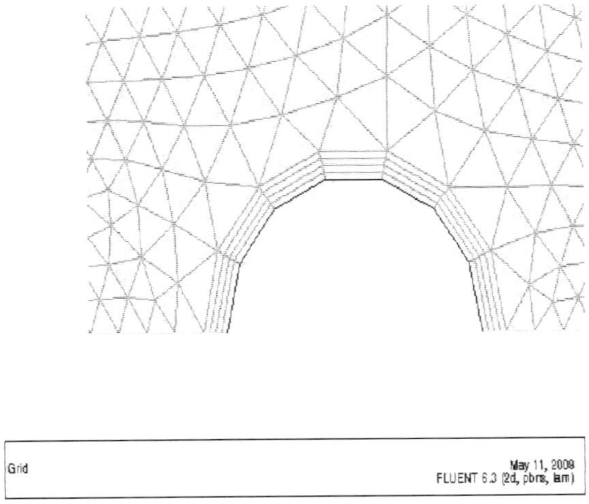

Figure 1b. Meshing scheme around axisymmetric particle surface

The control volume based technique consists of:

- The domain (as shown in Figure 1b) was divided into discrete control volumes using a computational grid;
- Integration of the governing equations over the individual control volumes was used to formulate non-linear algebraic equations for the discrete dependent variables of velocities, temperatures, species and reactions.

The solution algorithm for the non-linear algebraic equations involves:

- The linearization of the discretized equations and solution of the resultant linear equation system to yield updated values of the dependent variables starting with a given initial value and boundary conditions;
- The update continues until the scaled relative difference between two successive iterations has been reduced below a user-set threshold. In this study, the threshold value was set as

$$\frac{\left|\varphi^{k+1} - \varphi^{k}\right|}{\varphi^{k}} \leq 10^{-6}, \qquad (11)$$

where φ^{k} represents temperatures, species etc. of the kth iteration.

The computational domains and the partial differential equations describing the flow were discretized before being solved. Different meshing schemes were tested. It was found that mixed (quadrilateral and tri-geometrical) paving mesh converged quit fast although relatively larger number of control volumes is required.

Thus, an O-grid with a quadrilateral paving scheme was applied near the particle surface (boundary layer region) for the discretization of the computational domains, generating very fine resolution near the particle surface.

4.2 Grid dependency

The grid independence of the results, Figure 2, was achieved by changing the thickness of the boundary layer and comparing the average film temperature with the calculated value at selected time intervals. The influence of cell thickness and number of cells is shown in Figure 2. Three cases are compared namely, no boundary layer ($r_f/r_p = 1$); boundary layer less than film thickness; and boundary layer greater than film thickness. The latter is investigated for the optimal number of discrete. It is clearly shown from Figure 2 that improper meshing in the boundary layer results in alteration in the prediction of the film temperature which in turn affect the prediction of the amount of sulfur released.

In regard of the number of mesh lines in the boundary layer region, the estimation of average film temperature initially increases and then remains constant after about 4 lines of cells with $0.05r_p$ thickness. This indicates that increasing the number of cells (control volumes) beyond a certain value will not have any significant effect on the film temperature (see Figure 2), but will unnecessarily increase the computational time.

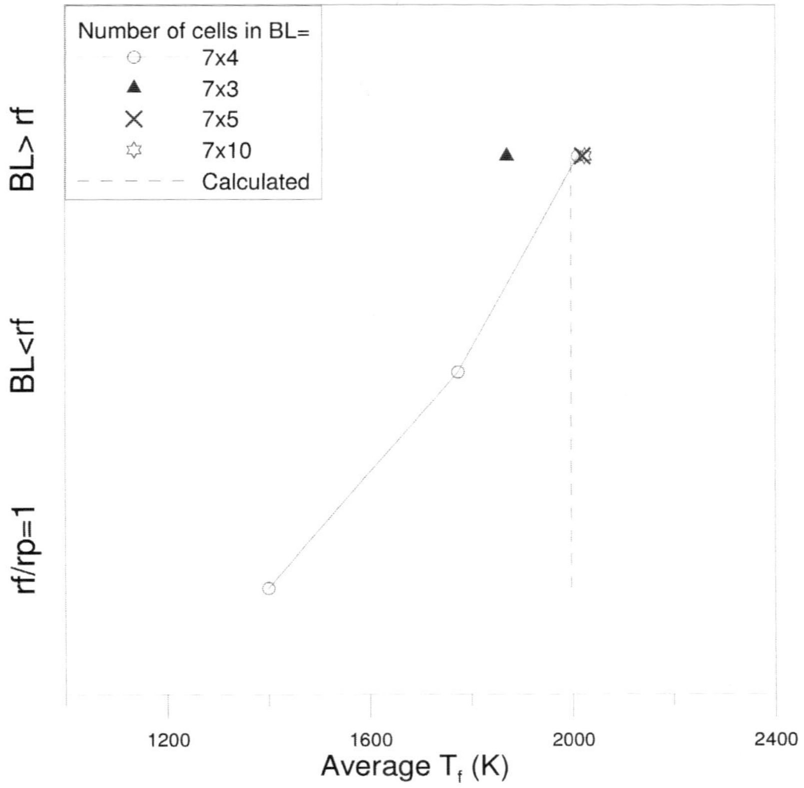

Figure 2. Grid dependency.

5. Results and Discussion

5.1 Temperature history

The computational fluid dynamics technique was developed to describe the transformation of a 45 μm diameter pyrite during combustion. Combustion conditions given in Table 1 are used as input parameters to the CFD software (FLUENT was used here). Figure 3 shows the profile of the particle temperature distribution for several time steps. The CFD simulation predictions clearly show a non-uniform temperature gradient at the particle surface that causes the particle to fragment. The average particle temperature of each time interval was estimated and compared with model prediction. Figure 4 illustrates the modeled thermal history of the pure pyrite particle initially injected in gas stream environment. The results indicate hat there is good agreement between the proposed models (Srinivasachar, 1989) and the CFD prediction, which gives an indication on the model reliability.

Table 1. Initial conditions for pyrite simulation.

Bulk temperature	1400 K
Wall temperature	1400 K
Pressure	101.325 Kpa
Oxygen mole fraction	0.21
Initial particle temperature	273
Particle emissivity	0.5

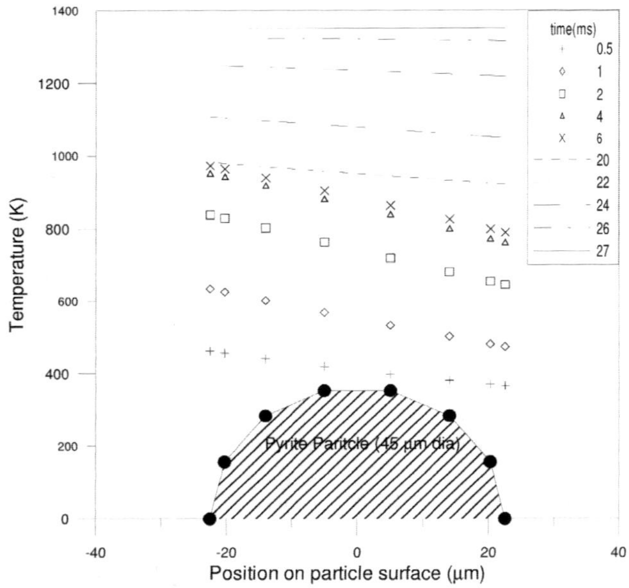

Figure 3. particle temperature distribution history predicted by CFD.

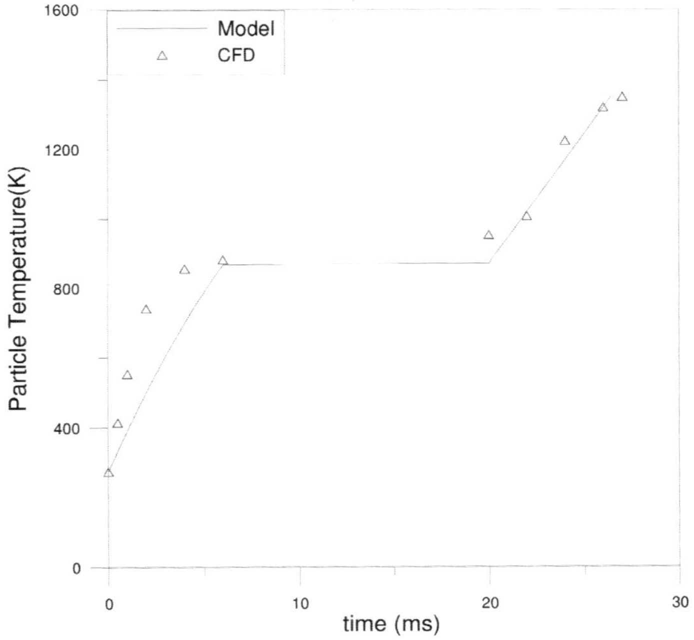

Figure 4. Comparison of particle temperature predicted by the proposed Model (Srinivasachar, 1989) and CFD.

5.2 Effect of fragmentation

The fragmentation behavior of pure pyrite during combustion is examined using the approach discussed earlier. The particle temperature with and without fragmentation is plotted in Figure 5. The figure shows that fragmentation results in an increase in the temperature of the fragmented particles decreases the duration of the melt phase during oxidation. Further, the fragmentation may increase the peak gas temperature perhaps to the level of melting point of magnetite and hematite causing sticking.

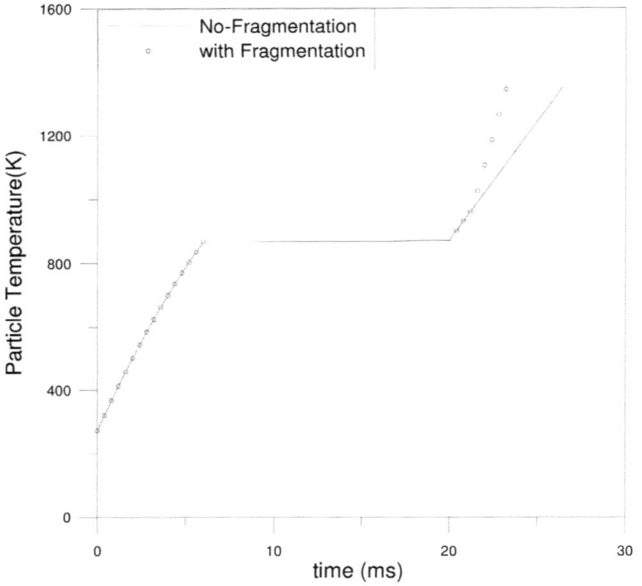

Figure 5. Particle thermal history with fragmentation.

When the particle temperature increases due to heat transfer from the bulk flow, sulfur begins to be released from the particle and form around the particle surface in the boundary layer region. Heat produced from the reaction of released sulfur with the bulk oxygen in the boundary layer region is fed back onto the particle surface, causing further increase in the particle temperature and devolatilization rate. The sulfur released is defined as the amount of sulfur removed from the particle, referenced by the original amount of sulfur contain in pyrite. The sulfur release rate is investigated and compared with CFD prediction. It was found that fragmentation increases the amount of sulfur released and that amount is greatest early in the oxidation process specifically at the formation and destruction of pyrrhotite. As shown by Figure 6, the data obtained from the model (Srinivasachar, 1989) is in excellent agreement with the CFD simulation.

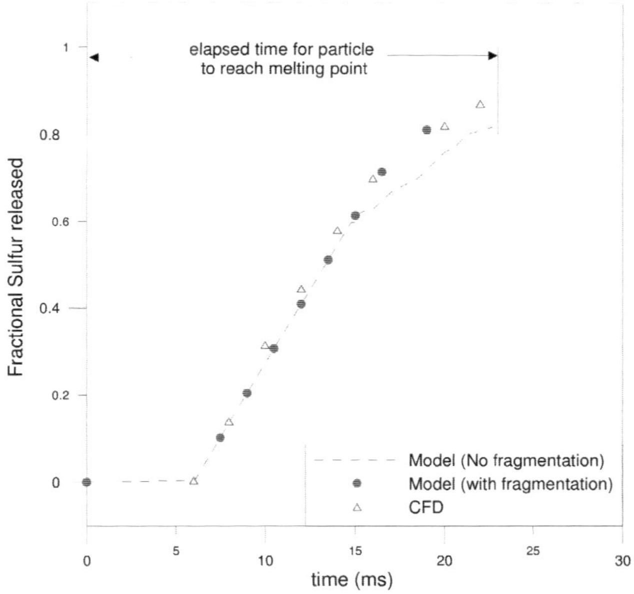

Figure 6. Percent of Sulfur released with fragmentation.

5.3 Effect of oxygen content

The concentration of oxygen in the bulk environment was examined to simulate gasification conditions. The oxygen level strongly influenced the amount of sulfur released from the pyrite particle. Figure 7 illustrates the influence of oxygen concentration on the releasing of sulfur. The time required for 60% sulfur release and almost completely released of sulfur for different concentration of oxygen namely 0.5%, 5%, and 21% are plotted and compared. As the bulk oxygen content decreases, the time required by the sulfur to completely release is significantly increases to 1000 ms as compared to 40 ms, for the 0.5 and 21% oxygen cases, respectively. This could be returned to two reasons; the first is low oxygen concentration means less penetration and diffusion of oxygen to the depth of particle to replace the sulfur released and the second reason refers to the quantitatively reduction in the exothermic sulfur/oxygen reaction due to low oxygen concentration results in decreasing the incremental rate in particle temperature. Hence it can be concluded that the time elapsed to remove sulfur is almost inversely proportion to the oxygen concentration in the bulk environment.

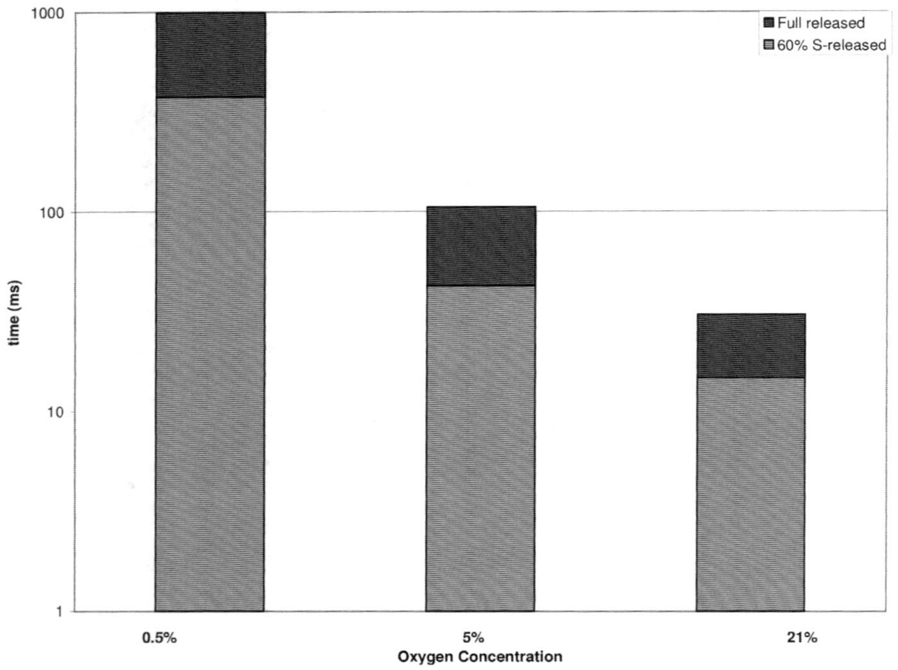

Figure 7. Time required to release sulfur as a function of oxygen content.

6. Conclusion

A mechanistic model of pyrite fragmentation during combustion and gasification processes was examined. The combustion and gasification processes influence the pyrite thermal history, fragmentation behavior, and rate of sulfur release. For example, an excluded pyrite particle exposed to combustion conditions results in particle fragmentation producing smaller size particles and releasing more sulfur. Under reducing conditions the rate of sulfur release is significantly reduced that will result in more low melting point sulfide and reduced iron phases. In addition, the degree of fragmentation was decreased.

The results of the CFD model were compared to other modeling efforts. The results of the comparisons showed good agreement.

Experimental data are needed to validate and enhance the model in the future work.

Acknowledgment

This material is based upon work supported by the Department of Energy office of Science under Award DE-FG02-06ER46292 (ND state grant, "coal combustion emission mechanisem").

References

Akan-Etuk, A. and Niksa, S., Paper No. 89-82 WSS/CI Fall Mtg., Livermore, CA, October 1989.

Akan-Etuk, A., "Pyrite Transformations During Pyrolysis," Report prepared under the Supervision of Prof. S. Niksa, Mechanical Engineering Department, Stanford University, 1990.

Asaki Z, Mori S, Ikeda M & Kondo Y "Oxidation of pyrrhotite particles falling through a vertical tube", Metallurgical Transaction B 16B: 627-638, 1985.

Benson, S.A.; Jones, M.L.; Harb, J.N. Ash Formation and Deposition. In Fundamentals of Coal Combustion for Clean and Efficient Use; Smoot, D.L., Ed.; Coal Science and Technology 20 Series; Elsevier: Amsterdam, 1993; Chapter 4, pp 299–373.

Finkelman, R.B., Modes of occurrence of potentially hazardous elements in coal: levels of confidence, Special Issue: Trace Elements Transformations in Coal-Fired Power Systems, Ed. S.A. Benson, E.N. Steadman, A.K. Mehta, and C.E. Schmidt, Fuel Processing Technology, 39, 21-24, 1994.

Groves S.J., Williamson J., and Sanyal A., "Decomposition of pyrite during pulverized coal combustion", Fuel 66, 461-466, 1987.

Helble J.J., Srinivasachar S., and Boni A., "factors influencing the transformation of minerals during pulverized coal combustion", Prog. Energy Combustion science, 16, 267-279, 1990.

Jorgensen F.R.A., "On Maximum Temperatures Attained During Single Particle Combustion of Pyrite," *Trans. Instn. Min. Metall.* C 90, pp. 10–16, 1981.

Lawrence E. Bool III, Thomas W., Jost O.L., "The partitioning of Iron during the combustion of pulverized coal", Combustion and Flame 100, 262-270, 1995.

Mitchell R.E., Ma L., Kim B., "On the behaviour of pulverized coal chars", Combustion and Flame 151, 426-436, 2007.

Mitchell R.E., "Mechanisms of Pyrite oxidation to Nonslagging Species", Report submitted to National Energy Technology Laboratory, The United States Department Of Energy,DE-FG22-94PC94205, Stanford University, 2002.

Rassak E., "Creation, Capture and Coalescence of Mineral Species in Coal Flame", Journal of the Institute of Energy, v57, 231-239, 1984.

Seames W.S., "An initial study of the fine fragmentation fly ash particle mode generated during pulverized coal combustion", Fuel process Technology, 81, 109-125, 2003.

Srinivasachar S., and Boni A., "A kinetic model for pyrite transformations in a combustion environment", Fuel 68, 829-836, 1989.

Srinivasachar S., Helble J.J., Boni A., Shah N., Huffman G., and Huggins F., "Mineral behaviour during coal combustion 2.Illite transformations", Progress Energy combustion Since 16, 293-302, 1990.

Ten Brink H.M., Eenkhoorn S., and Weeda M., "Flame transformations of coal-siderite", proceedings of engineering foundation conference on the impact of Ash Deposition on coal fired plants, Taylor & Francis, 373-383, 1993.

Ten Brink H.M., Eenkhoorn S., and Weeda M., "The behaviour of coal mineral carbonates in a simulated coal flame", Fuel Processing technology, v47, 233-243, 1996.

Tukey J.W., "Moments of Random Group Size Distribution", Annals of Mathematical Statistics, v20, 523-539, 1949.

Yan L., Gupta R.P., Wall T.F., "A mathematical model of ash formation during pulverized coal combustion", Fuel 81, 337-344, 2002.

Yan L., Gupta R.P., Wall T.F., "Fragmentation behaviour of pyrite and calcite during high-temperature processing and mathematical simulation", Energy &Fuels 15, 389-394, 2001.

Yan L. CCSEM analysis of minerals in pulverized coal and ash formation modeling, PhD thesis, The University of Newcastle, 2000.

Effects and Roles of Air Feed on the Noncatalytic Gasification of Jet Fuel in Supercritical Water

Jason Picou, Jonathan Wenzel, Michael Stever, Jared Bouquet, and Sunggyu Lee, Missouri University of Science and Technology, Rolla, MO

Abstract

The air oxidation of jet fuel can be beneficially incorporated in the novel supercritical water reformation. This paper experimentally and comparatively examines the roles of the concurrent oxidation reactions on the overall process chemistry, and specifically the effect of air feed position. The effect of air feed position, either at the inlet of the 48-inch long tubular reactor where the water/fuel mixture also enters, or a mid-reactor feed position 21.8" downstream from this point, on the noncatalytic reformation of jet fuel was evaluated using a tubular 0.4-L Haynes® Alloy 230 supercritical water reactor. The temperature, pressure and water-to-fuel ratio were kept nominally identical, while the air feed position and air flow rate were varied in order to elucidate its effect on reformation. Over the conditions examined, the mid-reactor air feed achieved higher hydrogen and carbon dioxide yields than the inlet air feed position for the same air feed rates, while the carbon conversion and methane yield for both feed positions was similar. The result indicates that the air feed position has little effect over the pyrolysis reaction while the mid-reactor air feed enhances the oxidation and water gas shift reactions more than the inlet air feed.

Introduction

The ability to efficiently produce hydrogen from a wide variety of hydrocarbon sources on demand from a small, portable apparatus is important to the future viability of the hydrogen economy. One of the more difficult to reform, but frequently used hydrocarbons, is jet fuel. The difficulty of reforming jet fuel is mostly due to the high concentration of sulfur, a variety of fuel additives used, the variety of hydrocarbon compounds and the longer chain and cyclic hydrocarbons that constitute jet fuel. Sulfur is a traditional poison for metallic catalysts, which are used in most reformation processes.

These inherent technological difficulties are overcome in a novel, non-catalytic approach using supercritical water partial oxidation for the reformation of jet fuel. In this process, supercritical water functions as a highly energized reforming agent, or a gasifying reactant, and also as a homogenizing reaction medium. Supercritical water is a non-polar solvent unlike its counterpart ambient water, allowing hydrocarbon molecules to be miscible with water molecules in all proportions, as well as oxygen.[1,2] The absence of a heterogeneous noble metal catalyst negates sulfur poisoning and coke fouling. The addition of oxygen in the form of air provides *in-situ* exothermic heat of combustion to supplement the endothermic reformation reaction. The purpose of these experiments is to determine if introducing air at a location farther inside the reactor, as opposed to the inlet of the reactor where the fuel and water mixture enters, will improve the process efficiency of the non-catalytic supercritical water reformation of jet fuel.

Apparatus & Chemicals

The water used in the reformer was deionized water, and the air used was Airgas Breathing Quality Grade D compressed air. The jet fuel used was both civilian jet fuel (Jet-A) and aviation logistic fuel (JP-8), both of which are an assortment of hydrocarbons including straight chain, branched and cyclic. An ASTM D2887 boiling range distribution analysis determined that the length of the carbon bonds varied from 7 to 17 carbons, with the average being 12 for both fuels. Therefore, the fuel was modeled as a single representative molecular species, n-dodecane, which has a chemical formula of $C_{12}H_{26}$. Both jet fuels were tested by Texas Oil Tech Laboratories for sulfur content by ASTM D4294 and found that the Jet-A contained 990 ppm sulfur and the JP-8 contained 810 ppm sulfur. Even though two different jet fuels were used due to difficulty in obtaining additional JP-8, the two samples were considered nominally the same from the standpoints of the fuel reformation and the specific process employed.

The supercritical water reformation system consists of a liquid feed system, feed preheater, reactor, reactor heaters, air feed system, sample collection system, and data acquisition and control system, of which a schematic process flow diagram is illustrated in Figure 1. The supercritical water reactor has a volume of 383 mL, and is constructed of Haynes® Alloy 230, which is an alloy of mostly nickel, chromium, tungsten, and molybdenum, among other elements.[3] The stress-strain properties, as well as creep and rupture properties, of Haynes® Alloy 230 allow the reactor to be operated at temperatures up to 800°C at a pressure of 36.2 MPa.

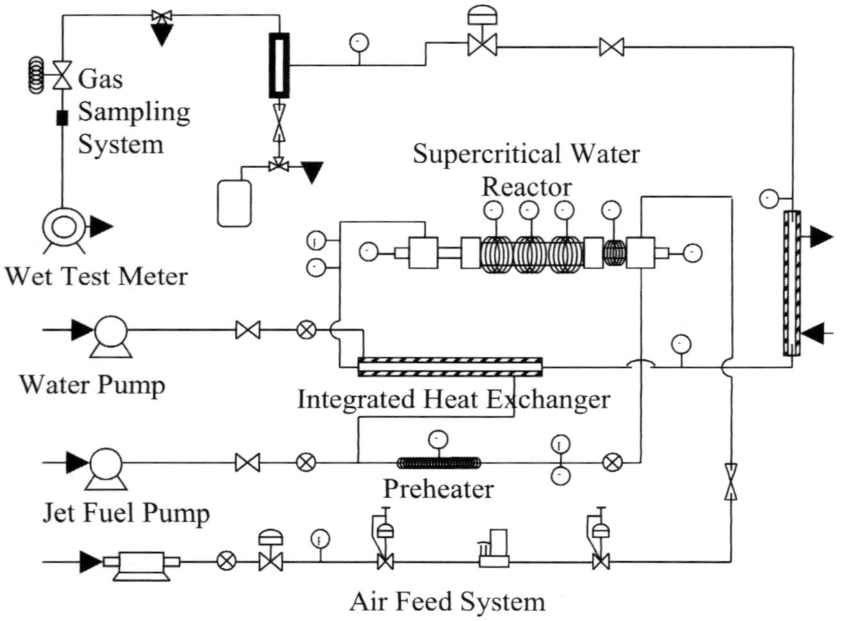

Figure 1. A schematic of the supercritical water reformation system at the Laboratory of Transportation Fuels and Polymer Processing, Missouri University of Science and Technology.

Three of the experiments were conducted with air fed into the inlet cross, directly across from where the fuel and water mixture enters, as shown in Figure 2. Three of the experiments were conducted with the air inlet inside the reactor 21.8" past the original feed spot. The first set initiates the partial oxidation of jet fuel at an entrance portion of the reactor, whereas the latter set brings about the partial oxidation of the reaction mixture in the mid-section of the reactor. Analysis of the gaseous effluent was performed using a HP 5890 Series A gas chromatograph with a thermal conductivity detector. The gas chromatograph is calibrated to detect hydrogen, nitrogen, carbon monoxide, methane, carbon dioxide, acetylene, ethylene and ethane. The carrier gas for the gas chromatograph was Airgas Ultra High Purity Argon with a purity of 99.9995%.

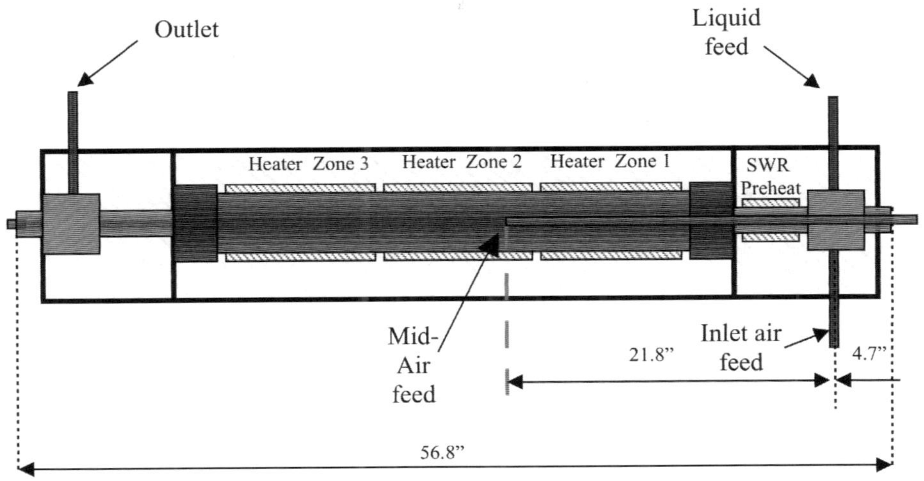

Figure 2. Air feed positions and lengths of the Haynes® Alloy 230 supercritical water reformation reactor at the Missouri University of Science and Technology.

Chemical Reactions

A variety of reactions are conceivable in the supercritical water reformation process, the most significant of which are listed below.[4] The overall reformation reaction of jet fuel may be written as:

$$C_{12}H_{26} + 12\ H_2O \rightarrow 12\ CO + 25\ H_2 \qquad \Delta H^{\circ}_{298} = 1866.2\ \text{kJ/mol} \qquad (1)$$

where jet fuel is represented by a model compound of n-dodecane for stoichiometric simplicity based on its average molecular weight. The above reaction is in competition and occurs in parallel with the pyrolysis reaction:

$$C_{12}H_{26} \rightarrow C_aH_b + C_xH_y + zH_2 \qquad (12 = a + x \text{ and } 26 = b + y + 2z) \qquad (2)$$

The endothermic pyrolysis reaction is thought to be responsible for any gaseous light hydrocarbons contained in the effluent gas. Repeated pyrolysis leaves hydrogen deficient fractions, which eventually become solid coke.[5]

In the presence of oxygen, another set of reactions occur. The first (reaction 3) is the partial oxidation of hydrocarbons, while the second (reaction 4) is the complete oxidation reaction:

$$C_{12}H_{26} + 6\ O_2 \rightarrow 12\ CO + 13\ H_2 \qquad \Delta H^{\circ}_{298} = -1035.6\ \text{kJ/mol} \qquad (3)$$

$$C_{12}H_{26} + 18\tfrac{1}{2}\ O_2 \rightarrow 12\ CO_2 + 13\ H_2O \qquad \Delta H^{\circ}_{298} = -7575.0\ \text{kJ/mol} \qquad (4)$$

In addition to these reactions, the water gas shift reaction also takes place. The forward reaction is a highly desirable reaction, since additional hydrogen is produced. The forward reaction is thermodynamically favored at temperatures of 815°C or below:[6]

$$CO + H_2O \leftrightarrow CO_2 + H_2 \qquad \Delta H^{\circ}_{298} = -41.2 \text{ kJ/mol} \qquad (5)$$

None of these reactions occur alone or independently as isolated events, nor are they mutually exclusive. For example, pyrolysis or oxidation could break down the original jet fuel hydrocarbons, then reformation occurs on the resulting products. Various other reactions, like methanation or the Boudouard reaction, are also possible, even though the thermodynamic equilibrium for the forward Boudouard reaction is not favorable for the process conditions of the current study. The discussion therefore is limited to the aforementioned reactions for simplicity and since they effectively describe all of the products observed.

Experimental

The air feed position and the feed rate of air was varied to investigate the effects upon the reformation of jet fuel in supercritical water. The temperature was held at 769.2 ± 0.9°C, while a constant pressure of 24.05 ± 0.02 MPa and a 15:1 water-to-fuel mass ratio was maintained. This corresponds to a water-to-carbon (H_2O/C) molar ratio of about 12, or a jet fuel concentration of 6.25 weight percent. Table 1 gives the experimental run number, temperature, pressure, fuel type, and water, fuel and air flow rates along with the air feed position for each experiment. The inlet air feed position is when the air feed is introduced directly across from the fuel-water mixture feed in the inlet cross, while mid or middle air feed is when the air is fed 21.8" downstream from this point. The experiments were conducted in the order given by the Run ID, from one to thirteen. Most of these experiments were performed without any air feed, in order to determine the effects of both normal and mid-reactor air feed on the reactions, as well as to determine the reproducibility of the experiments.

Table 1. Jet fuel reformation experiments in supercritical water. Water-to-fuel mass ratio is constant at 15:1, temperature at 769.3 ± 0.9°C, and pressure at 24.03 ± 0.02 MPa. The air flow rate and air feed position are varied.

Run No.	Temp (°C)	Pressure (MPa)	Water (g/min)	Fuel Type	Fuel Flow (g/min)	Air Flow (slpm)	Air feed position
13	768	23.9	14.5	JP-8	1.08	0.00	none
9	767	24.0	15.9	JP-8	1.01	0.00	none
8	769	24.0	14.1	JP-8	1.04	0.00	none
7	770	24.0	15.1	JP-8	0.93	0.00	none
3	770	24.1	15.4	JP-8	0.99	0.00	none
2	780	24.1	15.3	Jet-A	0.99	0.00	none
1	767	24.1	15.3	Jet-A	0.91	0.00	none
6	772	24.0	15.1	JP-8	0.95	0.50	inlet
12	766	24.0	15.1	JP-8	1.09	0.50	mid
5	772	24.1	15.1	JP-8	0.95	1.00	inlet
11	766	23.8	15.0	JP-8	1.08	0.99	mid
4	765	24.1	14.3	JP-8	1.00	3.01	inlet
10	769	24.1	15.7	JP-8	1.15	2.97	mid

Results and Discussion

The results of the experiments outlined in Table 1 are shown in Table 2, which lists the flow rate of gaseous components of the reactor effluent at ambient conditions (72°F and 1 atm) and the carbon gasification percentage. The carbon gasification percentage is defined as the amount of carbon in the gaseous reactor effluent divided by the carbon that entered as jet fuel.

Table 3. Gas effluent volumetric flow rates as measured at ambient conditions and the carbon gasification percentage for the experiments conducted.

Run #	Gas flow rate (L/min)							Carbon Gasification
	H_2	N_2	CO	CH_4	CO_2	C_2H_6	Total	
13	0.67	0.00	0.09	0.77	0.26	0.02	1.82	62.6%
9	0.59	0.00	0.13	0.71	0.19	0.01	1.64	60.5%
8	0.60	0.00	0.13	0.78	0.18	0.00	1.69	60.8%
7	0.57	0.00	0.12	0.76	0.21	0.02	1.68	71.0%
3	0.51	0.00	0.11	0.70	0.15	0.02	1.49	58.8%
2	0.57	0.00	0.10	0.70	0.21	0.01	1.60	60.8%
1	0.72	0.00	0.08	0.66	0.30	0.01	1.77	67.3%
6	0.58	0.45	0.15	0.69	0.25	0.02	2.13	68.2%
12	0.82	0.43	0.12	0.77	0.37	0.02	2.53	69.8%
5	0.56	0.88	0.18	0.66	0.31	0.01	2.60	71.4%
11	0.91	0.81	0.12	0.73	0.49	0.02	3.08	74.0%
4	0.70	2.44	0.25	0.60	0.65	0.00	4.64	87.2%
10	1.05	2.44	0.10	0.66	0.78	0.00	5.04	78.6%

The results of these experiments are shown in the following figures, for ease of comparison. Figure 3 is the carbon conversion as a function of increasing oxygen-to-carbon ratio. The oxygen-to-carbon ratio, since the fuel flow rate is kept constant, is a measure of air flow. Experiments with no air flow have a zero oxygen-to-carbon ratio.

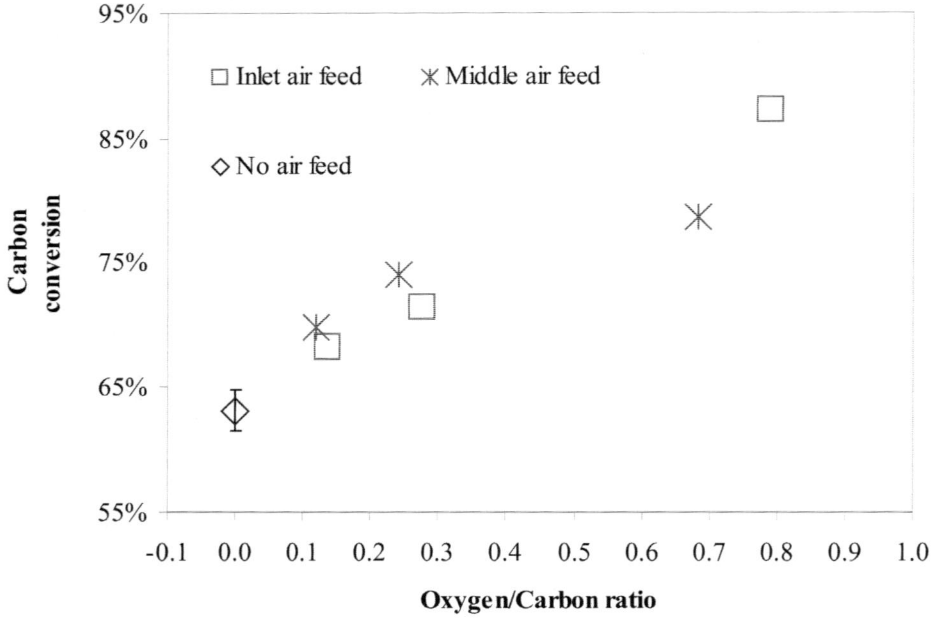

Figure 3. Carbon conversion as a function of oxygen-to-carbon ratio for both normal and middle reactor air feed.

As shown in Figure 3, the carbon conversion increases with increasing oxygen-to-carbon ratio for both air feed methods. The seven experiments performed without any air flow have a mean carbon conversion of 63.1%, with a standard error of 1.7%, which demonstrates the reproducibility of these experiments. When air is fed at the inlet position the carbon conversion is 68.2% at an oxygen-to-carbon ratio of 0.14, increases to 71.4% at 0.28, and reaches a maximum of 87.2% at a 0.79 oxygen/carbon ratio for the experiments conducted for this flow configuration. The middle air feed exhibits a similar progression in regards to carbon conversion as a function of oxygen feed. The carbon conversion is 69.8% at the lowest oxygen-to-carbon ratio, increases to 74.0% at 0.24, and is 78.6% at the highest oxygen-to-carbon ratio of 0.68 used for the mid-air feed experiments. This indicates that the air feed position has little effect on the carbon conversion for the conditions examined.

Figure 4 illustrates the hydrogen and carbon monoxide yields for the experiments conducted. The gas yield is calculated as the molar flow rate of that particular gaseous species out of the system divided by the molar flow rate of jet fuel into the system resulting in a dimensionless number that accounts for variations in the inlet fuel flow rate.

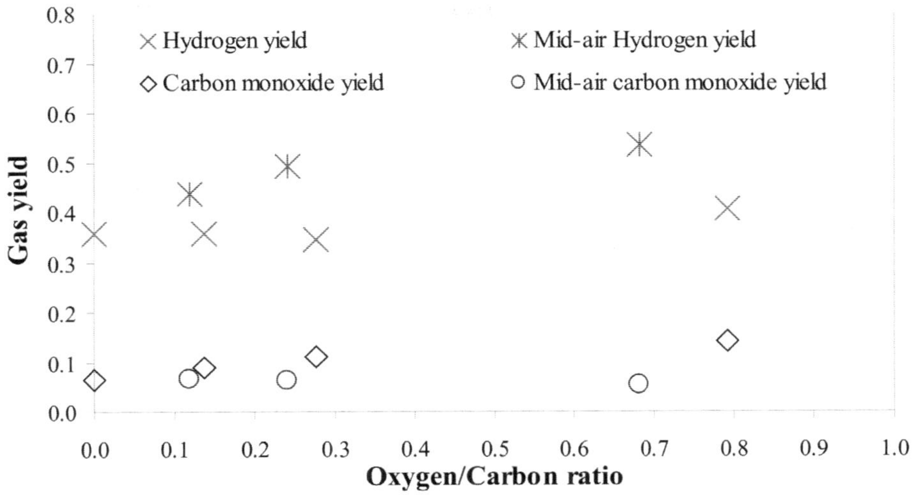

Figure 4. The hydrogen and carbon monoxide gas yield for changing oxygen to carbon ratios with normal and middle reactor air feed.

In Figure 4, the hydrogen and carbon monoxide yields for the normal air flow path and the experiments without air flow are combined for simplicity, the experiments with no air flow have a oxygen-to-carbon ratio of zero. The results for the air flow that began 21.8" from the inlet air entrance is denoted by the prefix "Mid air" while the inlet air feed has no prefix. Beginning with the baseline experiments with no air flow, the hydrogen yield was 0.36 ± 0.02 and 0.065 ± 0.004 for carbon monoxide, which is a 5% and 7% standard error respectively.

The hydrogen yield for inlet air flow is 0.36 at an oxygen-to-carbon ratio of 0.14 and 0.34 at a 0.27 O_2/C ratio. The hydrogen yield was 0.41 at the highest O_2/C ratio of 0.79, which is the highest hydrogen yield obtained at the inlet air feed position. When air is feed into the middle of the reactor, hydrogen yield increases with increasing oxygen-to-carbon ratio, with a yield of 0.44 at a 0.12 O_2/C ratio, 0.49 at a 0.24 O_2/C ratio, and a yield of 0.53 at a 0.68 O_2/C ratio. These mid-reactor hydrogen yields are 20% greater than the inlet hydrogen yields at similar oxygen-to-carbon ratios. While the feed location of air does not substantially affect the carbon conversion, as shown in Figure 3, it does affect hydrogen yield as shown in Figure 4.

The carbon monoxide yield, increases with increasing oxygen-to-carbon ratio for the normal air feed and decreases for the middle air feed. With normal air flow, the carbon monoxide yield goes from 0.09 at the lowest O_2/C ratio to a maximum at 0.14 at the highest O_2/C ratio. The middle air feed setup resulted in a carbon monoxide yield decrease from 0.07 at the lowest O_2/C ratio to 0.05 at the highest O_2/C ratio. These results indicate that either less carbon monoxide is produced yet more hydrocarbons are being converted to hydrogen in the mid-air feed system, or that the water gas shift reaction is more active, consuming carbon monoxide to produce more carbon dioxide and hydrogen gas.

The other major species present in the gaseous effluent are methane and carbon dioxide. Figure 5 illustrates the effects of the air feed position upon the methane and carbon dioxide yield for increasing O_2/C ratio, which complements the discussion of some of the effects presented in the Figure 4. It should be noted that the ethane yield for the different air feed positions is not compared in the figures due to its small yield as compared to the other gaseous species; however it is listed in Table 2.

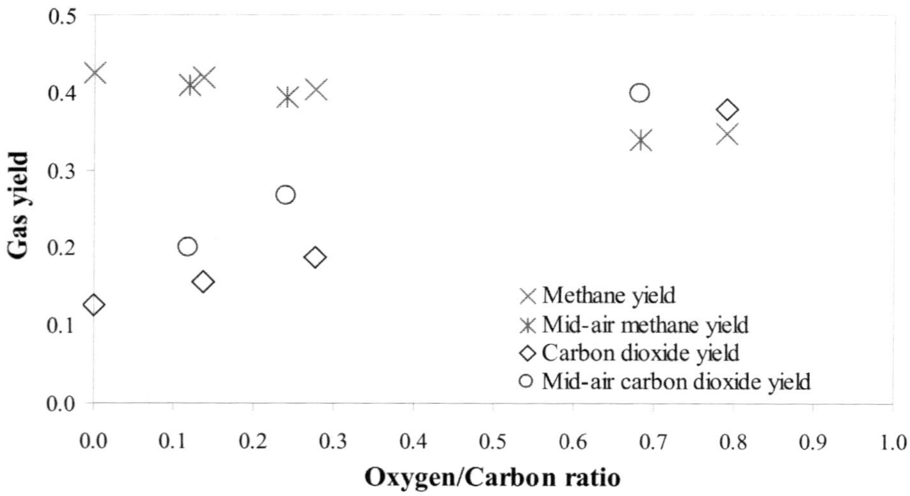

Figure 5. The methane and carbon dioxide gas yield for changing oxygen to carbon ratios with normal and middle reactor air feed.

As with Figure 4, the data points in Figure 5 for air fed into the middle of the reactor have "Mid-air" as a prefix while the data points for air fed at the inlet cross of the reactor do not have a prefix. The methane yield is 0.425 ± 0.009 and the carbon dioxide yield is 0.13 ± 0.01 with no air flow. Methane yields for both air feed systems decrease for increasing air feed, and the methane yield for any particular oxygen-to-carbon ratio are similar between the two feed systems. For the inlet air feed the methane yield decreases from 0.42 to 0.41 to 0.35 with increasing oxygen-to-carbon ratios. The middle reactor feed had a methane yield decreasing from 0.41 to 0.39 to 0.34 with increasing oxygen-to-carbon ratios. These results are nearly identical, and it may be concluded that the air feed position has limited impact on the production of methane.

In comparison to methane, carbon dioxide has a very different result. For both air feed positions the carbon dioxide yield increases with increasing oxygen-to-carbon ratio, which is expected since the oxidation reaction would increase with increasing oxygen feed. The normal air feed position increases from 0.16 to 0.19 to 0.38 for the highest O_2/C ratio, while the carbon dioxide yield for the mid-reactor air feed increases from 0.20 to 0.27 to 0.40 at the highest O_2/C ratio for this feed position. For the oxygen-to-carbon ratios of about 0.13 and 0.25, the middle reactor air feed had a carbon dioxide yield between 20% to 30% greater than the normal air feed. This indicates that either the oxidation reaction produces more carbon monoxide with air fed at the middle of the reactor, or that the forward-water gas shift reaction is converting more carbon monoxide into carbon dioxide and hydrogen, or both. The fact that the carbon monoxide yield remains relatively constant for air fed at the middle of the reactor, while hydrogen and carbon dioxide yields increase faster than normal air feed, indicates that the forward-water gas shift reaction contributes to the increased hydrogen yield.

Conclusion

The effect of air feed position, either at the point where the fuel and water mixture enters a tubular 0.4-L Haynes® Alloy 230 supercritical water reactor or 21.8" into the reactor from this point, on the supercritical water reformation and oxidation of jet fuel was studied using a 0.4-L Haynes® Alloy 230 supercritical water reactor. The temperature, pressure and water-to-fuel mass ratio were kept nominally identical at 769.3 ± 0.9°C, 24.03 ± 0.02 MPa and 15:1 respectively, while the air feed position and amount of air fed was varied. Seven duplicate experiments were conducted without air feed to determine the reproducibility of the experiments as well as to give a comparison to determine the effects of air flow. Carbon conversion, the amount of carbon that exits the reactor as gas divided by the amount of carbon in the fuel that entered, was not noticeably affected by air feed position, and for both feed positions carbon conversion increased with increasing air flow. The highest carbon conversion was 87%, which was achieved at a 0.79 oxygen-to-carbon ratio. The gas flow rate changed with increasing oxygen-to-carbon ratio, with hydrogen and carbon dioxide increasing and methane decreasing for both air feed positions. Hydrogen yield was 20% greater and carbon dioxide yield 30% greater when air was feed into the middle of the reactor when compared to inlet air flow. Carbon monoxide yield was less for mid-air feed, and methane yield in general was unchanged. The pyrolysis and methanation reactions are not affected by the air feed position, but the forward-water gas shift reaction and the oxidation reactions are affected and are more active with middle air feed in comparison to feeding air where the fuel and water enter the reactor.

Acknowledgements

This work was sponsored by the U.S. Army under PE Number 0602705A through a subcontract from DRS Technical Services, Inc. The authors are grateful for insightful technological guidance and supervision provided by Dr. Terry Dubois, CERDEC, U.S. Army, Ft. Belvoir, Virginia and Dr. H. Brian Lanterman of DRS Technical Services.

References

(1) Holliday, R.L.; Jong, B.Y,M.; Kolis, J.W. Organic synthesis in subcritical water: Oxidation of alkyl aromatics. *J. Supercrit. Fluids* **1998**, 12, 255-260.
(2) Klijn, J.E.; Engberts, J.B.F.N. Fast reactions 'on water'. *Nature* **2005**, 435, 746-747.
(3) *Haynes® 230 Alloy*, Haynes High-Temperature Alloys International, 2004 pp 2.
(4) Picou, J. W. Autothermal Non-Catalytic Reformation of Jet Fuel in a Supercritical Water Medium. M.S. Thesis, Missouri University of Science and Technology, Rolla, MO, June 2008.
(5) Lee, S., Lanterman, H., Wenzel, J., Edwards, N., Adams, A., Wootton, J., Garcia, A. Noncatalytic direct reformation of JP-8 fuel in supercritical water. *Prepr. Pap. Am. Chem. Soc., Div. Pet. Chem.* **2006**, *51*, 521-523
(6) Lee, S.; *Alternative Fuels*, Taylor & Francis, Philadelphia, 1996; pp.140-151.

A Framework for the Development of Dynamic Multi-Zonal Gasification Reactor Models

David C. Miller, Collaboratory for Process & Dynamic Systems Research, National Energy Technology Laboratory, U.S. Department of Energy, Morgantown, WV

Sara E. Zenner, Gasification Division, National Energy Technology Laboratory, U.S. Department of Energy, Morgantown, WV

Gasification of coal, coke and/or biomass offers a significant opportunity to cleanly produce energy, liquid fuels and/or chemicals. In order to understand how to best design gasification systems, accurate and computationally efficient models of the gasification reactor are required. Currently, most modeling approaches focus on either highly sophisticated computational fluid dynamics (CFD) models or overly simplified equilibrium reactors. While the former approach offers high accuracy, it suffers from long computation times that limit its ability to be used within a flowsheet process simulator. The latter approach improves calculation speed at the expense of detail and accuracy. Many simulations employing the equilibrium model approach make use of restricted equilibrium or approach temperature to better match data from experiments or more detailed models; however, these approaches typically limit the model applicability to a single operating point. Multi-zonal models offer a balance between the computational expense of a detailed CFD model and the speed of a simple equilibrium model. This paper will describe a framework for developing dynamic multi-zonal gasification reactor models. This approach can be implemented using CFD results, experimental data or a combination of the two. The effectiveness of the approach will be demonstrated using the equipment and process model development and simulation environment, Aspen Custom Modeler®.

See more of this Session: Advances in Gasification Research - II
See more of this Group/Topical: Energy and Transport Processes

Mathematical Modeling of a Transient 1D Entrained-Flow Texaco Gasifier Using Aspen Custom Modeler

Richard Turton, Chemical Engineering, West Virginia University, Morgantown, WV

Debangsu Bhattacharyya, Chemical Engineering, West Virginia University, Morgantown, WV

Stephen E. Zitney, Collaboratory for Process and Dynamic Systems Research, National Energy Technology Laboratory, U.S. Department of Energy, Morgantown, WV

Job Kasule, Chemical Engineering, West Virginia University, Morgantown, WV

Coal-fired integrated gasification combined cycle (IGCC) power plants produce synthesis gas that is subsequently converted into electricity through combustion in a gas turbine. Several mathematical models have been developed to study the processes taking place inside the gasifier. Such models range from simple one-dimensional steady-state models to sophisticated transient three dimensional CFD-based models that incorporate turbulence effects in the reactor. The practical operation of the gasifier is dynamic in nature but most one dimensional and some higher dimensional models are often steady state and although many higher order CFD based models are dynamic in nature, they are too complicated with many process parameters to be used directly in design and scale-up studies. Thus lower dimensional models are still useful in these types of studies.

In the current study, a one dimensional transient model for the entrained flow GE/Texaco type gasifier is developed using Aspen Custom Modeler ®, which is an equation-based industrial simulator for creating, editing and re-using models of process units.

The current model is based on mass, momentum and energy balances for the solid and gas phases that exist in the gasifier. The chemical reactions considered in the model are the devolatilization/pyrolysis, gasification and combustion, and homogeneous gas-gas reactions. This highly nonlinear system of equation is solved using the inbuilt Newton-based non-linear solvers.

The results to be presented are the transient and steady state solutions to our model equations giving the velocity, concentration (mass fractions), and temperature profiles in the gasifier. Parametric studies results for which the effect of the process parameters on the performance of the gasifier is investigated would be presented for different types of coals. Comparisons of the current model results to available Texaco pilot plant data and other published models will also be made.

See more of this Session: Advances in Gasification Research - II
See more of this Group/Topical: Energy and Transport Processes

Performance Prediction of Entrained Flow Coal Gasifier for IGCC System Using a Steady-State Process Model

Thanh D. B. Nguyen, Department of Chemical Engineering, Hankyong National University, Anseong-si, Gyonggi-do, 456-749, South Korea

Young-Il Lim, Department of Chemical Engineering, Hankyong National University, Anseong-si, Gyonggi-do, 456-749, South Korea

Byung-Ho Song, Department of Chemical Engineering, Kunsan National University, Gunsan, Jeonbuk 573 - 701, South Korea

Si-Moon Kim, IGCC Group, Korea Electric Power Research Institute, Daejeon, 305 - 380, South Korea

Yong-Jin Joo, IGCC Group, Korea Electric Power Research Institute, Daejeon, 305 - 380, South Korea

Dal-Hong Ahn, IGCC Group, Korea Electric Power Research Institute, Daejeon, 305 - 380, South Korea

In an Integrated-coal Gasification Combined-Cycle (IGCC) plant, the gasifier is one of the most important units, since it converts solid fossil fuels such as gaseous fuel, liquid fuel or chemical feedstocks into clean syngas. The entrained-flow coal gasifiers have been widely used in IGCC plants due to the high gasification rate achievement and the relatively clean gas production.

This study focuses on finding optimum operating conditions for the given entrained flow coal gasifiers using a steady-state process model to predict the process performance. The process model includes two-stage equilibrium submodels and a carbon burnout mechanism for which a kinetic model is used to calculate the amount of combusted coal and the heat released to provide the gasification process. The stoichiometric equilibrium model for char gasification and gas phase reactions is used to calculate the temperature and composition of the syngas at the gasifier exit. The parametric effects of operating pressure, oxygen/carbon ratio and steam/carbon ratio on the gasification performance are also studied.

See more of this Session: Advances in Gasification Research - II
See more of this Group/Topical: Energy and Transport Processes

Integrated Tool for IGCC Power Plants Design

Mar Perez-Fortes, Chemical Engineering Department, Universitat Politècnica de Catalunya - ETSEIB, Barcelona, Spain

Aaron D. Bojarski, Chemical Engineering Department, Universitat Politècnica de Catalunya - ETSEIB, Barcelona, Spain

Enric Velo, Universitat Politécnica de Catalunya, Barcelona, Spain

Jose M. Nougués, Chemical Engineering Department, Universitat Politècnica de Catalunya - ETSEIB, Barcelona, Spain

Luis Puigjaner, Chemical Engineering Department, Universitat Politècnica de Catalunya - ETSEIB, Barcelona, Spain

Abstract: Expanding economies and global warming are increasingly important and interrelated challenging issues which require cleaner and more efficient power plants designs to comply with sustainable energy demand are required. Integrated Gasification Combined Cycle (IGCC) power plants have an important role, because of their more efficient way of producing energy from fossil fuels. Actions are focused on clean power and H_2 from coal, and bioproducts. So, one of the future of coal-based power generation strategies should be CO_2 transport and storage in order to obtain a purer hydrogen stream. This paper proposes a methodology and supporting tool for estimating plant performance in terms of power effiency and environmental compliance and making an economic assessment for different scenarios. The technical performance has been modeled in Aspen Hysys and Aspen Plus, and models have been validated with real power plant data from ELCOGAS. Comparisons, in terms of power, emissions, efficiencies and costs between a wide variety of plant designs are presented: they enhance differences in raw material, purification units layout, and hydrogen obtention. Results obtained are examined and discussed towards future work.

Keywords: IGCC power plant, conceptual modelling, performance prediction Aspen Hysys modeling, plant layout.

1. MODELLING PERFORMANCE AND PREDICTIONS

In this paper, an integrated platform for different scenarios prediction has been done. On the one hand, Aspen Hysys is the software used as the main platform for building the conceptual model. On the other hand, Microsoft Excel is the tool used for the economic evaluation. The IGCC plant flowsheet can be changed (by the addition or the sustraction of units, or directly by changing the layout of the flowsheet). From the Aspen Hysys model, data such as power, emissions and efficiency can be obtained. From Microsoft Excel calculations it is possible to estimate preliminary costs: direct costs, the total capital requirement, operating costs and the cost of electricity (COE). The two platforms works in an integrated way.

1.1 Conceptual Modelling.

The model is implemented using two main chemical flowsheeting environments: Aspen Hysys and Aspen Plus, which are flexible tools for process simulation by providing thermodynamic models for the estimation of chemical properties and unit operation models for many processes. Aspen Hysys has been chosen as the platform for the overall process simulation because of its capability of accepting custom models as extensions, and its ease of coding. These models could represent complex chemical reactions (COS hydrolysis) or partial gasification steps (pyrolysis). The proposed platform also allows creating new chemical components not included in its database, such as non stoichiometric solids for raw material and char definition. Alternatively, Aspen Plus is used for calculations involving water systems and electrolytes. These ionic models are required to solve phase equilibrium problems for unit operation systems such as venturi scrubbers, SWS or MDEA absorbers. The aforementioned models have been integrated in Aspen Hysys by means of ANN extensions. Data required to train each of the ANN's comes from sensitivity analysis performed within Aspen Plus while the neural network training is carried out using the ANN package provided with Matlab 6.5. Matlab has been also used for gasifier model parameters estimation.

The global flowsheet studied in this work contemplates all components in a typical IGCC layout. It includes a PRENFLO gasifier and a series of purification units, i.e. a ceramic filter, a venturi scrubber, a COS hydrolysis reactor, a sour water stripper, a MDEA absorber and a Claus plant. An ASU is used to obtain oxygen at a purity of 85 wt%. Steam, oxygen and fuel raw materials enter the gasifier and they are converted into syngas, which is cooled down before entering the purification units. Heat is recovered by producing steam in the HRSG, which is used in a steam turbine. The ELCOGAS power plant is designed to produce around 335 MW of gross power with relatively low emissions. SO_2 emissions are controlled by removing sulfur species from the syngas before the combustion in the gas turbine cycle, and producing liquid sulphur. NO_x emissions are avoided by using preventive methods such as nitrogen saturation of the gas also before the combustion. It is worthwhile to mention that ashes are highly removed as slag in the gasification reactor due to high pressure and temperature conditions. Nevertheless, remaining particles are extracted from the syngas by means of filters. In relation to CO_2 and CO emissions, they are assumed to be produced in the gas turbine cycle. Other contaminants such as chlorine, mercury and heavy metals are distributed in the streams flowsheet by using experimental correlations, based on data developed within the AGAPUTE project (Advanced GAs Purification TEchnologies for co-gasification of coal, refinery by-products, biomass and waste, targeted to clean power produced from gas ans steam turbine-sets generator and fuel cells, supported by European Commission, project no. RFS-CR-04006, see Acknowledgements).

The model shows good agreement between simulated results and industrial data for outlet streams of purification gas units: venturi scrubber, sour water stripper, MDEA absorber, and Claus plant, as seen in Fig. 1. The error is a maximum of 18%, which is well within the accepted error margin at the preliminary design stage (25-40%).

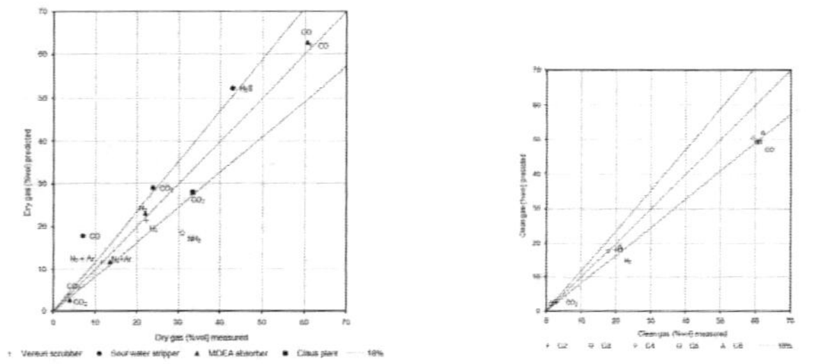

Fig. 1. Purification models validation with real power plant data.

Also, in orther to assure the model confidence, the result of the whole plant has been evaluated by means of the clean gas composition (where again a maximum error of 18% has been observed). See also Fig. 1, right side.

1.2 Best plant design

Election of costs, efficiency and emissions as KPI is directly related to the conception of a profitable gasification plant. The three authors' main criteria for IGCC best design are:

- Economic evaluation: costs à Cost Of Energy (COE).

- Current challenges: environmental impact à Emissions, like CO2, NOx and SO2.

- Process synergy à Energy efficiency.

1.2.1 Results: Scenarios comparison

The scenarios comparison enhance sixteen different cases:

- Eight with different raw materials (coal , coke, a mistureof them, and a mixture of them plus a waste biomass, orujillo)

- Six more cases with different purification gas units layout: by avoiding some units, and also by adding a Selexol absorber or a Dolomite unit.

- Two more by adding a CO_2 purification train (with water gas shift reactor, MEA absorber and PSA) and considering hydrogen to the market or to power production.

Pareto curves have been used to compare costs, environmental impact and efficiency of the different scenarios. A methodology to address the work main goal has been proposed: it consists of giving the tactic to find explicit and final solutions for concrete IGCC power plant flowsheet conditions. Costs, emissions and efficiency are considered as selection criteria. In this way, an exhaustive calculation of these parameters is developed in this work.

Acknowledgements: Data provided by partners (ELCOGAS, UPM and IJA-CSIC) of Agapute project (RFS-CR-04006,) and the support received from the European Commission are gratefully acknowledged.

REFERENCES

Christopher Frey H, Akunuri, N. (2001). Probabilistic Modeling and Evaluation of the Performance, Emissions, and Cost of Texaco Gasifier-Based Integrated Gasification Combined Cycle Systems Using ASPEN. Prepared by North Carolina State University for Carnegie Mellon University and U.S. Dept. of Energy, Pittsburgh.

See more of this Session: Advances in Gasification Research - II
See more of this Group/Topical: Energy and Transport Processes

Transport and Reaction Processes in Non-Homogeneous Porous Media for Fuel Cell Applications

Sesha H. Vemuri, Chemical Engineering, Carnegie Mellon University, Pittsburgh, PA

Parag Jain, Chemical Engineering, Carnegie Mellon University, Pittsburgh, PA

Dehee Kim, Chemical Engineering, Carnegie Mellon University, Pittsburgh, PA

Lorenz T. Biegler, Chemical Engineering, Carnegie Mellon University, Pittsburgh, PA

Myung S. Jhon, Chemical Engineering, Carnegie Mellon University, Pittsburgh, PA

The key technological issues in polymer electrolyte fuel cell (PEFC) technology stem from the performance inefficiencies within its membrane electrode assembly (MEA). A typical MEA comprises of the gas diffusion layer (GDL), the catalyst layer (CL), and the polymer electrolyte membrane, which are heterogeneous porous material sub-components, and through which heat & mass transport processes occur with or without electrochemical reaction. For overcoming the performance limitations within MEA, a detailed understanding of these transport processes is required. Traditional non-homogeneous media theories [1-3] estimate the effective molecular transport coefficients (without fluid flow information) through averaged phase hold-up and/or tortuosity parameters, therefore, do not account for the microscopic structural information within the media.

We recently adopted an agglomerate CL model, which captures the essential transport processes and accounts for species, proton, and electron transports within the GDL/CL sub-components as well as reaction/transport processes within the CL, and incorporated it into our in house state-of-the-art interior point optimization algorithm, IPOPT. [4] We examined the effect of various agglomerate shapes (sphere, cylinder, and plate) on simulations and optimizations via introducing agglomerate effectiveness factor and effective transport properties. We also investigated the agglomerate size effect via scaling analysis and obtained up to three-fold enhancement in the overall current density as the agglomerate size decreased by ten times.

We further incorporated mesoscopic structural information for estimating transport properties in non-homogeneous media via lattice Boltzmann methods (LBMs). The advantages of clear physical pictures, geometric flexibility, an inherently transient nature, and fully parallel algorithms, make LBM to be an attractive tool for multi-phase, multi-scale simulation for PEFC devices.[5-7] LBM has been employed previously for simulating individual transport processes (momentum, energy, or mass) however, simulating combination of these phenomena, present in a PEFC MEA requires a clever modification in the LBM scheme via equilibrium distribution function or relaxation time. We first developed the standard LBM with Bhatnagar-Gross-Krook approximation for porous media, and verified with the Maxwell-Eucken formulation, and accounted for local structural information via correlation functions. Prior to the real application to a PEFC MEA, we extend our LBM scheme to systematically estimate the effective transport properties in non-homogeneous media especially including fluid flow.

Our methodology will be used for addressing PEFC sub-component level issues through meso-/micro-scopic modeling, and is planned for integration with entire PEFC system with optimization.

References:

1. J. C. Maxwell, A Treatise on Electricity and Magnetism (Clarendon, Oxford, 1881), Vol. 1, p. 435.

2. D.A.G. Brugemann, Ann. Phys.(5) 24, 637 (1935).

3. Jianfeng Wang, James K. Carson, Mike F. North, Donald J. Cleland, International Journal of Heat and Mass Transfer 49 (2006) 3075-3083

4. P. Jain, L.T. Biegler, and M.S. Jhon, Electrochem Solid-State Lett, 11 (10), B193 (2008).

5. W.T. Kim, S.S. Ghai, Y. Zhou, I. Staroselsky, H.D. Chen, and M.S. Jhon, IEEE Trans. Magn., 41, 3016 (2005).

6. W.T. Kim, M.S. Jhon, Y. Zhou, I. Staroselsky, and H.D. Chen, J. Appl. Phys., 97, 10P304 (2005).

7. Y.S. Xu, Y. Liu, X.Z. Xu, and G.X. Huang, J. Electrochem. Soc. , 153, A607 (2006).

See more of this Session: Fundamental Research in Transport Processes II
See more of this Group/Topical: Engineering Sciences and Fundamentals

A Lagrangian Particle Method as An Alternative to Solute Lattice Boltzmann: Investigation of Mass Transfer in Biological Porous Media

Roman S. Voronov, School of Chemical, Biological, and Materials Engineering, University of Oklahoma, Norman, OK

Samuel VanGordon, Bioengineering, University of Oklahoma, Norman, OK

Bonnie Landy, Bioengineering, University of Oklahoma, Norman, OK

Vassilios I. Sikavitsas, Chemical, Biological, and Materials Engineering, University of Oklahoma, Norman, OK

Dimitrios V. Papavassiliou, Chemical, Biological, and Materials Engineering, University of Oklahoma, Norman, OK

An effective and inherently parallelizable numerical method for the simulation of microfluidic flows in complicated geometries is the Lattice Boltzmann Method (LBM). Although several methods exist for modeling solute transport with LBM (for example in multi-component LBM the second component can mimic a solute, when its non-local interaction with the bulk fluid is grossly reduced), these methods are set in the Eulerian framework. Useful statistical quantities, such as the solute survival distance, survival time, effective diffusion coefficient, collision frequency, etc. cannot be extracted directly from such simulations. Moreover, just as in the case of classical LBM, the range of solute diffusivity that can be modeled is limited by the numerical solute's relaxation time. As an alternative to these methods, our work presents a Lagrangian approach to simulating convective scalar transfer in conjunction with LBM.

A house hybrid MPI/Open MP parallelized scheme is employed in order to take advantage of the inherent LBM parallelizability. Macroscopic mass transfer is modeled using the Lagrangian scalar tracking (LST) method (Papavassiliou, Int. J. Heat Mass Transfer, 45(17), 3571-3583, 2002) in conjunction with the LBM algorithm. In LST, the motion of scalar markers is used to synthesize the scalar profile. The trajectories of these markers are composed by a convection part (obtained using the velocity field from the LBM simulations) and a diffusion part (i.e., Brownian motion obtained from a mesoscopic Monte-Carlo approach). This method is resourceful in terms of computational efficiency, in that it can be used to simulate various Schmidt number fluids with a single flow field resulting from an LBM simulation. In addition, modifications of the LST algorithm allow the simulation of a whole spectrum of solute reaction rates, also using just a single flow field obtained from LBM.

The presentation will include the description of the numerical methodology and the validation of the method for known cases of flow in porous media. The usefulness of LST will be demonstrated for the particular case of flow through biodegradable synthetic porous polymer scaffolds, which are used in flow perfusion bioreactors for the growth of bone tissue as potential replacement therapies for damaged or lost bone tissue. Knowledge of the mechanism of mass transfer of nutrients to osteoblast cells growing within the porous scaffold can be obtained from LST simulations, allowing the optimization of the 3D scaffold structure for enhanced tissue growth. Micro-Computed Tomography scans of the bone tissue growth along with computer simulation results are used to obtain insights into the tissue growth process with the ultimate goal of being able to predict where the tissue will grow a priori - based only on the plain scaffold geometry and the flow rates.

See more of this Session: Fundamental Research in Transport Processes II
See more of this Group/Topical: Engineering Sciences and Fundamentals

Separation of SWCNTs by Type Using Electric Field, Field-Flow Fractionation*

Frederick R. Phelan Jr.† and Barry J. Bauer

Polymers Division
National Institute of Standards and Technology
Gaithersburg, MD, USA

ABSTRACT

Modeling of the separation of single wall carbon nanotubes (SWCNTs) in Electric Field, Field-Flow Fractionation (EF-FFF) is examined using a Brownian dynamics method in which the forces and torques arising from an AC electric field acting in the field direction have been incorporated. Two mechanisms for type separation – *i.e.*, separation of metallic from semiconducting tubes – are elucidated. In a uniform electric field, E-field aligned rods can be induced to exhibit steric inversion, which leads to rapid type separation under steric mode conditions. In a non-uniform electric field, the dielectrophoretic (DEP) force can be used to induce type separation under normal mode conditions. Both methods rely on exploiting polarizability differences between metallic and semiconducting tubes. The relative efficiency of the two methods is compared and discussed.

Keywords: Brownian Dynamics; Dielectrophoresis; Field-Flow Fractionation; Nanotubes; Separations; SWCNTs; Polarizability

1 INTRODUCTION

The manufacturing of single wall carbon nanotubes (SWCNTs) results in mixtures that are polydisperse with respect to length, diameter and chirality, the latter of which strongly affects the tube electronic properties. For nanotubes to achieve their full potential in applications, it is necessary to be able to separate them according to these characteristics. Separations are particularly critical for nanotube applications in electronics, where the chirality determines whether the tubes are metallic or semiconducting in nature.

One possible technique for achieving this is field-flow fractionation (FFF), depicted in Figure 1. In FFF, a mixture to be separated (typically, a dilute solution of particulates) is driven through a channel while a field is applied in a direction perpendicular to the streamwise direction [1-3]. The perpendicular field may be a second flow field, an electric field, or a temperature gradient, amongst a number of possibilities. The perpendicular field is chosen so that the interaction between the field and the streamwise parabolic velocity profile, promotes a separation of components based on their mobility in the field direction.

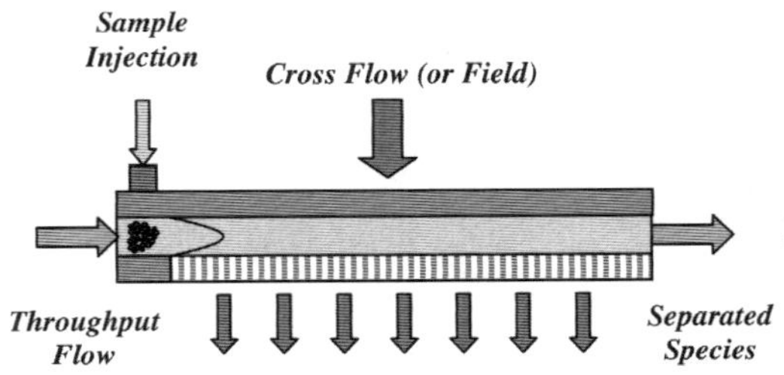

Figure 1: Montage of field-flow fractionation techniques. In the asymmetric variation of classical flow-FFF, the lower wall is porous.

In the family of FFF techniques, a number of variations can be fashioned to promote separations according to the various nanotube properties. Classical flow-FFF, in which the perpendicular field is a cross-flow, generates separations based on the difference in the diffusion coefficients of smaller and larger particles [1-3]. For SWCNTs, flow-FFF has been shown to promote length based separations [4-7]. Electric Field, Field-Flow Fractionation (EF-FFF), in which an electric field is applied in addition to the cross flow, can be expected to promote separations based on the different electronic properties of the particles, in particular, the differing electrophoretic mobility [8,9] and the polarizability [10,11] of metallic and semiconducting tubes. Sedimentation-FFF (Sd-FFF), in which a centripetal force is applied in the field direction [12,13], generates separations based on density differences. For SWCNTs this can be expected to promote diameter separations since SWCNT density varies mainly with diameter [14,15].

In EF-FFF with an AC electric field, the primary variable for controlling separations is the complex

† Corresponding author. Tel.: (301) 975-6761; E-mail address: frederick.phelan@nist.gov.

polarizability. The principal values for the complex polarizability of anisotropic particles are given by

$$\hat{\alpha}_{ii} = \varepsilon_m \frac{\left(\hat{\varepsilon}_p - \hat{\varepsilon}_m\right)}{\hat{\varepsilon}_m + L_{ii}\left(\hat{\varepsilon}_p - \hat{\varepsilon}_m\right)} \tag{1.1}$$

where $\hat{\varepsilon}_p$ and $\hat{\varepsilon}_m$ are the complex permittivities of the particle and medium, given by

$$\hat{\varepsilon}_p = \varepsilon_p - i\frac{\sigma_p}{\omega} \tag{1.2}$$

$$\hat{\varepsilon}_m = \varepsilon_m - i\frac{\sigma_m}{\omega} \tag{1.3}$$

ε_p and ε_m are the permittivity of the particle and medium, σ_p and σ_m are the conductivity of the particle and medium, L_{ii} are shape factors expressed in terms of elliptic integrals [16-18], ω is the angular frequency of the applied field and i is the imaginary number. Analytical expressions for shape factors for various cases of practical interest are shown in Table 1.

Figure 2 shows the real part of anisotropy of the polarizability based on the cylinder (slender body) model for 1 μm SWCNTS in an aqueous solution based on the physical properties for metallic and semiconducting tubes (Table 2). While there is some difference in the values across the entire range of frequencies, what is most pertinent is that the value for the semiconducting tubes drops off towards zero in the Megahertz range, meaning that metallics will be selectively active in this regime. The same hold true for the real part of the axial polarizability as shown in Figure 3. While the exact differences in real systems will depend on the specific solvent/surfactant combinations, in general the differences in polarizability between metallic and semiconducting SWCNTs offers a basis for type separation in both FFF and other liquid based separation schemes [10,11].

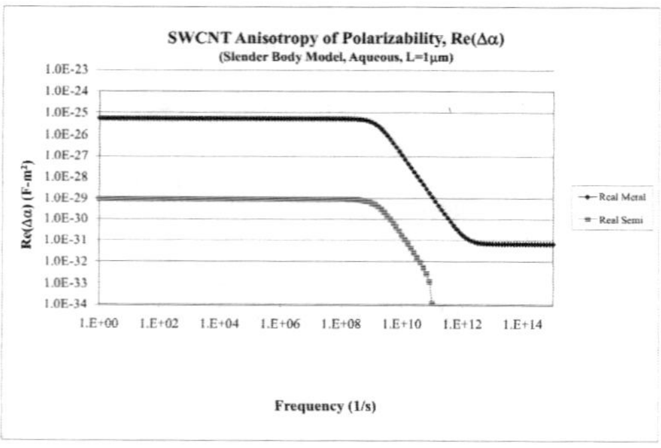

Figure 2: Real part of the anisotropy of the polarizability (F-m^2) as a function of frequency for metallic and semiconducting SWCNTs.

In what follows, two mechanisms for type separation of metallic from semiconducting tubes using EF-FFF are analyzed theoretically and numerically using a Brownian dynamics simulation. First, the case of separation using a uniform AC electric field applied in the gradient direction is examined. It is shown that E-field aligned rods can be induced to exhibit steric inversion, which leads to rapid type separation under steric mode conditions. Next, the case of a non-uniform AC electric field also acting in the gradient direction is examined. For this case it is shown that the dielectrophoretic (DEP) force can be used to induce type separation under normal mode conditions by selectively retaining metallic tubes. Both methods rely on exploiting polarizability differences between metallic and semiconducting tubes. The relative efficiency of the two methods is compared and discussed.

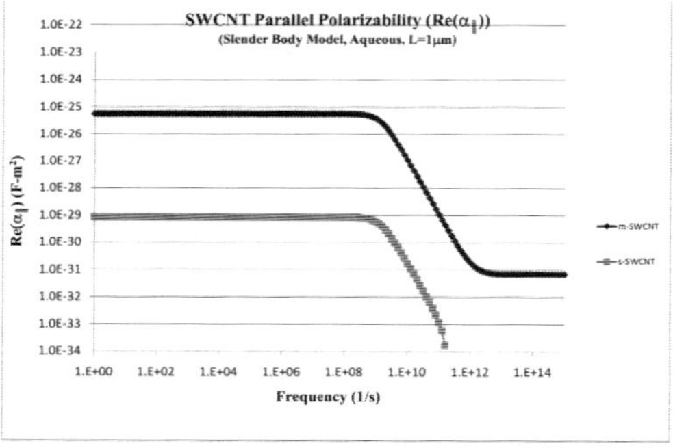

Figure 3: Real part of the axial polarizability (F-m^2) as a function of frequency for metallic and semiconducting SWCNTs.

2 SIMULATION

Full details of the Brownian dynamics simulation used to investigate the separation of rodlike particles in flow-FFF and EF-FFF are found in references [4,5,19-21]. A summary of pertinent details of the numerical method are discussed below.

2.1 EF-FFF Model Equations

The rodlike nanotubes are modeled as prolate ellipsoids, with major axis radius a and minor axis radius b. The governing equations for an ensemble of ellipsoids in EF-FFF, individually denoted by the superscript (i), can be written as [4,5,22-24]

$$\frac{d}{dt}\left(\underline{R}^{(i)}\right) = \underline{v}\left(\underline{R}^{(i)}\right) + \left[\underline{\underline{\zeta}}^{(i)}\right]^{-1} \cdot \left(\underline{F}_B^{(i)} + \underline{F}_{DEP}^{(i)} + \underline{F}_{EP}^{(i)}\right) \tag{1.4}$$

$$\frac{d}{dt}\left(\underline{p}^{(i)}\right) = -\underline{\underline{W}} \cdot \underline{p}^{(i)} + \lambda_p^{(i)}\left(\underline{\underline{D}} \cdot \underline{p}^{(i)} - \underline{\underline{D}} : \underline{p}^{(i)}\underline{p}^{(i)}\underline{p}^{(i)}\right)$$
$$-\underline{p}^{(i)} \times \left[\underline{\underline{\xi}}^{(i)}\right]^{-1} \cdot \left(\underline{T}_B^{(i)} + \underline{T}_E^{(i)}\right) \qquad (1.5)$$

where $\underline{R}$ and $\underline{p}$ are the position and orientation vectors of the particles, $\underline{F}_B$ and $\underline{T}_B$ are the Brownian force and torque, $\underline{F}_{DEP}$ is the dielectrophoretic force, $\underline{F}_{EP}$ is the electrophoretic force, $\underline{T}_E$ is the torque due to the electric field , $\underline{\underline{\zeta}}$ and $\underline{\underline{\xi}}$ are hydrodynamic resistance tensors, $\underline{\underline{D}}$ and $\underline{\underline{W}}$ are the stretching and vorticity tensors, respectively, and $\lambda_p = \frac{\Re^2 - 1}{\Re^2 + 1}$ with $\Re = \frac{a}{b}$. The electrophoretic force is zero in all AC fields, whether uniform or non-uniform, and thus, is not considered in the present analysis. The dielectrophoretic is zero for the case of a uniform AC electric field.

Eqs. (1.4) and (1.5) are integrated for an ensemble of particles with respect to initial conditions described by FFF flow-focusing operations; analytical kinematics are used for the flow field [4,5]. In order to integrate the time dependant equations for an AC field at the high frequencies used in these types of operations, time-averaged expressions for the DEP force and electric field torque must be used. The time-average DEP force is given by the expression [17,25,26]

$$\left\langle \underline{F}_{DEP} \right\rangle = \frac{1}{2} V \operatorname{Re}\left[\left(\underline{\underline{\hat{\alpha}}} \cdot \underline{\hat{E}}\right) \cdot \nabla \underline{\hat{E}}^*\right] \qquad (1.6)$$

where $\underline{\hat{E}}$ is the electric field phasor vector, $\underline{\hat{E}}^*$ is the complex conjugate, V is the particle volume, and $\underline{\underline{\hat{\alpha}}}$ is the complex polarizability tensor. The time-average torque in Eq. (1.5) due to an AC electric field is given by

$$\left\langle \underline{T}_E \right\rangle = \frac{1}{2} V^{(i)} \operatorname{Re}\left[\Delta\hat{\alpha}^{(i)}\left(\underline{p}^{(i)}\underline{p}^{(i)} \cdot \underline{\hat{E}}\right) \times \underline{\hat{E}}^* + \hat{\alpha}_\perp^{(i)} \underline{\hat{E}} \times \underline{\hat{E}}^*\right] \qquad (1.7)$$

where $\Delta\hat{\alpha} = \hat{\alpha}_\parallel - \hat{\alpha}_\perp$, and $\hat{\alpha}_\parallel$ and $\hat{\alpha}_\perp$ are the axial and perpendicular components of the polarizability.

2.2 Boundary Conditions

An important part of the scheme is the interaction of the particles with the boundaries. The cross flow velocity continually drives the particles towards the accumulation wall and the diffusion step being random may also cause the particles to collide with this boundary. To handle this, a no-penetration boundary condition is used, based on the rod size and 3-D orientation [5,19,20] as show in Figure 4, in which the particle velocity is set to its center of mass value with respect to its distance from the wall in the local velocity field. This distance is given by $\delta^{(i)} = a\left|p_y^{(i)}\right|$, where $p_y^{(i)}$ is the y-component of the orientation vector and the boundary condition for the velocity is $v_x = v_x\left(\delta^{(i)}\right)$.

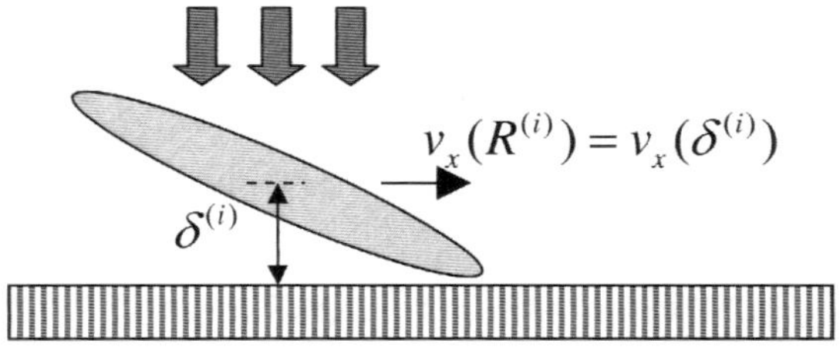

Figure 4: Boundary condition for ellipsoidal particles.

3 EF-FFF WITH UNIFORM AC FIELD

A number of different mechanisms can be exploited to achieve separation in FFF operations. Normal mode separation applies to particles which are small enough to undergo significant Brownian motion, and whose size is small compared to the cross flow gap size. In this case, smaller particles, which are more diffusive, have an average position closer to the centerline and elute faster than larger particles. Steric mode separation occurs when the particle layer in FFF is strongly compressed along the accumulation wall and due to size exclusion, larger particles are more highly entrained by the throughput flow and elute more quickly than smaller ones. This turnaround is called steric inversion.

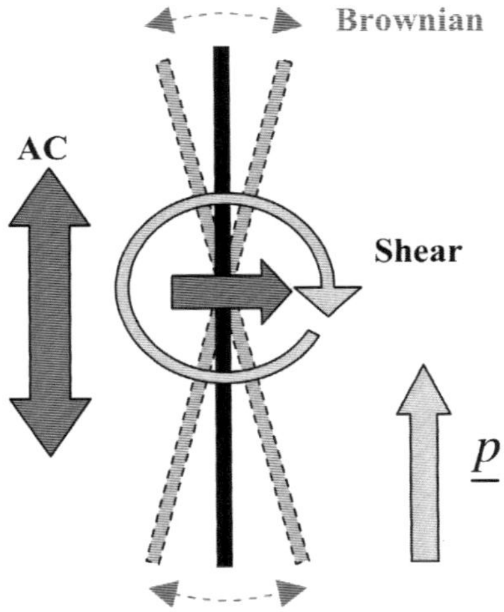

Figure 5: Forces on a rodlike particle subject to both shear flow, translational and rotational Brownian motion, and an AC electric field. Such a situation acts to induce wobbly alignment and modifies the FFF elution characteristics.

The steric inversion of spheres suggests that an even stronger steric effect might be obtained for rodlike particles due to their great length, by orienting them normal to the flow direction. This could be useful in the context of nanotube separation by preferentially orienting either metallic or semiconducting types relative to the other. This would create different elution rates for the two particle types. One way to induce an orientation of nanotube particles is by subjecting the particles to an electric field which acts in the gradient direction. When an anisotropic particle is subject to an electric field, a net torque which acts to align the particle in the field direction is produced [17]. In the present analysis, we consider the case of a uniform AC electric field acting in the gradient direction in EF-FFF. In such a case, the orientation torque produced by the electric field is in competition with both the shear field and Brownian motion as depicted in Figure 5. The degree of alignment depends on the polarizability of the particle, and the electric field strength. In this section, we develop an analytical expression to describe the Retention of oriented tubes, analyze the orientation as a function of electric field strength, shear rate and Brownian motion, and finally, analyze the elution of oriented particles.

3.1 Retention Theory

Particle separation in flow-FFF can be described by a theoretical variable called the retention, R, which represents the ratio of the average residence time of non-retained tracers, t_0, to the average retention time of the particles, t_r, *i.e.*, $t_r = t_0 / R$, where $0 \le R \le 1$. For the case of normal mode separation, the retention is given by

$$R = 6\lambda\left[\coth\left(\frac{1}{2\lambda}\right) - 2\lambda\right] \quad (1.8)$$

where λ is an inverse Peclet Number (Pe) given by

$$\lambda = \frac{D}{|v_c| H} = \frac{1}{Pe} \quad (1.9)$$

v_c is the cross flow velocity, D is the diffusion coefficient of the particle in the cross flow direction, and H is the cross-flow thickness.

A Retention model which takes into account both normal mode diffusion and steric effects for spheres in FFF was derived by Giddings [1]

$$R = 6\left(\alpha - \alpha^2\right) + 6\lambda\left(1 - 2\alpha\right)\left[\coth\left(\frac{1-2\alpha}{2\lambda}\right) - \frac{2\lambda}{1-2\alpha}\right] \quad (1.10)$$

where $\alpha = r / H$ is the ratio of the particle radius to gap thickness. For the case of negligible particle size, Eq. (1.10) reduces to Eq. (1.8). It can be seen that, theoretically, normal mode elution is independent of the particle shape and depends only on the particle diffusion coefficient. Steric mode elution becomes independent of diffusion, and depends mainly on the particle dimensions.

Little is known about how steric mode separations are effected by particle shapes other than spheres. For the case a nanotube oriented perpendicular to a surface, the boundary condition of the particle becomes as is shown in Figure 6, and the particle velocity will be at its maximum possible steric exclusion value, similar to the case for spheres. Under this condition, an appropriate expression for the Retention for the case of perfect orientation can be obtained by the modification of Eq. (1.10) with the substitutions $\alpha \to \alpha' = \frac{1}{2}\frac{L}{H}$, where L is the axial length of the rod, and $\lambda \to \lambda' = \frac{D_{\parallel}}{|v_c| H}$, where $D_{\parallel}$ is the axial diffusion coefficient of the rod (see Appendix), yielding the expression

$$R = 6\left(\alpha' - (\alpha')^2\right) + 6\lambda'\left(1 - 2\alpha'\right)\left[\coth\left(\frac{1-2\alpha'}{2\lambda'}\right) - \frac{2\lambda'}{1-2\alpha'}\right] \quad (1.11)$$

We compare this expression with simulation results for highly oriented tubes.

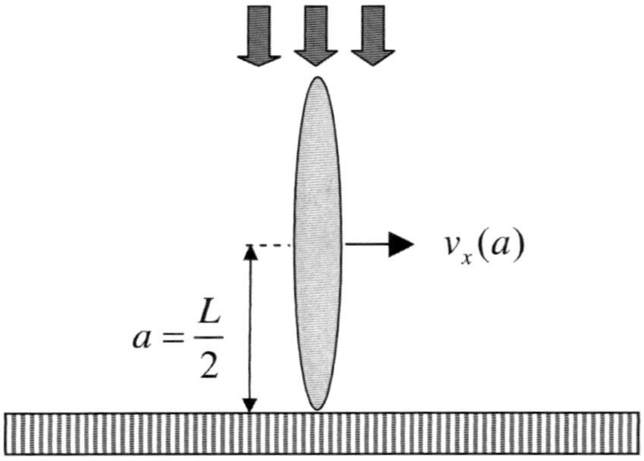

Figure 6: Boundary condition on a rodlike particle oriented normal to the flow direction.

3.2 Electro-Orientation

The orientation behavior of rodlike particles in a steady shear flow subject to a uniform AC electric field and Brownian motion is examined by solving Eq. (1.5) together with Eq. (1.7). The motivation for this calculation is to determine the effect of shear on particle orientation, and to determine range of electric field and polarizability needed to orient particles in EF-FFF. Theoretically, it is expected that the degree of particle orientation will depend on the ratio of the dimensionless groups $N_{DE} = \frac{V\,\mathrm{Re}[\Delta\hat{\alpha}]E_{rms}^2}{kT}$, which is the ratio of the dielectrophoretic potential energy to kT, and $Pe_r = \frac{\eta\dot{\gamma}Y_c^{(i)}}{kT}$ which is the Peclet number for rotational diffusion (where $Y_c^{(i)}$ is rotational friction coefficient, see Appendix). At lower values of Pe_r when the particles are more diffusive, higher values of N_{DE} are needed to orient the particles.

The particle orientation in a uniform electric AC field is characterized by an order parameter given by [27,28]

$$S = \left\langle \frac{3\cos^2\theta - 1}{2} \right\rangle \tag{1.12}$$

where θ is the angle between the particle orientation and the electric field vector, $\langle ... \rangle = \int_0^{\pi} (...) f(\theta) \sin\theta d\theta$, denotes an average over a potential, $f(\theta)$ is the potential

$$f(\theta) = \frac{e^{-\frac{U(\theta)}{kT}}}{\int_0^{\pi} e^{-\frac{U(\theta)}{kT}} \sin\theta d\theta} \tag{1.13}$$

and $U(\theta)$ is the interaction energy between the particle and the electric field

$$U(\theta) = -\frac{1}{2} V\,\mathrm{Re}[\Delta\hat{\alpha}] E_{rms}^2 \cos^2\theta \tag{1.14}$$

The value of S is equal to 1 when the particles are perfectly oriented with the field direction, 0 when they are oriented randomly with respect to the field direction, and a value of -½ when they are oriented normally to the field.

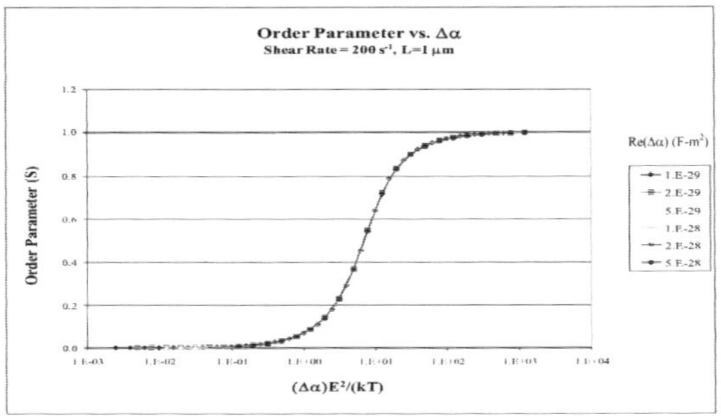

Figure 7: Order parameter at constant shear rate as a function of effective polarizability (*F-m*²).

Orientation results for a particle with a 1000 nm length and 1 nm diameter, at a constant shear rate of 200 are shown in Figure 7. The orientation is plotted as a function of the dimensionless group $N_{DE} = \frac{V\,\mathrm{Re}[\Delta\hat{\alpha}]E_{rms}^2}{kT}$, which is the ratio of the dielectrophoretic potential energy to kT. Effective polarizibility values, $\Delta\alpha = V \cdot \mathrm{Re}[\Delta\hat{\alpha}]$, were chosen in a range consistent with values obtained from analytical models and recent experimental investigations [28-31]. Shear rates were chosen in a range consistent with the maximum rates seen in experimental FFF studies [6,19]. The figure shows that the particle orientation data at constant shear rate collapses onto a single master curve.

The particle orientation at constant polarizabilty as a function of shear rate is shown in Figure 8. These results also collapse onto a single curve and indicate that at this size scale, the particle orientation is insensitive to the shear flow, and that the orientation is mainly a competition between the Brownian motion and the torque due to the electric field. Thus we expect that in electro-orientation EF-FFF the average orientation is constant in the gradient direction, and that high shear rates at the lower boundary are not a hinderance to generating a steric mode separation.

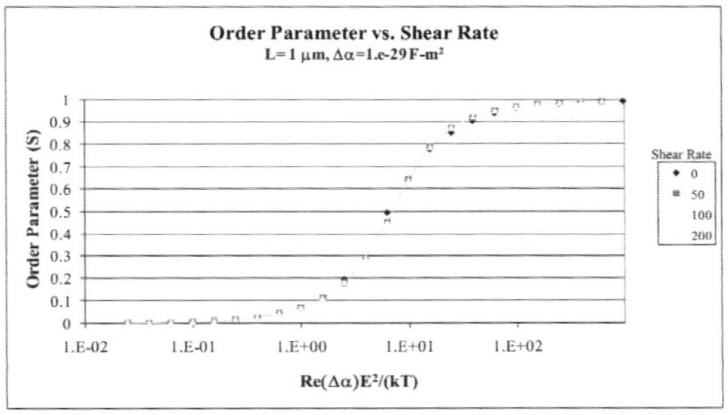

Figure 8: Order parameter at constant polarizability as a function of shear rate.

3.3 EF-FFF with Electro-Orientation

Simulation results are shown in Figure 9 for the processing parameters given in Table 2, and particle lengths ranging from 10 nm to 1000 nm, with a 1 nm diameter. An ensemble size of 1000 particles was used in all cases. Results for the steric inversion expression, Eq. (1.11), are plotted vs. EF-FFF results for $S = [0.1, 0.99, 1]$, and also the case of normal mode flow-FFF, which is for the most part is nearly random orientation or $S = 0$. The results for $S = 1$ were obtained by setting the particle initial condition at an orientation normal to the flow direction, and turning off all rotational motion.

A number of observations can be made from the Figure. First, for the oriented particles, steric inversion is clearly observed. All the EF-FFF curves diverge sharply from the flow-FFF curve (which are governed by normal mode elution) at a length of about 100 nm. For the given conditions, the steric transition occurs when the particle length is approximately 400 nm. The greatest degree of inversion occurs for the modified steric inversion theory of Giddings, *i.e.*, Eq. (1.11). This acts as a lower bound on the elution times while the flow-FFF curve acts as an upper bound. The simulation results for $S = 1$ (constrained rotational motion) are in close agreement with Eq. (1.11). The data for $S = 0.99$ and $S = 0.1$ show that as the average particle orientation decreases, there is still steric inversion, but the elution times begin to slow down. However, mergence with the flow-FFF curve is slow.

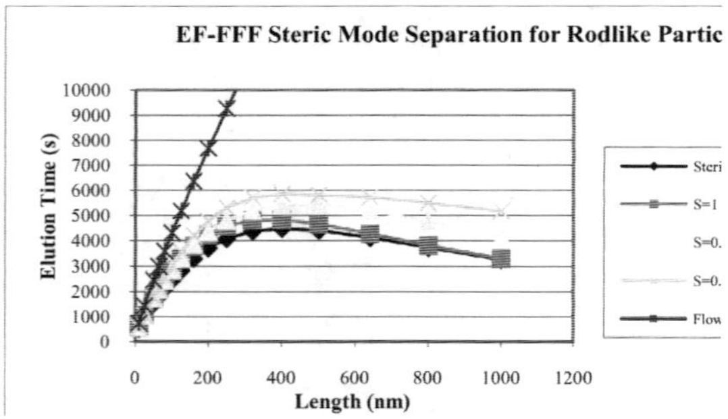

Figure 9: Comparison of elution time for rodlike particles under normal mode conditions, and for rods oriented perpendicular to flow direction.

These results can be used in the context of nanotube type separation by selectively orienting the more active metallic tubes relative to the semis by controlling the relative polarizability of the two different types by means of frequency, *cf.* Figure 2, and appropriate solvent/surfactant combinations. Orienting one type relative to the other will allow the orienting and non-orienting types to be separated into two fractions. The first fraction will contain oriented rods of all sizes, and non-oriented Brownian rods whose size is such that their elution time is less than the elution time corresponding to the steric inversion point for the oriented rods. The second fraction will contain all the longer non-oriented rods. For the case shown in Figure 9, the non-orienting cutoff size in the first fraction is about 150 nm; smaller cutoff sizes can be obtained by optimizing parameter conditions. Both fractions may then be further refined using size separation only.

4 DEP-FFF

Another means to separate particles in electric field, FFF is through the use of a non-uniform electric field, which generates a DEP force on particles. DEP has been used in conjunction with FFF in a number of studies [10,32-37] and has also been used as a means to separate nanotubes in non-FFF studies [11,38-40]. In practice, there are two ways to implement DEP-FFF depending on the size of the particles [35]. In levitation DEP-FFF, the DEP force is used to counterbalance the gravitational forces on particles, and is therefore used to lift particles away from the accumulation wall. In retention DEP-FFF, the DEP force is used in addition to the cross flow to counterbalance the Brownian force, and is therefore used to pull particles towards the retaining wall. In this case we analyze that latter which is the appropriate choice for SCWNTs which are highly Brownian.

4.1 Electrode Configuration

A typical configuration for DEP-FFF consists of an array of interdigitated electrodes along the lower wall [35]. A 2-D cross section depicting the electrodes and the approximate electric field lines generated by this type of arrangement is shown in Figure 10.

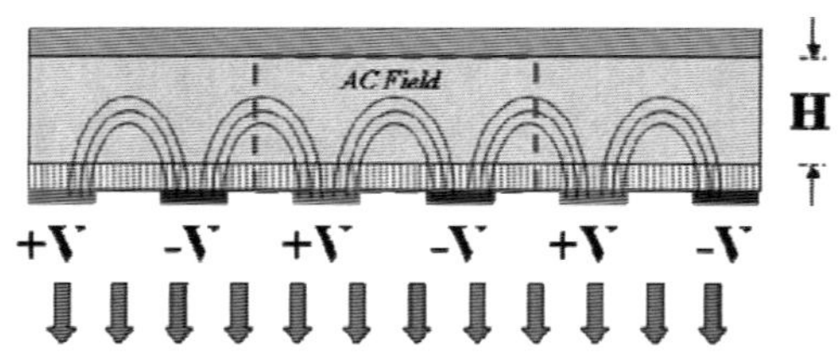

Figure 10: Electrode configuration in a typical retention DEP-FFF setup and approximate sketch of the electric field lines. The repeat unit for the voltage calculation is outlined. Particles which experience "positive" DEP move towards the accumulation wall.

The configuration is characterized by an electrode width, L_1, and an electrode spacing L_2, as depicted in Figure 11. Typical values for the electrode width and spacing are on the order of 50 μm. Since an FFF apparatus is on the order of 25 cm, there are on the order of 600 electrode pairs in a setup.

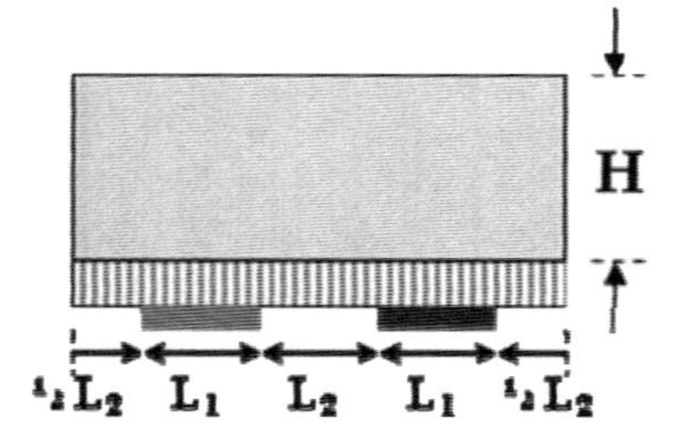

Figure 11: Repeat unit for voltage calculation in DEP-FFF electrode configuration.

4.2 Computation of Average Force

Because of the large number of electrode pairs in an actual FFF device, it is not practical to compute the detailed DEP force through all the electrode pairs over the entire length of the device. Thus, the average DEP forces over the space of a single repeat unit are used in the FFF calculation.

The net motion due to the DEP force in Eq. (1.4) for particles experiencing rotational motion during flow is given by the expression

$$\left[\underline{\underline{\zeta}}^{(i)}\right]^{-1}\cdot\left\langle \underline{F}^{(i)}_{DEP}\right\rangle = \underline{\underline{M}}^{(i)}\cdot\nabla\left|\underline{\hat{E}}\right|^{2} \quad (1.15)$$

where

$$\underline{\underline{M}}^{(i)} = \frac{1}{2}\frac{V}{\eta}\mathrm{Re}\left[\left(\frac{\hat{\alpha}_{\parallel}}{X^{(i)}_{A}} - \frac{\hat{\alpha}_{\perp}}{Y^{(i)}_{A}}\right)\underline{p}^{(i)}\underline{p}^{(i)} + \frac{\hat{\alpha}_{\perp}}{Y^{(i)}_{A}}\underline{\underline{I}}\right] \quad (1.16)$$

and $X^{(i)}_{A}$ and $Y^{(i)}_{A}$ are the axial and transverse drag coefficients for the particles defined in the Appendix. Since for nanotubes it is expected that the components of $\underline{\underline{\hat{\alpha}}}$ are positive and dominated by $\hat{\alpha}_{\parallel}$, according to Eq. (1.15) SWCNTs should experience "positive" DEP and move in the direction of the gradient $\nabla\left|\underline{\hat{E}}\right|^{2}$. Due to symmetry, the average force on the particles in the x-direction is zero, but is mainly downward and acts to move the particles towards the wall in the y-direction.

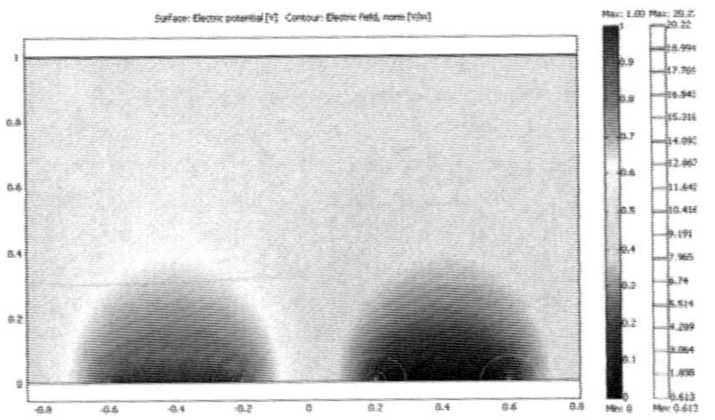

Figure 12: COMSOL solution for voltage showing electric field lines in dimensionless repeat unit.

To solve for the electric field needed to compute the DEP force, the electric potential in the fluid is computed by solving the equation

$$\nabla\cdot\left(\sigma_{m}\nabla V\right) = 0 \quad (1.17)$$

where σ_m is the conductivity of the fluid and V is the voltage. The solution is obtained using COMSOL Multiphysics[‡] [41]. The components of the electric field vector are then computed from the relation $\underline{E} = \nabla V$, and these values are used to calculated the electric field phasor vector $\underline{\hat{E}}$.

Figure 12 shows the COMSOL solution for the voltage and electric field lines. The problem is scaled to a unit cross-sectional thickness and voltage. The average (dimensionless) value of $\nabla_y\left|\underline{E}\right|^2$ in the gradient direction used to compute the DEP force is shown in Figure 13. The solution for $\nabla_y\left|\underline{E}\right|^2$ is converted to real units using the scale factor V_0^2/H^3, where V_0 is the applied voltage. The gradient is sharp and highly non-linear near the boundary.

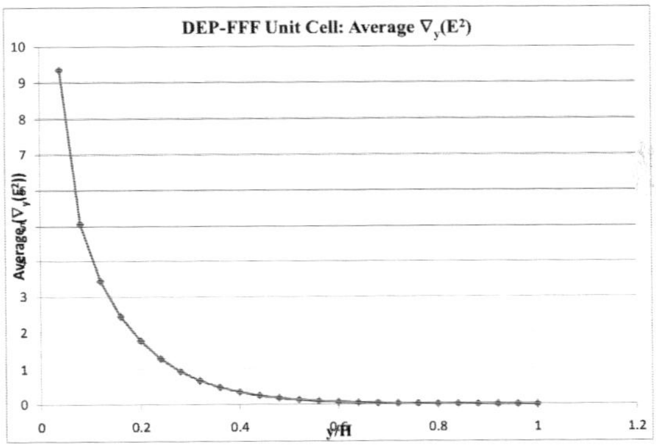

Figure 13: The average value of $\nabla_y\left|\underline{E}\right|^2$ over a repeat unit in the gradient direction.

4.3 Results for DEP-FFF

Simulation results in the form of elution profiles for DEP-FF are shown in Figs. 14-15 for 100 nm particles with a 1 nm diameter, for the processing parameters given in Table 4. An ensemble size of 1000 particles was used in all cases. The electrode width and spacing were set to 50 μm. The polarizability for the different particle types was computed using cylinder model with the property values listed in Table 2.

‡ Identification of a commercial product is made only to facilitate reproducibility and to adequately describe procedure. In no case does it imply endorsement by the National Institute of Standards and Technology (NIST) or imply that it is necessarily the best product for the procedure.

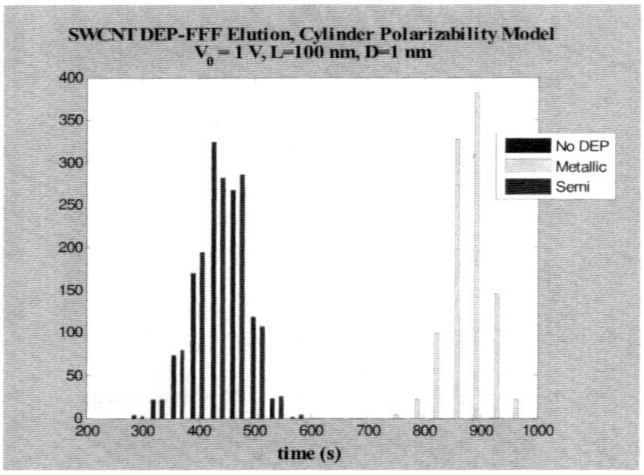

Figure 14: Results for DEP-FFF elution of m-SWCNTs and s-SWCNTs using the cylinder polarizability model.

Figure 14 shows results for DEP-FFF elution of m-SWCNTs and s-SWCNTs for an applied voltage of 1 V and at frequency of 10 Ghz, in comparison with the elution based on flow-FFF (which is not sensitive to type). The s-SWCNTs are slightly more retained, but the overall elution profile is not much different than that for flow-FFF (No DEP). By comparison, the m-SWCNTs are much more highly retained and the average elution time changes by over 400 s.

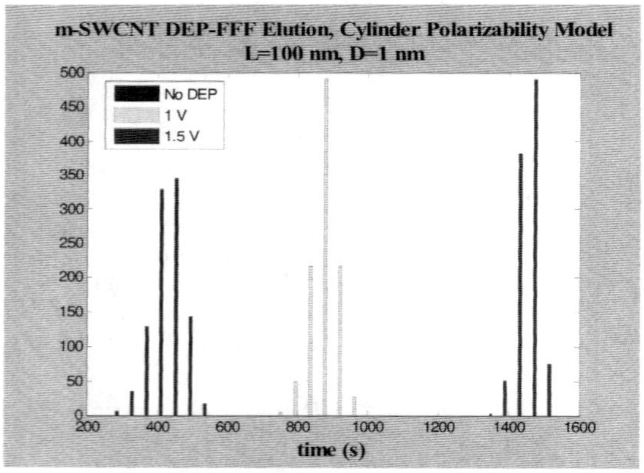

Figure 15: Results for DEP-FFF elution of m-SWCNTs using the cylinder polarizability model for different voltages.

The sensitivity of the elution of the m-SWCNTs is further shown in Figure 15, which compares flow-FFF elution with results for m-SWCNTs at 1 V and 1.5 V, again at 10 GHz. Due to the highly non-linear nature of the gradient $\nabla_y |E|^2$ near the boundary, even the small change in voltage produces a large change in the average elution time. This sensitivity suggests a scheme where an initially large voltage is used to effectively trap m-SWCNTs until adequate separation with s-SWCNTs is achieved, which is then tapered off to allow the m-SWCNTs to elute. Current work is focused on this.

5 SUMMARY AND CONCLUSIONS

We have investigated particle separation mechanisms in EF-FFF using a Brownian dynamics method. Two types of separation mechanisms have been investigated based on both uniform and non-uniform electric fields. Both methods exploit polarizability differences between metallic and semiconducting tubes. The application of the present results to real systems will depend on the specific solvent/surfactant combinations which certainly effect the polarizability, and possibly effect the hydrodynamic properties of the particles. A greater understanding of such effects are needed. Based on the best data currently available, the following conclusions can be drawn from the present work:

- In the range shear rates common for FFF, shear flow has little effect on the electo-orientation of rodlike particles in EF-FFF with a uniform electric field, as Brownian motion is the dominant opposition force.
- Rods oriented normal to the flow direction can be expected to exhibit steric inversion. The steric effect grows stronger with increasing particle orientation. Eq. (1.11) provides a useful expression for the Retention of perfectly oriented particles.
- The use of steric inversion could be useful in the context of nanotube separation by preferentially orienting metallic tubes relative to the semiconducting types. The first fraction can be obtained much more rapidly than for normal mode elution and will be mainly metallics. The second fraction will be purely semiconducting.
- Based on the theoretical values of the polarizability for SWCNTs, metallics will be much more highly retained in DEP-FFF than semiconducting tubes. This suggests that a separation scheme using programmable voltages (high to low) would be effective in sorting m-SWCNTs and s-SWCNTs into two fractions.
- The implementation of electro-orientation based EF-FFF with a uniform electric field is potentially simpler to implement and control, but based on available polarizability data, will require somewhat higher voltages than DEP-FFF.
- Properly programmed DEP-FFF potentially offers more control over type separation than electro-orientation EF-FFF, as the sterically separated

phase in the latter, will contain a small amount of the non-oriented phase.

REFERENCES

1. J. C. Giddings, Separation Science and Technology **13,** 241 (1978).
2. J. Janca, in *Field-Flow Fractionation*, (Marcel Dekker, New York, 1987).
3. in *Field-Flow Fractionation Handbook*, (Wiley-Interscience, 2000).
4. F. R. Phelan Jr. and B. J. Bauer, Chemical Engineering Science **62,** 4620 (2007).
5. F. R. Phelan Jr. and B. J. Bauer, Chemical Engineering Science **64,** 1747 (2009).
6. J. Chun, J. A. Fagan, E. K. Hobbie, and B. J. Bauer, Analytical Chemistry **80,** 2514 (2008).
7. M. H. Moon, D. J. Kang, J. H. Jung, and J. M. Kim, Journal of Separation Science **27,** 710 (2004).
8. S. K. Doorn *et al.*, Journal of Physical Chemistry B **107,** 6063 (2003).
9. B. Suarez, B. M. Simonet, S. Cardenas, and M. Valcarcel, Journal of Chromatography A **1128,** 282 (2006).
10. H. Q. Peng *et al.*, Journal of the American Chemical Society **128,** 8396 (2006).
11. M. J. Mendes, H. K. Schmidt, and M. Pasquali, Journal of Chemical Physics B **112,** 7467 (2008).
12. J. C. Giddings, M. N. Myers, M. H. Moon, and B. N. Barman, Acs Symposium Series **472,** 198 (1991).
13. J. C. Giddings, M. H. Moon, P. S. Williams, and M. N. Myers, Analytical Chemistry **63,** 1366 (1991).
14. M. S. Arnold *et al.*, Nature Nanotechnology **1,** 60 (2006).
15. J. A. Fagan, M. L. Becker, J. Chun, and E. K. Hobbie, Advanced Materials **20,** 1609 (2008).
16. H. A. Pohl, in *Dielectrophoresis*, (Cambridge University Press, London, 1978).
17. T. B. Jones, in *Electromechanics of Particles*, (Cambridge University Press, New York, 1995).
18. A. Sihvola, in *Electromagnetic Mixing Formulas and Applications*, (INSPEC Inc., 2000).
19. F. R. Phelan Jr. and B. J. Bauer, Chemical Engineering Communications **in preparation,** (2008).
20. F. R. Phelan Jr., S. Kuruvilla, and B. J. Bauer, AICHE Journal **in preparation,** (2008).
21. F. R. Phelan Jr., S. Kuruvilla, and B. J. Bauer, *Steric Mode Separation of Nanotubes Using Electric Field, Field-Flow Fractionation* , (AICHE, AIChE 2008 Annual Meeting, 2008).
22. A. Satoh, in *Introduction to Molecular-Microsimulation of Colloidal Dispersions*, (Elsevier Science B.V., Amsterdam, The Netherlands, 2003).
23. S. Kim and S. J. Karrila, in *Microhydrodynamics: Principles and selected applications*, (Butterworth-Heinemann, Stoneham, 1991).
24. F. R. Phelan Jr., S. Kuruvilla, and B. J. Bauer, Microfluidics and Nanofluidics **submitted,** (2009).
25. T. B. Jones, IEEE Engineering in Medicine and Biology Magazine **22,** 33 (2003).
26. T. B. Jones, Biophysical Journal **64,** 1588 (1993).
27. C. T. O'Konski, K. Yoshioka, and W. H. Orttung, Journal of Chemical Physics **63,** 1558 (1959).
28. J. A. Fagan, V. Bajpai, B. J. Bauer, and E. K. Hobbie, Applied Physics Letters **91,** 213105-213105-3 (2007).
29. K. Bubke *et al.*, Applied Physics Letters **71,** 1906 (1997).
30. M. S. Brown, J. W. Shan, C. Lin, and F. M. Zimmerman, Applied Physics Letters **90,** 203108 (2007).
31. K. Donovan and K. Scott, Physical Review B **72,** 195432-195432-8 (2005).
32. G. H. Markx, J. Rousselet, and R. Pethig, Journal of Liquid Chromatography & Related Technologies **20,** 2857 (1997).
33. Y. Huang *et al.*, Journal of Hematotherapy & Stem Cell Research **8,** 481 (1999).
34. J. Yang *et al.*, Biophysical Journal **78,** 2680 (2000).
35. X. B. Wang *et al.*, Analytical Chemistry **72,** 832 (2000).
36. J. Vykoukal *et al.*, Lab on A Chip **8,** 1386 (2008).
37. T. S. Leu and C. Y. Weng, Journal of Micro-Nanolithography Mems and Moems **8,** (2009).
38. F. Hennrich, R. Krupke, M. M. Kappes, and H. V. Lohneysen, Journal of Nanoscience and Nanotechnology **5,** 1166 (2005).
39. S. Banerjee, T. Hemraj-Benny, and S. S. Wong, Journal of Nanoscience and Nanotechnology **5,** 841 (2005).
40. Z. Chen *et al.*, Analytical Chemistry **78,** 8069 (2006).
41. in *COMSOL MP 3.5 User's Guide*, (COMSOL AB, Burlington, MA, 2009).

6 TABLES

Case	L_{11}	L_{22}	L_{33}
Sphere	⅓	⅓	⅓
Cylinder	0	½	½
Ellipsoid	$\frac{1-s^2}{2s^3}\left(\ln\frac{1+s}{1-s}-2s\right)$	$\frac{1}{2}(1-L_{11})$	$\frac{1}{2}(1-L_{11})$

Table 1 – Polarizability shape factors.

Component	Relative Dielectric Constant	Conductivity (A/V-m)
m-SWCNT	1.E+04	1.E+08
s-SWCNT	2.5	1.6E+04
Water	80.1	1.0

Table 2 – Properties used for computing complex polarizability in Figures 2-3.

Processing Parameters	
Length, L (m)	0.25
Thickness, H (m)	0.00025
Width, W (m)	0.02
Focus Point, x_f (m)	0.01
T (K)	293
Viscosity (Pa-s)	0.001
Q_x (m^3/s)	2.5 x 10^{-8}
Q_y (m^3/s)	2.0 x 10^{-7}
Q_f (m^3/s)	2.0 x 10^{-8}

Table 3 – Processing parameters used in the EF-FFF with electro-orientation simulations.

Processing Parameters	
Length, L (m)	0.25
Thickness, H (m)	0.00025
Width, W (m)	0.02
Focus Point, x_f (m)	0.01
T (K)	293
Viscosity (Pa-s)	0.001
Q_x (m^3/s)	2.5 x 10^{-8}
Q_y (m^3/s)	2.0 x 10^{-8}
Q_f (m^3/s)	2.0 x 10^{-8}

Table 4 – Processing parameters used in the DEP-FFF simulations.

7 ACKNOWLEDGEMENTS

FRPJr thanks Prof. T.B. Jones of the University of Rochester for his guidance in deriving the appropriate torque expressions for AC fields. Sibu Kuruvilla of the University of Illinois Urbana Champaign did the electro-orientation calculations and the some of the EF-FFF calculations. Milena R. Bobea Rodriguez of the University of Puerto Rico did the DEP calculations.

8 APPENDIX – ELLIPSOID FRICTION COEFFICIENTS

$D_{\|}$, the axial diffusion coefficient of the ellipsoid, is given by

$$D_{\|} = \frac{kT}{\eta X_A} \tag{1.18}$$

where X_A, the axial friction coefficient of an ellipsoidal particle is given by

$$\frac{X_A}{6\pi a} = \frac{8}{3} \cdot \frac{s^3}{-2s + (1+s^2)Z} \tag{1.19}$$

with

$$s = \sqrt{1 - \Re^2} \tag{1.20}$$

$$Z = \ln \frac{(1+s)}{(1-s)} \tag{1.21}$$

$D_{\perp}$, the transverse diffusion coefficient of the ellipsoid, is given by

$$D_{\perp} = \frac{kT}{\eta Y_A} \tag{1.22}$$

where Y_A, the transverse friction coefficient of an ellipsoidal particle is given by

$$\frac{Y_A}{6\pi a} = \frac{16}{3} \cdot \frac{s^3}{2s + (3s^2 - 1)Z} \tag{1.23}$$

D_r, the rotational diffusion coefficient of the ellipsoid, is given by

$$D_r = \frac{kT}{\eta Y_C} \tag{1.24}$$

where Y_C, the rotational friction coefficient of an ellipsoidal particle is given by

$$\frac{Y_C}{8\pi a^3} = \frac{4}{3} \cdot \frac{s^3(2-s^2)}{-2s + (1+s^2)Z} \tag{1.25}$$

EXPERIMENTAL AND NUMERICAL STUDY OF HYDRODYNAMIC DISPERSION AND COLLOIDAL TRANSPORT IN SOLID CELLULAR FOAM

Tyler R. Brosten[1], Einar O. Fridjonsson[1], Joseph D. Seymour[1], Robert S. Maier[2] and Sarah L. Codd[1]

[1]Magnetic Resonance Lab, College of Engineering, Montana State University, Bozeman MT USA
[2]U.S. Army Engineer Research and Development Center, Vicksburg MS USA

Introduction

The transport of a colloidal suspension through a porous material is an important process found within a broad range of disciplines; e.g. geophysics, engineering, and medicine. The importance of colloidal transport (e.g. microbe, radio-nuclide, etc.) in the subsurface has motivated much theoretical and experimental work focused upon understanding the mechanisms of colloid transport and deposition in porous media; e.g. [1,2]. The practical challenges associated with experimentally observing the transport of colloids in opaque media have limited past techniques to classic breakthrough methods [2-4]. This work presents results from nuclear magnetic resonance (NMR) measurement of moderate concentration (volume fraction $\Phi = 0.15$), neutrally charged, Brownian (hard sphere particle radius of 1.25μm), colloidal suspension transport through a high porosity open cell polymer foam. The recent development of core shell oil filled particles [5] has allowed for the separation of liquid and solid-phase dynamics in NMR measurements of colloidal suspensions [6-8]. The proven application of NMR in the study of single phase transport in porous media, [9], coupled with recent developments in NMR active colloidal suspensions have established all the experimental details necessary to directly probe the transport of colloidal suspensions in porous media. The experimental observations of colloid transport in this work are limited to the long time limit of transport; times much greater than the transient so-called "clean-bed removal" filtration phase [3]. In this "stationary" regime the interplay of colloid attachment and flow has equilibrated so that the solute and colloid dynamics are stationary.

Materials and Methods

Polyurethane open-cell foam (Foamex, Inc.) was used as the porous media. The foam has a porosity of 97% and an approximate average pore size of 110PPI (pores per inch). Note that the 110PPI designation given by the manufacturer is only an approximate characterization of the pore structure. The NMR transport measurements employed in this work consisted of standard pulsed gradient spin echo (PGSE) methods. The reader is referred to the text of Callaghan [10] for a detailed discussion of these methods and related NMR fundamentals. The NMR transport experiments began by first flushing the sample with de-ionized water (Millipore, Inc.) for about 12 hours at 20mm/s using a peristaltic pump (Masterflex, Inc.). Transport measurements of pure water flow from a Pharmacia P500 pump were then taken for comparative purposes. After the pure water measurements were finished the sample was pre-equilibrated with the colloidal suspension through 6 hours of continuous up-flow at a mean axial flow velocity of $\langle u \rangle = 1$ mm/s. The flow of colloidal suspension was driven by a programmable syringe pump (KDS Scientific, Inc.) interfaced to a 50 ml glass syringe (Hamilton, Inc.).

The core-shell oil particles were manufactured using the method of Loxley and Vincent [5]. The microspheres are short range repulsive due to short chain PVA molecules which adhere to the surface and exhibit hard sphere behavior [11]. The core of these particles are liquid oil and therefore are "NMR active". The suspension dynamics studied in this work are limited to a colloidal volume fraction of $\Phi = 0.15$ with a particle diameter of $d = 2.5 \pm 0.9 \mu m$.

Results and Discussion

Displacement propagator dynamics

The temporal evolution of the displacement propagator, $P(\mathbf{R},\Delta)$, was measured as a function of observation time for both pure water and the colloidal suspension. The phase separated propagator of the colloidal suspension was obtained by integrating the spectral dimension of the two dimensional Fourier transformed signal over the water and oil peaks. The phase separated axial and transverse displacement propagators of the colloidal suspension are compared in Figure 1 for $<u> = 1.0$ mm/s at several observation times.

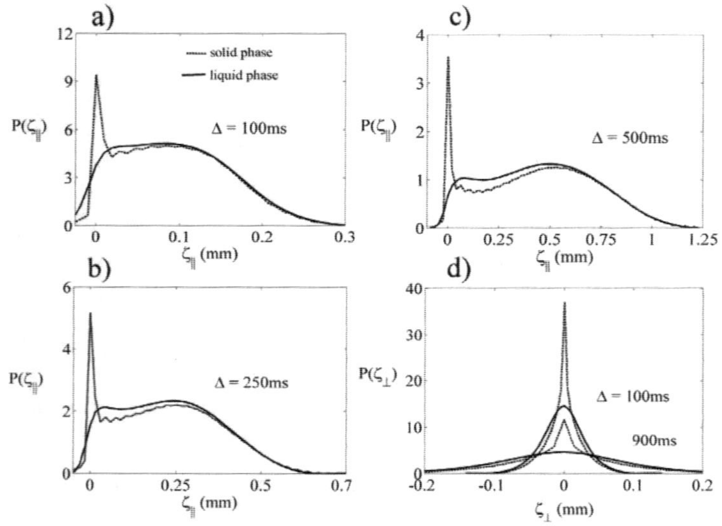

Figure 1: Axial and transverse propagator temporal evolution of the phase separated colloidal suspension at a mean flow velocity of 1 mm/s; a) - c) colloidal suspension axial propagators, d) colloidal suspension transverse propagators

The axial propagator of the colloidal suspension solid-phase, Figs. 1 a) through c) contains a significant portion of the population near axial displacement length $\zeta_{||} = <u>\Delta$ 0, indicating hold-up of colloidal particles. The variance or second moment, σ^2, of the displacement propagator can be determined from a low q fit to the PGSE signal amplitude. A time dependent dispersion coefficient is defined as the variance divided by the observation time, $D = \sigma^2/(\Delta)$. The two quantities share the same information content but in different form. The time dependent phase-separated variance and dispersion coefficient are shown in Figure 2 for the transverse and axial direction. 2 a) & c) contain the axial dispersion dynamics as a function of mean displacement and 2 b) & d) contain the transverse dynamics. The data shown in Figures 1 and 2 was obtained following at least 6 hours of continuous flow of suspension.

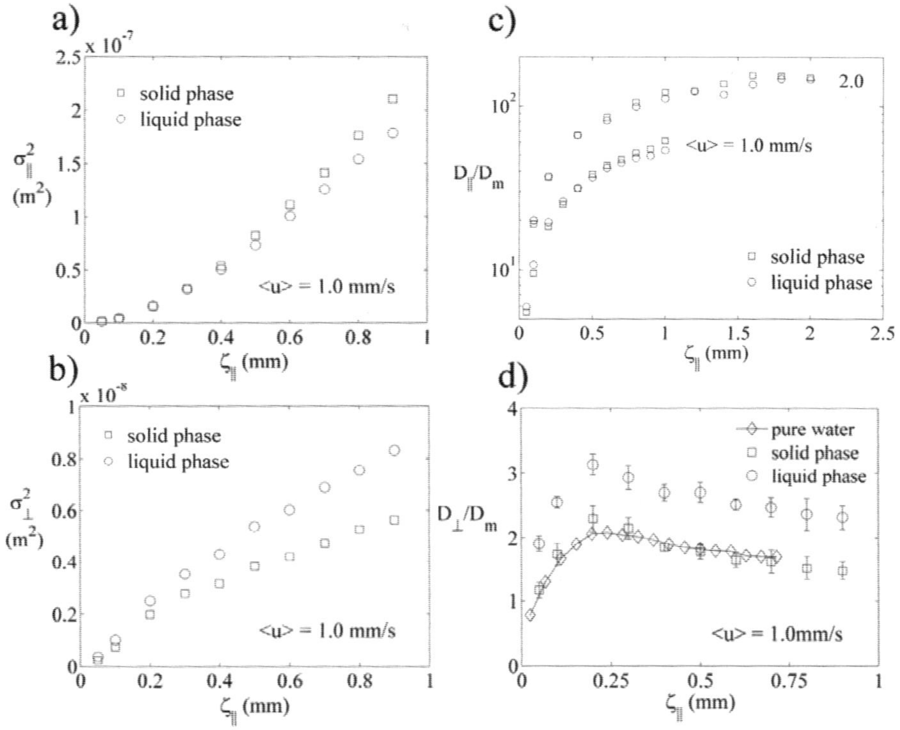

Figure 2: Phase separated axial and transverse dispersion dynamics of the colloidal suspension within the foam pore structure, a) & b) axial and transverse displacement variance as a function of mean displacement for a mean flow velocity of 1 mm/s, c) & d) axial and transverse effective dispersion coefficient as a function of mean displacement, D_m = 2x10^{-9} m^2/s, also shown in d) are the dynamics of pure water at a mean flow velocity of 1mm/s

The linear growth of the axial and transverse variance, Fig. 2 a) & b), at long observation time, $\zeta_{||} > 0.6$ mm, suggests the existence of a unique asymptotic dispersion coefficient for both phases. To our knowledge these are the first measurements directly demonstrating this behavior of a colloidal suspension in a porous media. At long times the axial variance of the colloidal suspension solid-phase grows faster than the liquid-phase variance. There is a transition point in the axial dynamics before which the liquid-phase axial dispersion is greater than the solid-phase, Fig. 2 c). This short time difference is the result of a number of potential factors including the significantly greater molecular mobility of the liquid-phase and also perhaps a greater scattering of the liquid-phase throughout the convective velocity field. In contrast the liquid-phase transverse dispersion is consistently greater than the solid-phase dispersion. The *backscattering* (increase then decrease) effect seen in the transverse dispersion dynamics of both phases, Fig. 2 d), is the result of oscillatory streamlines within the structure. Numerical lattice Boltzmann simulations of single phase transport in a 3D foam structure show this equivalent behavior. The experimental and numerical data of the backscattering effect and a nonequilibrium statistical mechanics description will be discussed in detail in a separate work. For pure water, at high Peclet number, the maximum of the transverse dispersion curves occurs at a fixed mean displacement of *l* = 250μm; a number which is on the order of the wavelength of the axial oscillations of streamlines within the pore space. This fixed displacement is a characteristic length of the structure and will be used in the Peclet number definition of the foam structure.

Asymptotic dispersion dynamics

The phase-separated axial and transverse asymptotic dispersion coefficients are plotted in Figure 3 as function of Peclet number. The Peclet number was varied by adjusting the mean flow velocity. In the convection dominated regime, i.e. Pe > 100, classical dispersion theory predicts a power law scaling of the asymptotic axial dispersion coefficient with Pe [12]. Power law fits to the axial asymptotic dispersion coefficients are shown in Fig. 3 a). The asymptotic coefficients shown in Figure 3 were determined from the long time behavior of $D(\Delta)$.

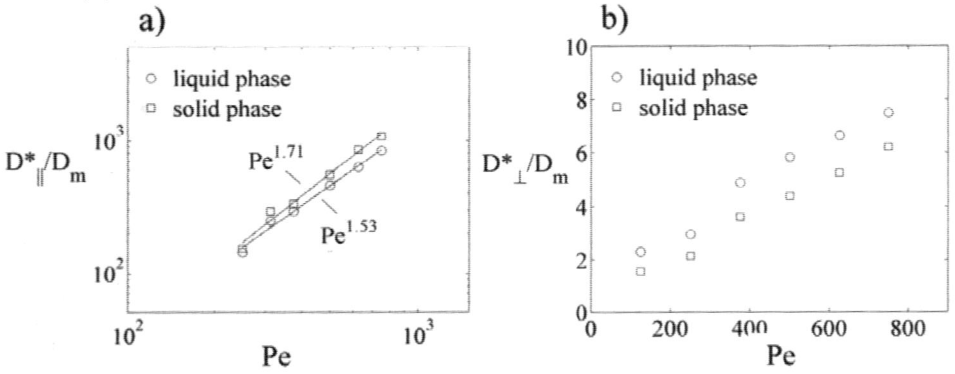

Figure 3: Phase separated a) axial and b) transverse asymptotic dispersion coefficients of the colloidal suspension as a function of Peclet number and non-dimensionalized by the molecular diffusion $D_m = 2x10^{-9}$ m^2/s

The asymptotic axial dispersion coefficient of the colloidal solution solid-phase was found to scale like $Pe^{1.71}$ while the liquid-phase was found to scale like $Pe^{1.53}$; for pure water the coefficient scales like $Pe^{1.81}$ (determined from a linear fit to the long time variance). Therefore, as mentioned in the earlier transverse dispersion discussion, the dynamics of the colloidal solution liquid-phase are more significantly driven by mechanical effects then either the solid-phase or pure water dynamics. The solid-phase dynamics however appear more like Taylor dispersion, where scaling is Pe^2.

Summary

The active transport of a colloidal suspension within cellular foam was probed using NMR. The dynamics of both the liquid and solid-phase of the suspension were presented over a range of observation times and mean flow velocities. The deposition of colloidal particles onto the solid pore structure were shown to cause a transformation of the solid pore structure which in turn alters the macro-scale dynamics of both phases relative to single phase transport. Asymptotic dispersion coefficients of both phases are measured and scaling with Peclet number examined.

References

1. T. R. Ginn, B. D. Wood, K. E. Nelson, T. D. Scheibe, E. M. Murphy, and T. P. Clement (2002), "Processes in microbial transport in the natural subsurface," *Advances in Water Resources,* vol. 25, pp. 1017-1042
2. A. B. Kersting, D. W. Efurd, D. L. Finnegan, D. J. Rokop, D. K. Smith, and J. L. Thompson (1999), "Migration of plutonium in ground water at the Nevada Test Site," *Nature,* vol. 397, pp. 56-59
3. J. N. Ryan and M. Elimelech (1996), "Colloid mobilization and transport in groundwater," *Colloids and Surfaces a-Physicochemical and Engineering Aspects,* vol. 107, pp. 1-56
4. X. Q. Li, T. D. Scheibe, and W. P. Johnson (2004), "Apparent decreases in colloid deposition rate coefficients with distance of transport under unfavorable deposition conditions: A general phenomenon," *Environmental Science & Technology,* vol. 38, pp. 5616-5625
5. A. Loxley and B. Vincent (1998), "Preparation of poly(methylmethacrylate) microcapsules with liquid cores," *Journal of Colloid and Interface Science,* vol. 208, pp. 49-62
6. J. R. Brown, J. D. Seymour, S. L. Codd, E. O. Fridjonsson, G. R. Cokelet, and M. Nyden, "Dynamics of the solid and liquid phases in dilute sheared Brownian suspensions: Irreversibility and particle migration," *Physical Review Letters,* vol. 99, pp. -, Dec 14 2007.
7. P. Stilbs, "Fourier transform pulsed-gradient spin-echo studies of molecular diffusion," *Progress in Nuclear Magnetic Resonance Spectroscopy,* vol. 19, pp. 1-45, 1987.
8. H. Wassenius, M. Nyden, and B. Vincent, "NMR diffusion studies of translational properties of oil inside core-shell latex particles," *Journal of Colloid and Interface Science,* vol. 264, pp. 538-547, Aug 15 2003.
9. J. D. Seymour and P. T. Callaghan, "Generalized approach to NMR analysis of flow and dispersion in porous media," *Aiche Journal,* vol. 43, pp. 2096-2111, Aug 1997.
10. P. T. Callaghan, *Principles of nuclear magnetic resonance microscopy*. Oxford [England] New York: Clarendon Press; Oxford University Press, 1991.
11. H. Wassenius, M. Nyden, and B. Vincent (2003), "NMR diffusion studies of translational properties of oil inside core-shell latex particles," *Journal of Colloid and Interface Science,* vol. 264, pp. 538-547
12. J. Salles, J. F. Thovert, R. Delannay, L. Prevors, J. L. Auriault, and P. M. Adler (1993),

Reaction-Diffusion Based Measurement of Protein-Ligand Affinity

Paul J. Wesson, Chemical and Biological Engineering, Northwestern University, Evanston, IL

Yanhu Wei, Chemical and Biological Engineering, Northwestern University, Evanston, IL

Bartosz A. Grzybowski, Department of Chemical and Biological Engineering and Department of Chemistry, Northwestern University, Evanston, IL

Many methods to measure protein ligand binding/dissociation constants require an intrinsic shift in a "signal" from the protein (i.e. fluorescence, NMR spectra, etc.), thus they require some optimization for each protein measured. However, by using reaction-diffusion (RD), one can accurately measure protein-ligand dissociation constants, Kd, for arbitrary protein-ligand combinations.

In this talk, we show that Kd may be measured using a general RD process. Here, we first "ink" a patterned agarose stamp with a protein, then apply the stamp to a poly(acryl amide) gel which contains immobilized ligand. As the protein diffuses from the stamp into the gel, it intereacts with the ligand: when the protein is bound to the ligand it is stationary, however, when it is not bound to the ligand it freely diffuses. After a prescribed period of time, the stamp is removed from the gel, and the gel is stained with either Coomassie blue or a combined Coomassie blue/silver staining method. The resulting pattern of stained protein may then be fit using the RD equations to determine the dissociation constant for that protein/ligand pair.

See more of this Session: Fundamental Research in Transport Processes II
See more of this Group/Topical: Engineering Sciences and Fundamentals

Measurement of Temperature Distribution Over 2000°C in Premixed Flame by Using Laser Speckle Method

Ken-ichiro Tanoue, Tokiwadai 2-16-1, Department of Mechanical Engineering, Graduate School of Science and Engineering, Yamaguchi University, Ube, Japan

Makoto Takayanagi, Tokiwadai 2-16-1, Department of Mechanical Engineering, Graduate School of Science and Engineering, Yamaguchi University, Ube, Japan

Yuuto Ogura, Tokiwadai 2-16-1, Department of Mechanical Engineering, Graduate School of Science and Engineering, Yamaguchi University, Ube, Japan

Tatsuo Nishimura, Tokiwadai 2-16-1, Department of Mechanical Engineering, Graduate School of Science and Engineering, Yamaguchi University, Ube, Japan

Temperature distribution of premixed flame over 2000°C between methane gas and air has been developed using laser speckle method. The inner diameter of the pipe exit was 13mm. The gas flow rate of methane changed from 0.7 L/min to 2L/min. The flow rate of air was ten times of the methane flow rate. The premixed flame consists of two different kinds of flames. One is the internal flame which the maximum temperature is greater than 2000°C. The other is the diffusion flame around the internal flame. From the movie of the speckle patterns, it was found that the motion of the premixed flame was unsteady and oscillatory. The more methane flow rate made not only the longer the period of the oscillation but also the more the unsteady domain of the diffusion flame. Furthermore, from Fermat's principle and Gradstone-Dale's equation, the temperature distribution of the unsteady and oscillatory premixed flame was predicted.

See more of this Session: Fundamental Research in Transport Processes II
See more of this Group/Topical: Engineering Sciences and Fundamentals

Comment On Coal-Based Chemical Looping Combustion Process

Yan Cao, Department of Chemistry, Institute for Combustion Science & Environmental Technology, Bowling Green, KY

Wei-Ping Pan, Chemistry, Institute for Combustion Science & Environmental Technology, Bowling Green, KY

There is a urgent push focusing on quick ways to curb CO2 emission based on wide-accepted scientific evidences on global warming. Under President Obama's economic stimulus and recovery plan, carbon managements will be the priority in fossil-fuel energy utilization, including coal-fired power plants to reduce CO2 emissions. The United States of America has abundant coal resources. More than 50% of U.S. electricity is derived from coal-based power. Consequently, the U.S. will have to either deal with or lead these upcoming carbon management regulations. However, carbon management in existing coal-fired power plants is burdened with high capital and operational costs based on post combustion capture processes.

This manuscript is to make technical comments on efforts made to develop coal-based chemical looping combustion process, which is an innovative coal-based combustion process dedicated to low cost carbon management and enhanced electricity generation efficiency. Oxygen carriers are tecnical cores of chemical looping combustion process. Selection and preparation of different oxygen carriers directly associated process economics and efficiencies. We are trying to conduct the feasibility study of coal-based chemical looing combustion, and thus point out the opportunity oxygen carriers, that lead to the successful development of coal-based chemical looing combustion process.

Based on selected oxygen carriers, a novel coal-based chemical looping combustion process are proposed. It could be retrofitted to existing coal fired power plants resulting in efficiency enhancement and capacity enlargement as well as 30%-70% CO2 capture under constraint of cost increase. This novel combustion process is the combination of chemical looping combustion (CLC) using novel oxygen carriers and subsequent combustion of residue char in existing, aged, low-efficiency, pulverized coal boilers. The potential impact of this process is the realization of affordable and timely CO2 capture technology that can be introduced into the existing coal-fired power generation fleet. The recovery of coal energy using different combustion technologies will be rationally optimized without energy penalty. Attrition-resisting and thermally stable oxygen carriers will be developed to achieve auto-thermal heat balance of the process, a high purity of CO2, favorable kinetics and participation of free-oxygen. The strategy of controlling multiple major air pollutants (SO2, NOx, and Hg) will be addressed.

See more of this Session: Chemical Looping Processes - II
See more of this Group/Topical: Energy and Transport Processes

Synfuel Production in a Carbon Constrained Scenario Using the Calcium Looping Process

Shwetha Ramkumar, Chemical and Biomolecular Engineering, The Ohio State University, Columbus, OH

Nihar Phalak, Chemical and Biomolecular Engineering, The Ohio State University, Columbus, OH

Liang-Shih Fan, Chemical and Biomolecular Engineering, The Ohio State University, Columbus, OH

The increasing energy demand and the necessity for energy independence have brought coal to liquid (CTL) technologies, which have the capability of near term implementation due to their compatibility with the existing infrastructure, to the forefront. Currently, the indirect production of coal derived liquid fuels is through the coal gasification and Fischer Tropsch (FT) process. It has been estimated that the carbon foot print of the coal to liquids process is at least 150 - 175% higher than petroleum based fuels. By the implementation of carbon capture and sequestration, the life cycle CO2 emissions for the coal to liquids process can be reduced by 20% compared to conventional fuel.

The Calcium Looping Process (CLP) is capable of producing a sequestration ready CO2 stream by capturing all the CO2 emitted during the coal to liquids process. In addition to achieving carbon capture, the CLP improves the efficiency of the coal to liquids process by conversion of the Fischer Tropsch reactor's off gases to hydrogen which is used to adjust the H2:CO ratio of the FT feed and for the product upgrader. The CLP integrates the reforming of hydrocarbons, water gas shift (WGS) reaction and in-situ carbon dioxide (CO2) removal at high temperatures in a single reactor and reduces the overall foot print of the hydrogen production process. The exothermic carbonation and WGSR convert the highly endothermic reforming of hydrocarbons into a heat neutral process thus simplifying the reforming process and reducing the temperature of reforming from >900C to 650C. In this paper, the reaction chemistry, thermodynamics, lab and bench scale experimental studies, process intensification strategies and energy conversion efficiencies will be presented.

See more of this Session: Chemical Looping Processes - II
See more of this Group/Topical: Energy and Transport Processes

Chemical Looping Combustion Process applied to liquid fuels

Ann Forret, Ali Hoteit*, Thierry Gauthier

IFP-Lyon, Solaize, France

Abstract

CLC is a new promising combustion process for CO_2 capture with less or even no energy penalty compared to other processes. Indeed, CLC does not require specific O_2 or CO_2 separation equipment and energy penalty appears very limited. CLC process is an oxy-combustion technology in which the oxygen required for combustion is supplied by metal oxides known as oxygen carriers.

The majority of the work performed to date on CLC has been performed using methane and coal and/or pet coke as the fuel. Up to now, there is no information reported in the literature concerning the use of liquid fuels in CLC. The objective of the present work is therefore to investigate the feasibility of liquid feed injection and contact with oxygen carrier to conduct controlled combustion in CLC conditions. A batch experimental fluidized bed set-up was developed to contact alternatively oxygen carrier with liquid fuels or air. $NiAl_{0.44}O_{1.67}$ was used as oxygen carriers in a fluidized bed reactor and dodecane as the combustible. The fluidized bed experiment results showed that during the reduction period, all the dodecane was converted. A high concentration of CO_2 (> 99%) could be achieved. No formation of carbon was evidenced on the oxygen carrier during the reduction step. Therefore, it was concluded in this study that complete combustion of dodecane is achievable in CLC mode , producing CO_2 and H_2O with stable operation.

Keywords: Chemical-looping combustion; Liquid fuels; CO_2 capture; Fluidization; Oxygen carriers; Clean Combustion

1. Introduction

Over the last 10-15 years, chemical looping routes have started to be investigated in order to provide clean and efficient solutions for several applications such as combustion and hydrogen production. CLC is one of the promising route for CO_2 capture since no specific separation of O_2 from air nor CO_2 from flue gas is required, resulting in much lower energy expected penalties. The chemical looping concept is based on the use of materials with oxidation-reduction cycles that can either oxidize or reduce feedstocks. In combustion, CO_2 rich effluents can be directly produced, enabling subsequent CO_2 transportation and storage. Chemical looping combustion (CLC) is a flameless combustion technology where there is no direct contact between air and fuel. Theoretical analyses of the technology indicate that its use results in increased power efficiency compared to regular combustion [1]. The important feature in the chemical looping combustion is that the carbon dioxide produced from the reduction step is inherently separated from the inert nitrogen and unreacted oxygen from the combustion air. Carbon dioxide is then easily separated from the water by condensation of the water resulting in carbon dioxide available for sequestration. The combustion process does not produce NO_x. The production of a

* Corresponding author: Tel. (+33) 4 78 02 55 71, Email: ali.hoteit@ifp.fr

sequestration ready CO_2 stream does not require any additional separation units and there is no energy penalty or reduction in power plant efficiency. This combustion process is divided into two steps (see Figure 1) : reduction and oxidation.

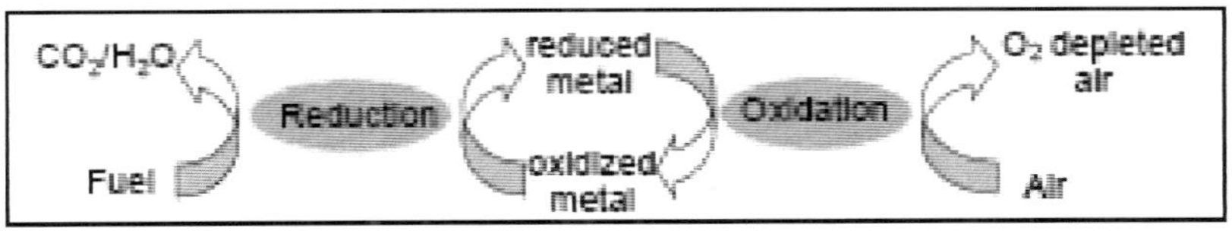

Figure 1: Principle of chemical looping combustion

In the reduction step, fuel gases are oxidized by reduction of a metal oxide via the following reactions:

$$(2n+m)Me_xO_y+C_nH_{2m}\rightarrow(2n+m)Me_xO_{y-1}+mH_2O+nCO_2$$
$$Me_xO_y + CO \rightarrow Me_xO_{y-1} + CO_2$$
$$Me_xO_y + H_2 \rightarrow Me_xO_{y-1} + H_2O$$

(where Me_xO_y is the metal oxide, and $MexOy\text{-}_1$ is the reduced metal oxide)
In the oxidation step, the reduced metal oxide is re-oxidized by contact with air via the following reaction:

$$Me_xO_{y-1} + \frac{1}{2} O_2 \rightarrow Me_xO_y$$

The net result of these two steps is the direct combustion of the fuel gases with oxygen:

$$C_nH_{2m} + \frac{1}{2} (2n+m)\, O_2 \rightarrow nCO_2 + mH_2O$$
$$CO + \frac{1}{2} O_2 \rightarrow CO_2$$
$$H_2 + \frac{1}{2} O_2 \rightarrow H_2O$$

Most of the work around chemical-looping combustion has focused on the use of solid fuel ([2], [3]; [4]; [5]) and on gaseous fuel ([6], [7]). The solid–solid reaction between coal and metal oxide is not very likely to occur at any appreciable rate. Instead the solid fuel needs to be gasified and then the oxygen carrying particles react with the gas produced, making the gasification the time limiting step [8].

2. The aim of the study

Liquid fuels such as heavy fuels resulting from oil distillation or conversion may also be interesting feed stocks to consider. Liquid fuels are challenging feedstock to deal with in fluidized beds. The objective of the present work is therefore to investigate the feasibility of liquid feed injection and contact with oxygen carrier in CLC conditions in order to conduct partial or complete combustion of hydrocarbons in these conditions.

3. Experimental

3.1 Oxygen-carrier materials

The oxygen carrier used in the present study is $NiO/NiAl_2O_4$ which consisted of 60 wt% active phase and 40% support material. This Ni-based Oxygen Carrier was manufactured by Marion Technologies (France). Its average arithmetic and harmonic diameters were 171 and 125 µm, respectively. The bulk density of the particles was

2200 kg/m^3. The BET surface area of the particles is 7 m^2/g. The maximum oxygen-carrying capacity of $NiO/NiAl_2O_4$ was measured as 14.5 wt% with thermogravimatric tests according to literature [9]. Most of the oxygen transfer capacity is accounted for by reduction of NiO to Ni (12.8%), and the remainder by the slower reduction of $NiAl_2O_4$ to Ni and Al_2O_3.

3.2 Reactivity tests in batch fluidized bed reactor

The scheme hereafter (see Figure 2) presents the experimental pilot scale. It consists of a fluidized bed reactor made of quartz with 20-mm inner diameter. The reactor was electrically heated up to 950°C and was operated at atmospheric pressure. K-type chrome/alumel thermocouples was used to measure continuously temperature of the oxygen carrier bed. 45g of oxygen carrier were preheated in an inert atmosphere (N_2) to the desired experimental temperature and then exposed to air to ensure that the particles were fully oxidized before the reduction period.

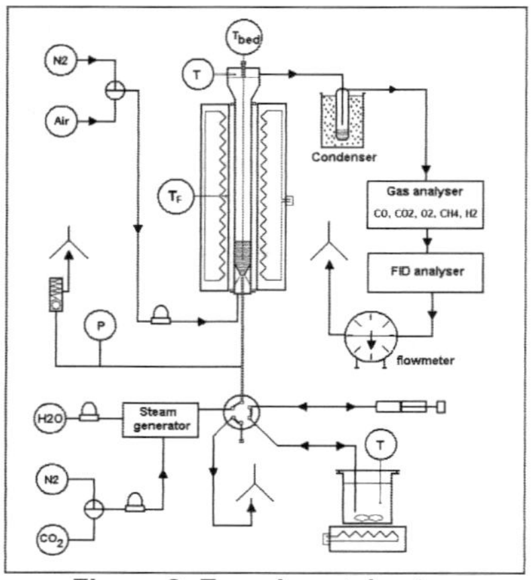

Figure 2: Experimental setup

During reduction step, a pulse of liquid fuel is injected at the bottom of the dense bed. In order to simplify flue gas analyses process of these tests, oxygen carriers are fluidized by nitrogen (instead of CO_2 or steam) during reduction steps. The combustion gas sample was pumped to a water condenser. The H_2, CO, CO_2, O_2, HC and CH_4 gas concentrations were then measured. CH_4, CO and CO_2 were measured by non-dispersive infrared analysis (NDIR); the O_2 concentration was measured in a paramagnetic analyzer and the H_2 concentration by gas conductivity. Hydrocarbons were also measured by Flame Ionization Detector (FID). Data are recorded using a data acquisition system connected to a personal computer running commercial software.

3.3 Evaluation of Data

The conversion of oxygen carrier or the degree of oxidation (X) was defined by the equation (1)

$$X = \frac{m - m_{red}}{m_{ox} - m_{red}} \qquad (1)$$

m, m_{ox}, m_{red} are respectively actual mass of oxygen carrier, mass of fully oxidized and mass of fully reduced oxygen carrier respectively. The conversions of oxygen carrier during reduction and oxidation phases of the cycle are evaluated by using the flue gas concentration as given by equations (2) and (3) respectively:

$$X_{i,red} = X_{i-1} - \int_{t_0}^{t_{red}} \frac{Q_{out}}{n_0 P_{tot}} \left(2p_{CO_2,out} + p_{CO,out} + p_{H_2O,out}\right) dt \quad (2)$$

$$X_{i,oxi} = X_{i-1} + \int_{t_0}^{t_{oxi}} \frac{2}{n_0 P_{tot}} \left(Q_{in} p_{O_2,in} - Q_{out}(p_{O_2,out} - 1/2\, p_{CO,out} - p_{CO_2,out})\right) dt \quad (3)$$

where, n_O is the moles of oxygen that can be removed from fully oxidized oxygen carrier. Q_{in}, Q_{out} are the molar flow of the incoming gas and the gas leaving the reactor respectively. P_{tot} is the total pressure. P_{in} is the partial pressure of incoming gas i and $P_{i,out}$ is the partial pressure of gas i leaving the reactor. and t is the time. The complete conversion equation for oxidation phase takes into account the formation of CO and CO_2 during the oxidation period due to the oxidation of carbon formed in the reduction period.

4. Results

In a first experiment, dodecane was injected on inert solids (carborundum SiC) in order to identify specific phenomena linked to vaporization and pyrolysis. In a second step, dodecane was injected on metal oxide particles $NiAl_{0.44}O_{1.67}$.

4.1 Dodecane injection on SiC

Figure 3 shows the conversion of dodecane on inert carborundum SiC at 800°C. 1 ml of dodecane was injected during 2 seconds at t = 1 minute. The oxygen carrier bed is then fluidized by nitrogen. The product stream of dodecane reaction on SiC consists of a mixture of gases, H_2, CH_4, unreacted dodecane (Hydrocarbons) and traces of CO. After 16 minutes of fluidization, nitrogen is then replaced by air. During the oxidation period, after O_2 introduction in the reactor, CO and CO_2 immediately appear at the reactor outlet stream, indicating that some carbon formation occurs during reduction and remains on the oxygen carrier .

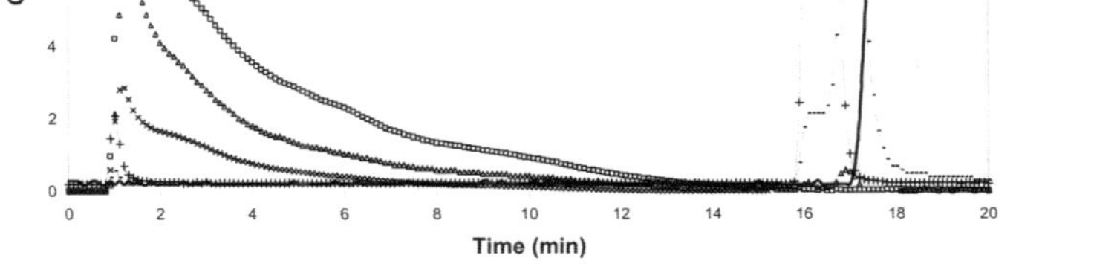

Figure 3: Pyrolysis of dodecane on SiC at 800°C

4.2 Reduction experiments with Dodecane

The oxygen present in 45g of nickel based oxygen carrier ($NiAl_{0.44}O_{1.67}$) totally oxidized is 12.3g. There is therefore in the bed a large amount of oxygen in excess to fully convert 1 mL of dodecane (2.6g) into CO_2 and H_2O.
As shown on Fig.4, after injection of dodecane into the reactor, CO_2 and H_2O are formed and no CH_4, CO nor H_2 can be observed during the carrier reduction time, thus indicating that dodecane conversion is complete .This figure shows also a decrease of the bed temperature due to the endothermic reduction of metal oxide. In the following reoxidation step, upon O_2 introduction to the reactor, CO and CO_2 is not detected at the reactor outlet stream. This result indicate that there is no significant carbon deposit on oxygen carrier at the end of the reduction step. It must be noted that the slight peak of hydrogen measured during the increase of CO_2 concentration ($1<t<1.5$ min) is an artifact: it comes from internal compensation calculation of interference of CO_2 on hydrogen by the analyzer.

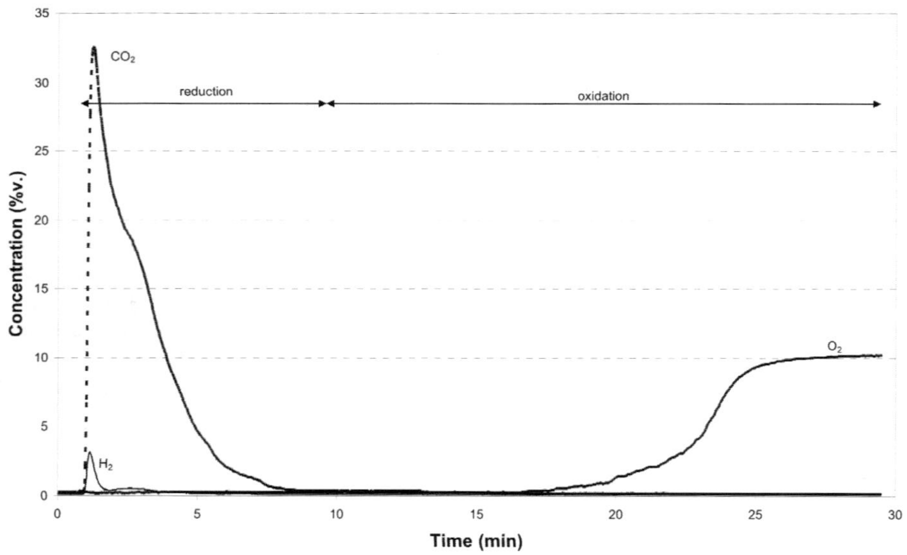

Figure 4: Outlet gas concentration after the injection of dodecane on $NiAl_{0.44}O_{1.67}$ at 800°C

4.3 Series of dodecane injections on $NiAl_{0.44}O_{1.67}$ at 800°C and 900°C

Figure 5 shows flue gas concentration for five consecutive injections of dodecane on a bed of 45g of $NiAl_{0.44}O_{1.67}$ at 800°C. Five injections of 1 mL of dodecane are performed at t = 2.0; 10.7; 21.1; 32.7 et 42.6 min, without intermediate oxidation.
During the two first injections, the conversion of dodecane is complete. This result indicates that, during the second injection time, the amount of available oxygen stored in the oxygen carrier still sufficient to convert dodecane into CO_2 and steam.

Figure 5 shows also the selectivities to synthesis gas. During the third injection, CO and H_2 and CH_4 are detected at the reactor outlet stream. The shift in the reaction is due to the quantity of O_2 present in the oxygen carrier during dodecane injection.

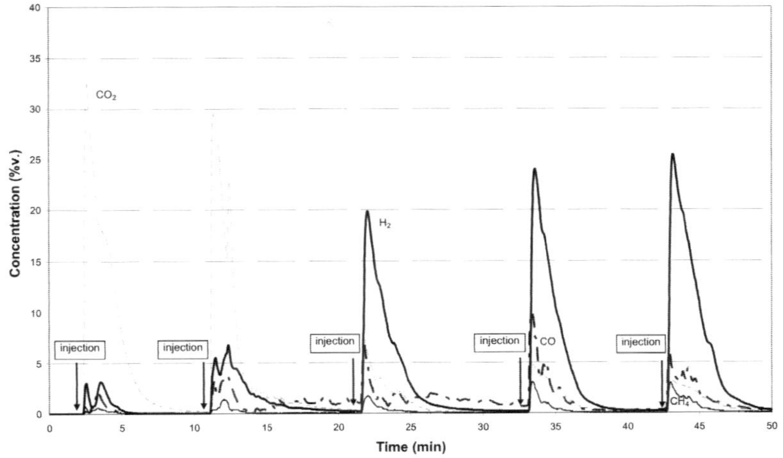

Figure 5: outlet gas concentration for five consecutive injections of 1 mL of dodecane on $NiAl_{0.44}O_{1.67}$ at 800°C (injections at t = 2.0; 10.7; 21.1; 32.7 et 42.6 min).

A similar experiment is performed with the same bed mass at 900°C. Figure 6 shows the flue gas concentrations for eight consecutive injections of dodecane on $NiAl_{0.44}O_{1.67}$. The first four injections of dodecane are converted into CO_2 and steam. That suggests that for the same amount of oxygen stored in the oxygen carriers bed, the quantity of oxygen transferred at 900°C from the oxygen carrier to the fuel is more important than that at 800°C. After 60 minutes of oxidation, an important release of CO is observed, resulting from carbon deposits combustion which was formed during the last metal reduction.

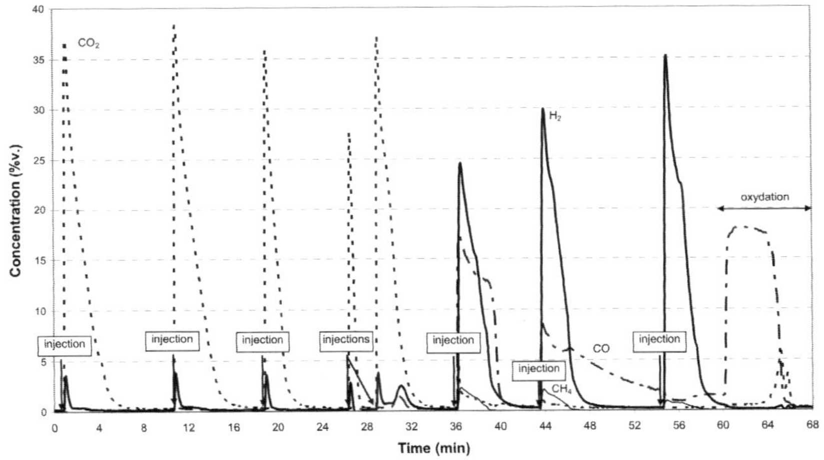

Figure 6: outlet gas concentration for eight consecutive injections of 1 mL of dodecane on $NiAl_{0.44}O_{1.67}$ at 900°C (injections at t= 0.7; 10.6; 18.7; 26.2; 28.7; 36.0; 43.4 et 54.5 min). Oxidation for t>60min

5. Conclusions

Experiments presented in this paper extend the feasibility of the Chemical Looping Combustion (CLC) process to liquid fuels. It was shown that dodecane is totally converted into CO_2 and steam using $NiAl_{0.44}O_{1.67}$ oxygen carrier. No carbon deposits was detected during the reduction step, resulting in an optimal CO_2 capture efficiency. CLC process can then be used to produce energy with a pure flow of CO_2 than can be easy to sequestrate. It is also possible to consider Chemical Looping Reforming for the production of synthesis gas or for cogeneration of H_2 and power.

6. Acknowledgements

The authors are thankful to French National Research Agency for its financial support and to William Pelletant for his co-operation.

References

[1] M. Anheden and G. Svedberg, Energy Convers. Mgmt. 39 (16-18) (1998) 1967-1980.
[2] Cao, Y., Pan, W.-P., 2006. Investigation of chemical looping combustion by solid fuels. 1. Process Analysis. Energy Fuels 20 (5), 1836–1844.
[3] Cao, Y., Casenas, B., Pan, W.-P., 2006. Investigation of chemical looping combustion by solid fuels. 2. Redox reaction kinetics and product characterization with coal, biomass, and solid waste as solid fuels and CuO as an oxygen carrier. Energy Fuels 20 (5), 1845–1854.
[4] Dennis, J.S., Scott, S.A., Hayhurst, A.N., 2006. In situ gasification of coal using steam with chemical looping: a technique for isolating CO2 from burning a solid fuel. J. Energy Inst. 79 (3), 187–190.
[5] Leion, H., Mattisson, T., Lyngfelt, A., 2007. The use of petroleum coke as fuel in chemical-looping combustion. Fuel 86 (12- 13), 1947–1958.
[6] Johansson, M., Mattisson, T., Lyngfelt, A., 2004. Investigation of Fe2O3 with MgAl2O4 for chemical-looping combustion. Indus. Eng. Chem. Res. 43 (22), 6978–6987
[7] Cho, P., Mattisson, T., Lyngfelt, A., 2004. Comparison of iron-, nickel-, copper- and manganese-based oxygen carriers for chemical-looping combustion. Fuel 83 (9), 1215–1225.
[8] Scott, S.A., Dennis, J.S., ., Hayhurst, A.N., Brown T., 2006. In situ gasification of a solid fuel and CO2 separation using chemical looping. AICHE J.52 (9), 3325 – 3328.
[9] Ishida M, Yamamoto M, Ohba T. Experimental results of chemical-looping combustion with NiO/NiAl2O4 particle circulation at 1200 C. Energ Convers Manage 2002;43:1469–78.

Gasification of Biomass and CO_2 Capture Using Chemical-Looping Combustion: Thermodynamic Analysis

Mohammad M. Hossain, Chemical and Biochemical Engineering, The University of Western Ontario, London, ON, Canada

Mohammad R. Quddus, Chemical and Biochemical Engineering, The University of Western Ontario, London, ON, Canada

Hugo de Lasa, Chemical and Biochemical Engineering, The University of Western Ontario, London, ON, Canada

In recent years, biomass including agricultural and municipal waste, has been received increasing attention considering a major source of renewable energy with zero net CO_2 emissions. The use of biomass as fuel, in conjunction with chemical-looping combustion (CLC), offers the additional potential of delivering negative CO_2 emissions. To accomplish this, a possible approach is to gasify the solid fuel producing syngas which can be burned inside the same fuel reactor of a CLC process. The overall gasification and combustion combined process can be described as follows:

Gasification step:

$$C_xH_yO_z \rightarrow H_2+CO+CO_2+CH_4+C$$

$$C+H_2O \rightarrow CO+H_2$$

$$CO+H_2O \rightarrow CO_2+H_2$$

$$C+CO_2 \rightarrow 2CO$$

Combustion step (with oxygen carrier)

$$M_yO_x+H_2 \rightarrow M_yO_{x-1}+ H_2$$

$$M_yO_x \rightarrow M_yO_{x-1} + CO_2$$

The above reactions are heavily influenced by the type of solid fuel, the oxygen carrier, the reaction temperature, and the partial pressure of steam in the fuel reactor (Leion et al., 2007; Cao et al., 2006). This approach has the following main advantages: (i) eliminates extra units for solid gasification and (ii) circumvents the dilution of the syngas with nitrogen (Leion et al. 2007; Scott et al. 2006).

An equilibrium thermodynamic model is reported in this communication able to simulate an integrated Gasifier-Combustor unit using glucose as a feedstock. Glucose is one of the building blocks of cellulose, a main biomass component. Alumina supported nickel based oxygen carrier is considered as the oxygen carrier inthe calculations. The effects of fuel composition, temperature, pressure, and extent of gasification are studied using the Gibbs free energy minimization to determine the product gas composition. This analysis shows that CO_2 can be concentrated and purified up to 99.0% in such a process during the reduction cycle by the gasification products (CO and H_2).

References:

Cao, Y., Pan, W., (2006), Energy & Fuels, 20, 1836-1844.

Scott, S.A., Dennis, J.S., Hayhurst A.N. (2006), AIChE Journal, 52 (9), 3325-3328.

Leion, H., Mattisson, T., Lyngfelt, A. (2007), Fuel 86, 1947-1958.

See more of this Session: Chemical Looping Processes - II
See more of this Group/Topical: Energy and Transport Processes

Insights From Process Engineering Simulations for Envisioning Design of Chemical Looping Combustion Process for Solid Fuels

Asad H. Sahir, Department of Chemical Engineering, University of Utah, Salt Lake City, UT

Adel. F. Sarofim, Dept. of Chemical Engineering, University of Utah, Salt Lake City, UT

JoAnn Slama Lighty, Chemical Engineering, University of Utah, Salt Lake City, UT

Chemical Looping Combustion is currently being investigated as a novel technique for coal combustion with carbon capture. The research community is exploring options to extend the promising results of laboratory studies on chemical looping combustion of coal into an economically viable combustion process.

Process Engineering Simulations for Chemical Looping Combustion of coal have been developed in ASPEN PLUS. A simplified case for chemical looping combustion of coal with copper oxide into gaseous components has been studied using equilibrium reactor models. The fluidized bed fuel and air reactors have been modeled as a series of equilibrium reactors. The temperature of the fluidized bed has been studied as a function of coal, copper oxide flow rates and heat input. It has been observed that the temperature of the conceptualized fuel equilibrium reactor was sensitive to the mass ratio of copper (II) oxide to coal. An inference which could be drawn from the study is that a proper distribution of coal and copper oxide particles is important in maintaining the temperature throughout the length of the fuel fluidized bed reactor. A similar study has been done for analyzing the effect of distribution of copper (I) oxide and air on the temperature throughout the length of the conceptualized air fluidized bed reactor. The analysis suggests possibilities to conceptualize novel process configurations for combustion of coal, taking into consideration that the combustion process for coal occurs over a wide temperature range.

The talk would also feature results of a techno-economic analysis conducted based on the simulations with kinetics of coal combustion incorporated in the process models.

See more of this Session: Chemical Looping Processes - II
See more of this Group/Topical: Energy and Transport Processes

Modeling and Sensitivity Analysis of the Chemical Looping Combustion Process

Balaji Sukumar, Chemical Engineering, Carnegie Mellon University, Pittsburgh, PA

Jovan Ilic, Electrical and Computer Engineering, Carnegie Mellon University, Pittsburgh, PA

Erik Ydstie, Chemical Engineering, Carnegie Mellon University, Pittsburgh, PA

Bruce Krogh, Electrical and Computer Engineering, Carnegie Mellon University, Pittsburgh, PA

Chemical looping is believed to a promising technology in clean combustion of gas or solid fuels for future power generation. In this process, oxygen required for combusting the fuel is provided through metal oxides rather than using air/oxygen. This prevents the production of pollutants like oxides of nitrogen as there is no nitrogen present in the feed. In addition, it has been proved that almost pure carbon dioxide can be produced as a result of this process which makes it easy for carbon sequestration. Noteworthy experimental works have been carried out in the past few years in determining the better metal oxide for these systems. However, the characteristics of metal oxides vary quite significantly with respect to fuel types. Thus, a universal metal oxide which suits for all kinds of fuel is yet to be determined. Therefore, devising a mathematical model to study the effectiveness of various important physical and operating parameters will be beneficial.

The process consists of two fluidized bed reactors connected through loop seals to maintain the pressure drop in the reactors. The inflow and outflow of the solid metal oxide and the fuel should be monitored appropriately for enhanced performance. The process is still at the nascent level and being controlled manually or by simple control algorithms. Once commercialized, there will be a definite need of advanced control strategies for controlled power output. Thus, in this study, we focus on developing a model for the entire CLC system with a view point of determining the optimal operating conditions and the sensitivity of important operating parameters. The model is based on system invariants under thermodynamic equilibrium coupled with the shrinking core mechanism to represent the gas-solid reactions. The model results are validated based on the experimental data available in the literature. To the authors' knowledge, this work is the first in proposing a control oriented model for the CLC system.

Keywords: Modeling, thermodynamic equilibrium, invariants, chemical looping combustion, control.

See more of this Session: Chemical Looping Processes - II
See more of this Group/Topical: Energy and Transport Processes

Integrations for Oxyfuel Power Production Process Utilizing Novel CO2 Purification System

Jeffrey Kloosterman, Tonnage Gases, Equipment and Energy - Technology, Air Products, Allentown, PA

Kevin B. Fogash, Tonnage Gases, Equipment and Energy - Technology, Air Products and Chemicals, Inc., Allentown, PA

Vince White, Tonnage Gases, Equipment and Energy - Technology, Air Products PLC, Walton-on-Thames, United Kingdom

Two key aspects of oxyfuel combustion of coal in power boilers are the process concepts for the oxy-combustion within the boiler scope and the backend CO2 purification and compression. Thoughtful integration of various unit operations can improve the overall efficiency of the oxycoal combustion process for power production. As for the CO2 capture process, special attention must be paid to the contaminants within the raw CO2 product beyond the typical H2O, O2, N2, and Ar. In particular, acid gases such as SO3, SO2, HCl and NOX produced as products of combustion represent both an opportunity and a challenge. These acidic impurities will need to be removed cost effectively from the CO2 stream before it is introduced into the pipeline to prevent corrosion and comply with possible regulations. The keys to providing a viable CO2 capture technology relies on two principals: 1) advanced technology that reduces the penalty associated with capture and 2) integration of capture technology within an oxyfuel power boiler flowsheet. Air Products' novel compression and purification technology provides a viable and unique solution to the CO2 capture problem which can be readily integrated into retrofit and new build markets.

Air Products has advanced this technology via many projects including a U.S. Department of Energy (DOE) project to design and construct its proprietary CO2 purification system for a slip stream (~1/3 MWth) pilot-scale tests at the site of a 15 MWth tangentially-fired (T-fired) boiler oxycombustion test facility. Air Products is also developing other plants for larger pilots of the technology that will demonstrate the sour compression technology at around the 1 MWth scale, plus purification with inerts and oxygen removal, together with other aspects of Air Products Oxyfuel Purification Technology, such as the use of a membrane to recover CO2 and O2 normally lost with the inerts stream. In this paper we review the latest process concepts/integrations for oxyfuel combustion of coal for CO2 capture as well as an integrated process for CO2 compression and simultaneous purification to produce CO2 product streams at purities up to essentially pure CO2.

Acknowledgment: "A portion of this material is based upon work supported by the Department of Energy's National Energy Technology Laboratory under Award Number DE-NT0005309."

See more of this Session: Oxycombustion of Coal - Needs, Opportunities, and Challenges - I
See more of this Group/Topical: Energy and Transport Processes

A Numerical Method for Optimizing Oxyfuel Power Plants

Xinxin Li and Klaus Lackner,

Earth and Environmental Engineering, Columbia University, New York, NY

1.Introduction

Choosing between near-term oxyfuel combustion technologies and novel next-generation technologies is difficult and requires a means of making comparisons across vastly different technologies. This paper proposes a computational methodology to evaluate different technologies and to identify optimal technologies based on a user supplied set of evaluation criteria. The tool makes it possible to compare different modules of oxyfuel power plants, different oxyfuel combustion power plant designs, and different technology pathways leading from today's power plants to the power plants of the future.

2.Different oxyfuel combustion technologies

Oxyfuel combustion technologies involve individual modules or components (for example air separation units, oxyfuel boilers, flue gas processing units, etc). The fossil fuel, usually coal, is combusted in pure oxygen that may be diluted with combustion products. The output is a nearly pure stream of carbon dioxide (Croiset, 2000). Oxyfuel combustion technologies are among the most promising low-emission power plant generation technologies.

The need for a fundamental redesign of today's power plant is driven by the rapidly growing consensus that excess carbon dioxide will cause a significant change in climate and will have repercussions on a wide variety of human activities. It is also driven by the prospect of increasingly stringent environmental regulations, which will eventually demand zero emission power plants that do not emit any gaseous effluent to the atmosphere (Lackner 2001). Oxyfuel combustion technologies, among all low-emission power generation technologies, offer a unique opportunity for improved environmental control performance. The new combustion environment without nitrogen makes it possible to design power plants where the entire off-gas stream is disposed of and kept out of the atmosphere.

A spectrum of competing oxyfuel combustion power plant designs already exists. On one end of the spectrum is a class of plant designs that simply modify existing pulverized coal plants. Such designs are suitable for a retrofit and they consist of adding a conventional air separation unit for producing oxygen and sealing the boiler system so as to avoid the accidental introduction of nitrogen rich outside air (Wilkinson et al., 2003). On the other end of the spectrum are plants that are based on entirely novel technologies. This includes new oxygen production methods, for example by separation with high temperature ion transport membranes (Sinner and Kilner 2003; Bouwmeester and Van Laar, 2002; Dyer et al,. 2000; Bredesen et al., 2004). One might even consider as advanced oxyfuel designs chemical looping systems that oxidize a metal in contact with air and carry the oxygen to the fuel bound to the metal (Zafar et al., 2005).

Choosing between different oxyfuel combustion technologies has a profound impact not only on the cost-efficiencies of an individual power plant, but more importantly on the pathway that will lead from today's power plant technology to future technologies. In order to choose from a wide range of technologies, and more importantly from different technological pathways, one needs to make comparisons between different plants and different pathways.

This paper aims to create conceptual plant designs and conduct engineering assessments of the component modules of an oxyfuel power plant; evaluate different power plant designs under various energetic, economic, environmental and infrastructural constraints; and perform optimization not just of the individual power plants, but also of pathways connecting current power generation technologies to future technologies.

3.Novel evaluation criteria based on a penalty model

Power plant development leading to a zero emission plant design could move through a set of new plants, each designed to the best available knowledge at the time and with little regard to the long term goal, or to the basic knowledge that is embedded in previous designs. In such a strategy one may introduce technologies even though it is clear from the outset that they do not lend themselves to further advances. For example, post combustion technology, may well be in this category. Any R&D investment into flue gas scrubbing is most likely made obsolete by the next generation of power plants. Alternatively, the goal could be achieved by a set of incremental improvements that are introduced in each new plant or in each upgraded plant, where changes are designed to build upon each other. Oxyfuel combustion designs are likely to fit into this category.

A consideration of the intermediate plant designs can reduce the long-term cost of power plant designs. However, a rational implementation of such an approach requires the means of making comparisons across different technologies and across different times. We propose a methodology by which we can make such an assessment. The method introduces a penalty function that can be applied to modules, plants, and sequences of plant designs. In optimizing the design, one varies design parameters so as to minimize the penalty function. The penalty function is zero for some perfect state of the system which is typically not attainable, and the penalty function is optimized by varying all the available design parameters. The penalty can be thought of as a sum of penalties for specific aspects of the plant, for example its efficiency, its cost or its environmental impact. Individual modules may have component penalties. Some aspect of the penalty will depend on properties that can only be defined for the entire plant, or even for a sequence of plants.

The relative weights of these penalties can be chosen appropriately by a user, who has specific goals. For examples, penalties one may associate with having to build a new plant on a new site may vary for users in different countries. For example, building new plants in China is likely to introduce a relatively small penalty for greenfield plants. The same decision in the West is likely to introduce a much bigger weight, because the political difficulties of opening up new sites are much larger. The different weights may result in alternative development pathways.

It is also recognized that the availability and maturity of novel technologies, as well as environmental thresholds for existing and potential criteria pollutants are likely to change over time, thus the weights for these penalties should not be considered as static. The dynamic nature of these penalty specifications, allow users to choose, instead of the best possible plant at a specific time ti, the best possible pathway connecting technologies from time t1 to time tn. Pursuing the optimal pathways on the basis of the minimum total pathway penalties helps users lower the cost of achieving the specific goals, even if it results in seemingly sub-optimal outcomes for individual plants.

Much of the work we present is in defining an appropriate set of penalties on which the optimization rests. The underlying algorithms have been studied for other applications. The approach outlined here has taken its inspiration from the typesetting software TeX, in which Donald Knuth demonstrated the power of these penalty based optimization algorithms for trading off between very different and very difficult to quantify properties of text. We will show how very similar algorithms can be used to select appropriate modules, and power plants to arrive at a sequence of plant designs that provides an advantageous technology pathway from today's power plant designs to a future design that has far higher efficiency, avoids all emissions to the air, and provides the CO_2 produced in a concentrated stream ready for disposal.

Enhanced CO_2 Sequestration From a Combined Cycle Power Plant

Sivaji Seepana, Department of Chemical Engineering, IIT Madras, Chennai, India

Sreenivas Jayanti, Department of Chemical Engineering, IIT Madras, Chennai, India

At present 80 % [1] of world energy is being generated by the combustion of fossil fuels because of their easy availability and technical maturity in transforming raw material to final useful products. The products of the burning of the fossil fuel carbondioxide (CO_2), which is a greenhouse gas, and nitrogen and sulphur oxides, which have long been recognized as pollutants. In order to avoid continuous accumulation of CO_2 in the atmosphere, efforts on several fronts are being carried out worldwide to tackle CO_2 build-up in the atmosphere. CO_2 capture and storage (CCS) is one of the most promising of such measures, which can reduce emissions from fossil fuel-fired power stations by 80-90% [2]; here, the CO_2 released from the combustion of hydrocarbon fossil fuels is captured at the source itself, for example, in a natural gas fired combined cycle power plant, and the flue gas is prevented from being released into the atmosphere, by storing it in liquid form in an underground geological formation. However, the difficulty with these conventional plants is the very low CO_2 concentration (~3.5% by volume) in the flue gas. Hence, the capture process is difficult, and powerful solvents are required to remove CO_2 in a conventional air-fired power plants. In addition, the presence of NOx and other pollutants also is a problem in the monoethanolamine (MEA)-based CO_2 absorption process [3] leading to a low CO_2 removal efficiency. These problems can be overcome if the flue gas is produced under oxy-fuel combustion mode. Here, combustion occurs in a CO_2/ recycled flue gas environment rather than in nitrogen environment (as in conventional combustion) and a CO_2-rich flue gas (up to 85% by volume) can be achieved. After the removal of water vapour, the remaining gases, mostly CO_2, will be sent for direct compression and then for storage.

In the present paper, a variation to the oxy-fuel combustion, namely, enriched CO_2 flue gas recycling is proposed [4] which is based on the concept of intermediate-temperature diluted oxygen carbon sequestration (INDOCS) [5] method in which preheated oxidant with diluted oxygen content is used to moderate the flame temperature. The oxidant is made up of recirculated exhaust gas enriched in CO_2 (obtained by partial removal of water vapour) and nearly pure oxygen from an air separation unit. It is shown here that this mode of operation is particularly suitable for a combined cycle, gas-fired power plant.

In a normal air combustion plant, air is drawn from the atmosphere, and is sent for compression. The compressed air at high velocity passes through to the combustion chamber where it mixes with fuel in desired proportion to generate sustained flames continuously. Usually, gas turbine combustion process occurs in the pressure range of 12-15 bar and gas turbine inlet temperature is in the range of 1200-1400 ^{o}C. The exhaust gas temperature from gas turbine is at 700-500 ^{o}C depending upon the downstream applications, the energy in the flue gas from gas turbine can be recovered using water/steam. This steam is sent for additional power generation in high pressure steam turbine and low pressure steam turbine. After recovering the energy, the gases will be sent through the stack to the atmosphere. The flue gases at the exit are in the temperature range of 125 to 175 ^{o}C. In the new variant proposed here, the flue gases at the exit from the gas turbine and steam generation are sent through the secondary energy recovery system to be brought down to a temperature in the range of 150 ^{o}C, and are then sent through a flue gas conditioner. Here a part of the flue gas is condensed and is drained out. The resulting mixture has now less moisture and is thus CO_2-enriched; it is preheated to a temperature in such way that, after compression, enriched CO_2 flue gas is in the range of 327 ^{o}C to 577 ^{o}C and is mixed with oxygen in such a proportion as to have a decreased O_2 concentration in the oxidizer of 16% by mass [5]. This will increase the CO_2

concentration in the exhaust gas which enhances the CO_2 sequestration efficiency. The advantages of the proposed CO_2-enriched flue gas recirculation over the conventional flue gas recycling for oxy-fuel are (i) reduction in the overall volumetric flow rate of the gases through the gas turbine resulting in decreased size of the turbine, and (ii) decreased amount of non-condensable gases in the exhaust leading to a more efficient sequestration of CO_2.

References

[1] IEA, Key World energy statistics, International Energy Agency, Paris, 2008

[2] Holloway S. Carbon dioxide capture and geological storage, Phil. Trans. R. Soc. A 2007;(365):1095-1107, (doi:10.1098/rsta.2006.1953)

[3] Yang H, Xu Z, Fa, M, Gupt, R, Sliman, RB, Bland AE, Wright I, 2008 Progress in carbon dioxide separation and capture: A review, Journal of Environmental Sciences;2008 (20): 14-27.

[4] Seepana S. and Jayanti S. Optimized enriched CO_2 recycle oxy-fuel combustion for high ash coals, Fuel 2009 (Accepted for publication).

[5] Seepana S. and Jayanti S., A new stable operating regime for oxy-fuel combustion, 2008, ASME Conference, Boston. (#IMECE2008-67091)

See more of this Session: Oxycombustion of Coal - Needs, Opportunities, and Challenges - I
See more of this Group/Topical: Energy and Transport Processes

Oxy Combustion of Coal - Needs, Opportunities, and Challenges

2009 AIChE Annual Meeting

Results of basic study on Oxy-fuel Combustion System for CO_2 Capture

Toru Ishii[a], Toshihiko Yamada[a], Makoto Takafuji[b], Yoshiyuki Iso[b], Nobuhiro Misawa[c], Takashi Kiga[d] and Chris Spero[e]

a IHI Corporatio , Toyosu IHI Building 1-1, Toyosu 3-chome, Koto-ku, Tokyo 135-8733, Japan
b IHI Corporation, 1 Shinnakahara-cho Isogo-ku, Yokohama-shi, Kanagawa 235-8501, Japan
c Electric Power Development Co., Ltd., 15-1, Ginza 6-chomem Chuo-ku, Tokyo 104-8165, Japan
d Japan Coal Energy Center, 9F Meiji Yasuda Seimei Mita Building, 3-14-10 Mita, Minato-ku, Tokyo 108-0073, Japan
e CS Energy Ltd., Level 21, 66 Eagle St., Brisbane, GPO Box 769, Qld 4001, Australia

1. Introduction

One of the main topics at Hokkaido Toyako summit held in Japan, July, 2008, was "Environment and Climate Change". As the long-tern Goal, the G8 leaders agreed to seek to share and adopt it with all Parties to the United Nations Framework Convention on Climate Change, with respect to the goal of achieving at least 50% reduction of global emissions by 2050. Concerning the global warming, Kyoto Protocol entered into force in February 2005, and the intension of Japanese government towards a low-carbon society, that means CO_2 emission reduction of 60 – 80 % as compared with the present amount of CO_2 emission by 2050, was expressed in the Fukuda vision. From a global viewpoint, however, thermal power plants are still releasing CO_2 in large quantity at present, which indicates the necessity for a power generation system with Carbon Dioxide Capture and Storage (CCS) in more effective and economical manner.

Among all the fossil fuels used at thermal power plants, coal produces the greatest amount of CO_2 per unit calorific value. Thus, emissions from power plants using coal seem to be one of the significant sources in capturing CO_2 effectively.

Recently many CO_2 capturing process are under development. They are generally divided into three types. Oxy-fuel combustion is one of the types. The others are pre-combustion and post-combustion. The comparison between the three types and the outline of oxy-fuel combustion system are shown in the next section.

After that some basic study results toward realization of oxy-fuel combustion system are introduced and then the progress report of the oxy-fuel demonstration project, the Callide Oxyfuel Project, is also introduced.

2. Outline of the oxy-fuel combustion system

In the flue gas of typical coal fired power plant CO_2 concentration is only about 13%. The main ingredient of the flue gas is nitrogen. This nitrogen originates in combustion air which has about 80% of nitrogen. Removing nitrogen from combustion air makes the flue gas composition theoretically no nitrogen and over 90 percents of CO_2 [1][2]. This higher CO_2 flue gas makes it possible to easily separate CO_2 from other gas components with compression and cooling. This is a concept of oxy-fuel combustion.

As mentioned above, there are other types of CO_2 capturing system. Fig. 1 shows three types of CO2 capturing system. First, numbering (1), is oxy-fuel combustion system. Second, numbering (2), is an example of pre-combustion type applied to coal gasification system. Third, numbering (3), is an example of post-combustion type applied to a conventional boiler system. The second type and the third type are CO_2 capturing system from fuel gas or flue gas containing CO_2. CO_2 is captured with some method such as chemical absorption, physical absorption, adsorption, or membrane permeation. These two types are based on the similar mechanism. As compared with them oxy-fuel combustion system is based on the different mechanism.

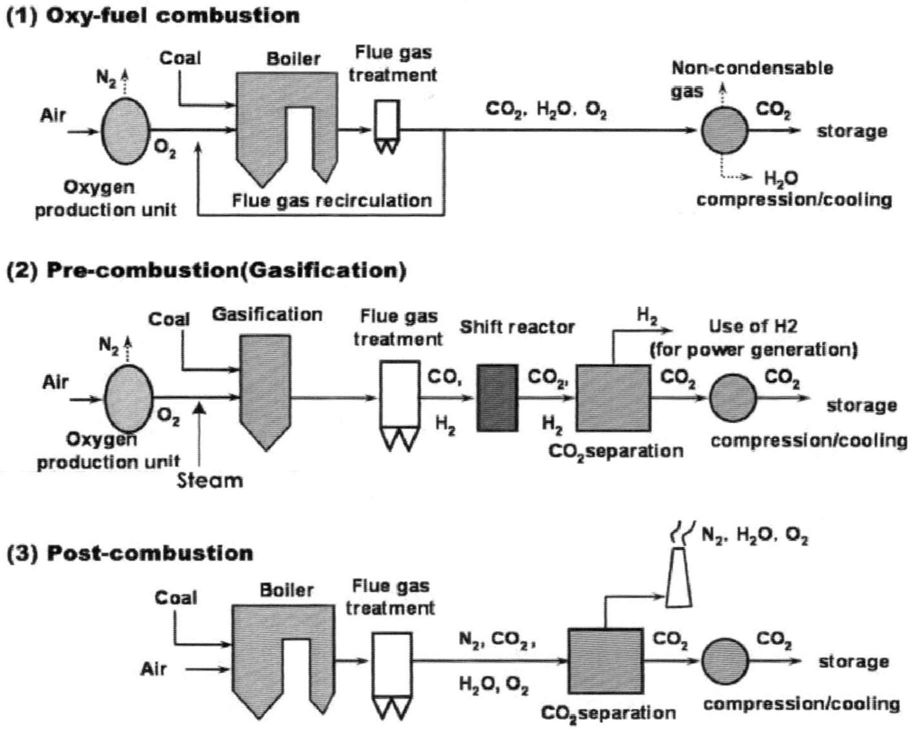

Fig. 1 Typical CO_2 Capturing systems

3. System flow and the subjects to be solved

In the actual oxy-fuel combustion system oxygen is diluted with recycled flue gas because of controlling combustion temperature. Fig. 2 shows simplified oxy-fuel combustion system flow on Callide Oxyfuel Project. Oxygen is separated from air and supplied toward boiler. Oxygen is mixed with recycled flue gas on the way of the boiler combustor inlet, marking as "Mixing point".

Boiler outlet flue gas is filtered with the fabric filter, marking as "FF" in fig. 2. The greater part of flue gas is recycled to the air heater, marking as "AH" in fig. 2. Then the recycled flue gas is heated up, mixed with oxygen, and supplied to the burners. A part of flue gas is conducted with the CO_2 purification unit, marking as "CPU", in fig. 2, and sent to storage system. In the air combustion mode, for example in startup and shutdown operation, flue gas is not recycled and air from atmosphere, marking "Air" in fig. 2, is used for combustion gas.

The fly ash passed through the filter comes to the air heater, marking as "AH" with recycled flue gas. Then the fly ash accumulates at the horizontal duct at the outlet of air heater. Here is the oxygen mixing point and accumulated fly ash is contacted with high concentration oxygen. Generally fly ash from coal combustion system includes several percents of carbon on weight. There is some risk as possibility of ignition of fly ash at the high temperature and high oxygen concentration.

Concerned with oxygen mixing there is another risk. Oxygen and recycled flue gas are mixed in the flow during the mixing point and burner inlet. In order to control all the burner of boiler furnace stably oxygen concentration has to be uniform at the inlet of burner wind box. In the limited area during the mixing point and burner wind box optimization of flow pattern is needed for good mixing of oxygen and recycled flue gas.

Mentioned above there are many subjects toward the realization of oxy-fuel combustion system. We will solve these subjects step by step.

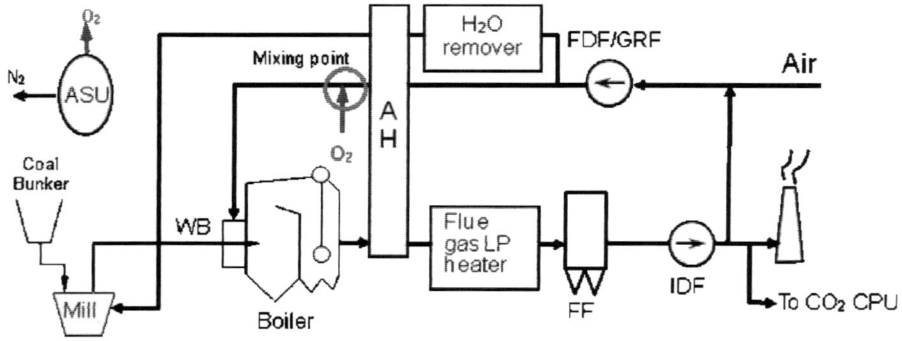

Fig. 2 Oxy-fuel combustion system flow on Callide Oxyfuel Project

4. Results of basic study towards the realization of oxy-fuel combustion technology

As mentioned above, there are some subjects on handling of pure oxygen. Oxygen is mixed with recycled flue gas to control oxygen concentration in the combustion gas. This operation is very important to keep combustion condition steady and safe. The mixing operation is carried out by a simple line mixing because of making the pressure drop at the flue gas recycle line minimize. Pure oxygen is put into the flue gas recycle duct with an injection nozzle therefore near by the injection nozzle oxygen concentration is locally very high. When the recycling flue gas flow is not uniform, there is every possibility of oxygen accumulation at the local area. Accumulation of oxygen is very hazardous, but that does not always lead to failure. It is necessary to confirm the safety condition on flammability of fly ash and material of duct. And it is also necessary to study the mixing performance. In this paper some basic study results are introduced. The first is behaviour of fly ash in the high oxygen condition, and the second is the results of gas mixing simulations. In addition to the previous two items, the test results on performance of flame detector that is a important sensor to control the combustion condition.

4.1 Behaviour of fly ash in the high oxygen concentration atmosphere

In this section we introduce the test results which are concerned with behaviour of fly ash under high oxygen concentration and high temperature. The tests were carried out using the small tube reactor with electric heater. System configuration of test apparatus is shown in Fig. 3. In the test, O_2 concentration in the supply gas to the furnace is controlled by the mass flow meter for O_2 and N_2 gas, and N_2 is used as the balanced gas in the test. Gas temperatures around the sample cell are monitored and gas temperature is increased controlling the rate of 3 degree C per a minute up to the setting temperature. Carbon content in the sample based on fly ash is adjusted by mixing the activated carbon and is set in the cell which is approximately 1cm^3. In the gold furnace, supply gas is heated at the lower section of the furnace filled with the ceramic balls by electrical heater.

Pattern diagrams of the test results are summarized in Fig. 4. Test results show the characteristics of ignition and weight loss depending on the O_2 concentration and carbon content in the sample and are classified into 3 types of results. First is the area of ignition in the left diagram of Fig. 4. This means that the temperature of the sample is rapidly increased at red heat at a certain temperature. Second is the area of no ignition with weight loss shown in the both diagram of Fig. 4. Temperature increasing was not detected however weight loss is measured after the test. This is because the carbon content in the sample was oxidized in the high temperature. Third is the un-reacted area shown in the right diagram of Fig. 4. Temperature increase was not detected and weight loss was not measured.

From now, many tests would be performed to evaluate the safety condition during mixing of O_2 with recycled flue gas. From these results, we can review the safety atmosphere and select the appropriate mixing point with O_2.

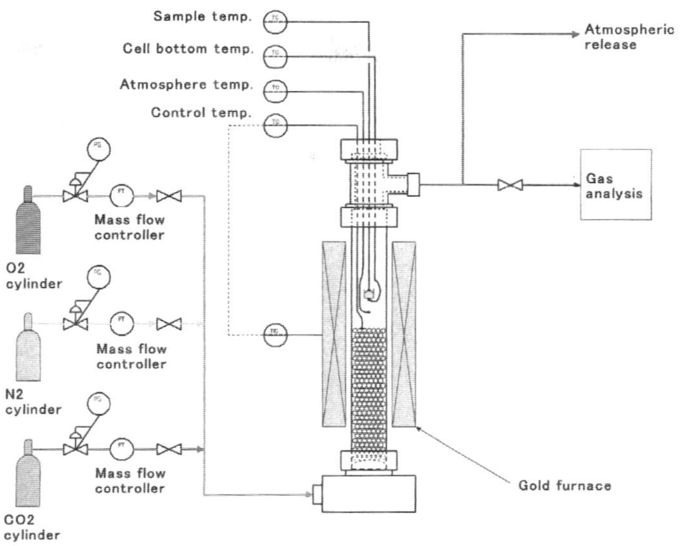

Fig.3 System configuration of confirmation test of the behaviour of fly ash under the high temperature and high O_2 concentration atmosphere

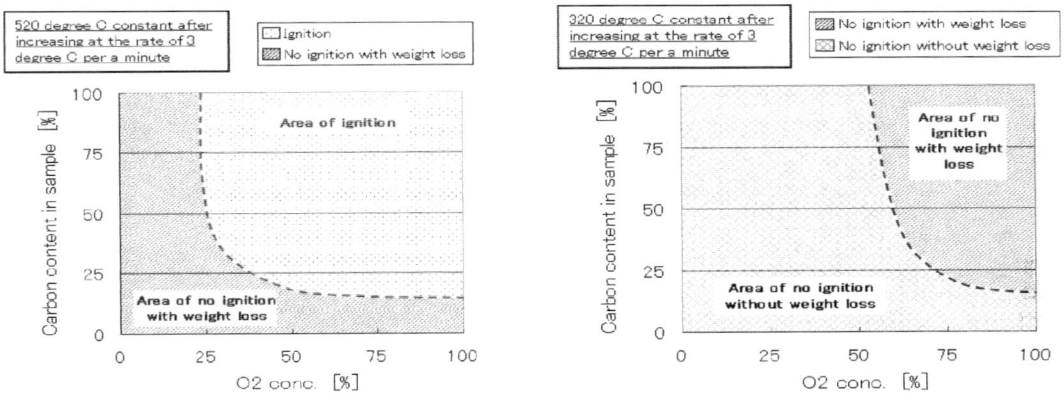

Fig. 4 Pattern diagrams of the behaviour of fly ash under high temperature and high O_2 concentration atmosphere

4.2 Mixing of O_2 with recycled flue gas

In the oxy-fuel combustion system, oxygen is mixed with recycled flue gas in order to keep the flame temperature before the combustion in the furnace. In the normal oxy-fuel process, oxygen is mixed with recycled flue gas at the outlet of air-heater which is upstream of burner wind-box with considering of avoidance the oxygen leakage at the air-heater. This means that mixing area is only duct to the burner after the mixing point. If there is the higher distribution gradient of O_2 concentration at the burner wind-box, flame temperature and combustion characteristics are different among the burners. This higher distribution gradient like this should be avoided in order to maintain the safety condition and the stable operation of the boiler. In addition, considering the lifetime and the safety for the duct, the O_2 concentration with higher temperature region should be as low as possible on the duct wall, as described in section 4.1. Therefore, mixing simulation study was performed to confirm the distribution gradient of O_2 concentration at the burner inlet.

Temperature of recycled flue gas and O_2 are approximately 300 degree C and normal temperature for each. Flow rate of O_2 is a quarter of flow rate of recycled flue gas.

Fig. 5 shows the mesh structure of the model that has 600,000 points from the air-heater to burner inlet (wind-box). STAR-CD is used as the program for thermo-fluid analysis in this study.

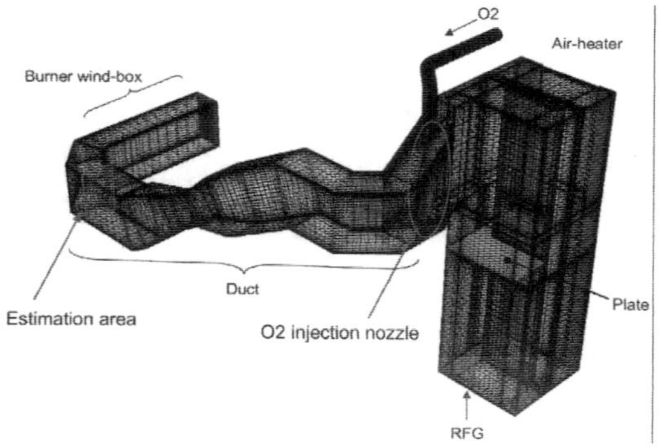

Fig. 5 Simulation model

Results of 2 types of the O_2 injection nozzle are introduced in this study. The first is the normal (non optimized structure and direction) injection nozzle type (Case 1) and the second is the final improved (optimum) injection nozzle type (Case 2) obtained by evaluating the results of any types of nozzles. Results of mixing for 2 types of nozzle are shown in Fig. 6 and Fig. 7.

From the results of case 1 in Fig. 6 and Fig. 7, O_2 flow more than 50% O_2 concentration pass along the bottom on the duct wall, because the direction and the distribution of O_2 injection nozzle is not optimum, it can be seen that there is the unbalance of O_2 concentration in the range of ±3%.

On the other hand, from the results of case 2, it is found that the area of higher O_2 concentration on the duct wall around the mixing point such like case 1 is improved and the distribution gradient of O_2 concentration at the inlet of burner wind-box is also improved to be in the range of less than ±1% as shown in Fig. 7.

These results obtained in this study will be reflected with the design of the demonstration plant and the mixing simulation like this will also be put into practical use at the commercialization stage of oxyfuel combustion system.

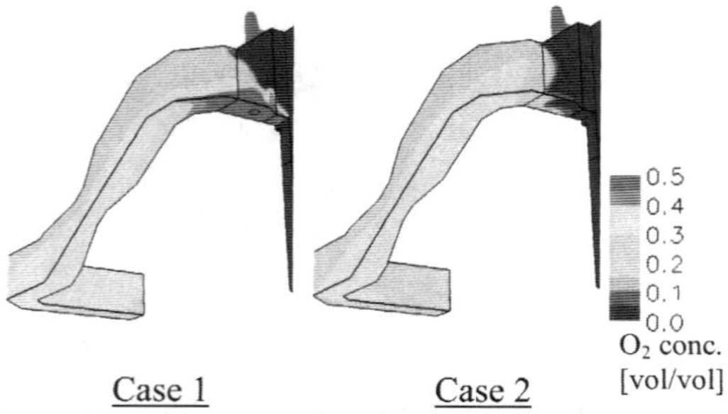

Fig. 6 Simulation results of oxygen concentration on the duct wall after the mixing

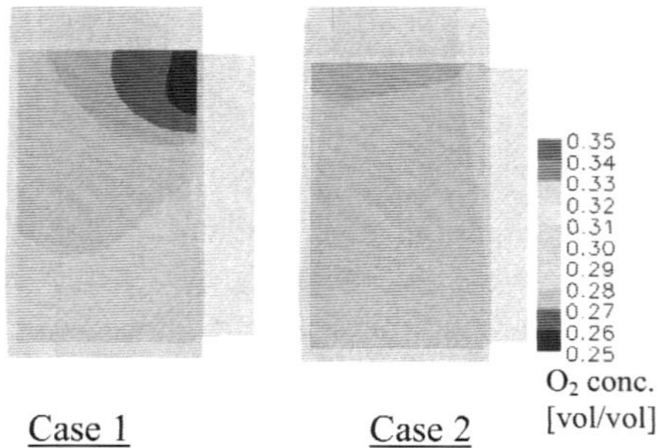

Fig. 7 Simulation results of oxygen concentration at the inlet of burner wind-box

4.3 Performance of flame detector

In the oxy-fuel combustion, the atmosphere for the combustion is different from the air combustion and is consist of the gases with higher concentration of CO_2 and H_2O. There is no example to check the performance of flame detector in such atmosphere. Flame loss signal of flame detector can be connected with mill trip or MFT (Master Fuel Trip) in some situations. To detect the flame properly by flame detector is key issue to operate of the power plant in safety. Therefore, performance of flame detector during oxy-fuel combustion was confirmed using the pilot combustion test facilities in Aioi works of IHI

Regarding the test facilities, the furnace is a vertical and cylindrical furnace with a capacity of 1.2 MWt, an inside diameter of 1.3 m, and a length of 7.5 m. In the oxy-fuel combustion, the flue gas taken out after the bag filter is supplied to the burner by the GRF (Gas Recirculation Fan) and is also used as gases for transporting pulverized coal as well as for combustion. The combustion gas is mixed with O_2 for adjusting the O_2 concentration at the inlet of the furnace. Combustion test under both air and oxy-fuel conditions combustion was performed monitoring the detection level of flame detector. Forney-IDD flame detector was used in this test.

Test condition is simulated with the operation condition at demonstration plant as shown in Table 1 and flow diagram of test facilities is also shown in Fig. 8. In the oxy-fuel combustion test, the concentration of flue gas CO_2 was about 77dry% as compared with 16dry% in air combustion.

Table 1 Test operation condition

Items	Air combustion	Oxy-fuel combustion
PC feed rate [kg/h]	100	100
Burner wind-box O_2 [wet%]	21(Air)	36
Total O_2 to the furnace [wet%, calc.]	21(Air)	27
Flue gas O_2 [dry%]	3.5	3.5
Flue gas CO_2 [dry%]	16	77

Fig. 9 shows the photos of the flame at air and oxy-fuel combustion. Both flame is almost same shape, however flame temperature of oxy-fuel combustion measured by radiation thermometer is lower than that of air combustion. This is because CO_2 and H_2O have higher emissivity at oxy-fuel combustion and the flame temperature is adjusted by recirculation gas flow to obtain the same heat flux on the furnace wall. In such situation, Fig. 10 shows the result of flame detection signal after the adjustment by Forney-IDD in the test. Signal output is in the range between -5V and +5V, and more than 0V of signal output means "Flame on". Both air and oxy-fuel combustion is performed with the continuous operation during one day. At the gain adjustment, gain level is decreased in order to check the behaviour of signal level at both combustions. From the result, it can be found of the slightly lower level of signal output in oxy-fuel combustion as compared with the level in air combustion, however it is confirmed that the signal level is high enough to detect the flame and adjustable signal level by gain in order to detect the flame signal stably.

Based on the test results, flame detector will be considered to be adjusted at the demonstration and commercial stage.

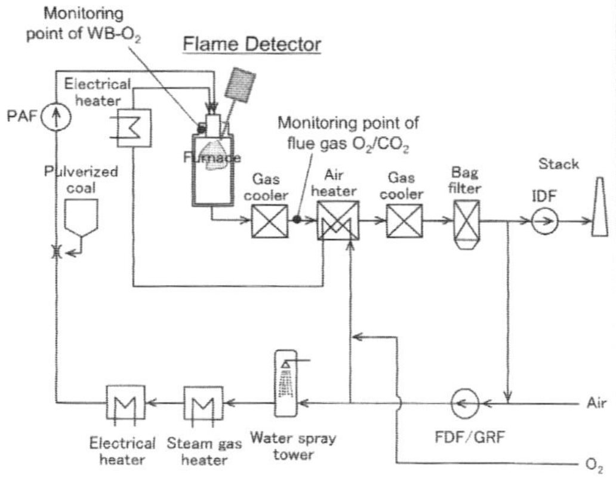

Fig. 8 Flow diagram of combustion test facilities in Aioi

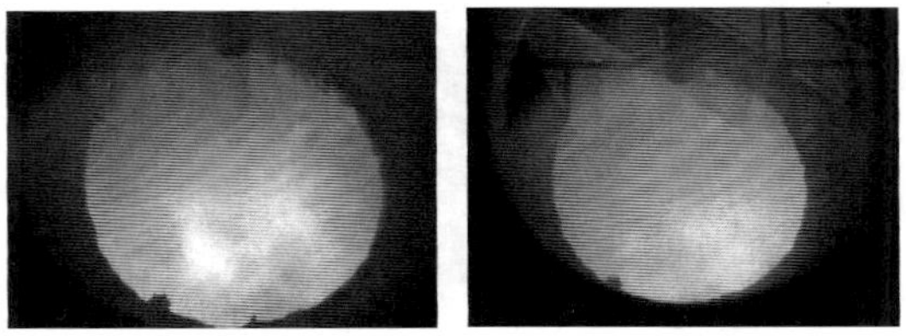

(PFT : 1400degree C) Air combustion

(PFT : 1330degree C) Oxy-fuel combustion

*PFT: Peak flame temperature measured by radiation thermometer

Fig. 9 Flame at both air and oxy-fuel combustion

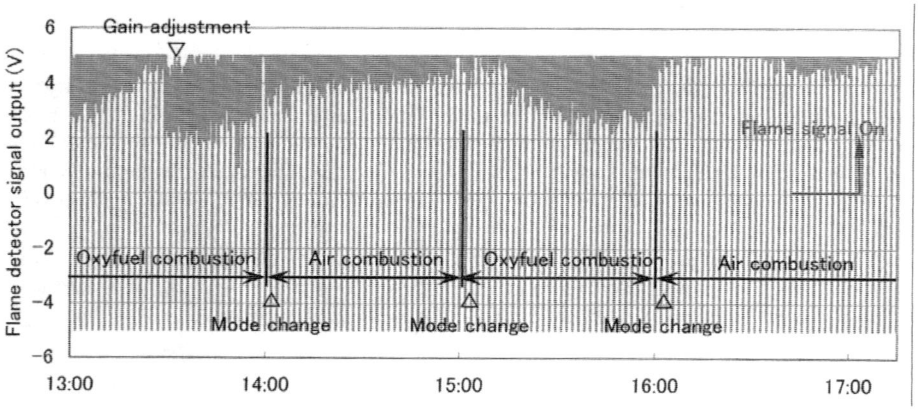

Fig. 10 Result of detection signal of the flame at both air and oxy-fuel combustion

5. Demonstration Project in Australia

On the basis of the study results described above and previous study results[2][3], Joint Venture for the demonstration project of oxy-fuel combustion was established in 2008, that is called "Callide Oxyfuel Project" [4], and the works are now under way for applying oxy-fuel combustion to an existing power plant by way of demonstration. This project aims at recovering and storage CO_2 from an actual existing power plant and also obtaining the economical and technical data or knowledge for the commercialization. The outline of the project is described as follows:

Fig.11 shows the site location of the existing power plant and the CO_2 storage of the demonstration project. Specifically, the project is implemented at the power generation system in Callide-A power plant No.4 unit owned by CS Energy on the east coast of Australia [5]. The capacity of this unit is 30MWe without reheat cycle, main steam flow at 30MWe is approximately 130t/h and coal feed rate is approximately 30t/h.

Regarding the application of oxy-fuel combustion to the existing power plant, 2 of air separation units with the capacity of 330t/h and O_2 purity of 98% will be installed near the power plant, air/gas flow system of the boiler will mainly be modified and CO_2 purification unit with the compression and cooling will also installed. It will be expected to have the CO_2 capture rate of 50 - 100 ton per a day at the condition of the liquefied CO_2.

The candidate CO_2 storage site is in the process of the site selection. At this stage, the area of CO_2 storage is planning to be about 250 km to the west of the power plant site. This area was selected because it is not far away from the power plant site, the estimated CO_2 storage capacity is sufficient, and the reservoir characteristics such as permeability and porosity are adequate for CO_2 storage. After the decision of the site and the layer, trial drilling at storage site will be implemented.

The entire schedule of the demonstration project is shown in Table 2. Before the moving to demonstration project, Australia and Japan jointly conducted a feasibility study of oxyfuel combustion technology in FY 2004 [6] and FY 2005 [7]. Based on its results, detailed studies have started in 2006, aiming of the application of this technology to the existing power plant and the integration with CO_2 storage. And simultaneously, the clarification of the project scheme towards the demonstration and the acquirement of the project budget were reviewed in 2006 and 2007. And, in the beginning of 2008, Joint Venture for the demonstration project was established under the supporting of both Australian and Japanese government. Electric Power Development Co., Ltd. (J-POWER), Mitsui & Co., Ltd. (Mitsui) and IHI Corporation (IHI) are participating in Joint Venture and Japan Coal Energy Center (JCOAL) is the supporting collaborator on technical issues from Japanese side. In November 2008, launch was held in the site to celebrate the start of site works [4].

Project mainly consists of stage 1 and 2. In stage 1, demonstration operation of oxy-fuel combustion with CO_2 capture will be performed for four years after the plant completion at the middle of 2011. In stage

2, the injection of captured CO_2 from the power plant will start in 2011 simultaneously, and the demonstration of CO_2 injection into the underground and the monitoring of CO_2 storage will be performed for three years. This demonstration project will be summarized at 2016.

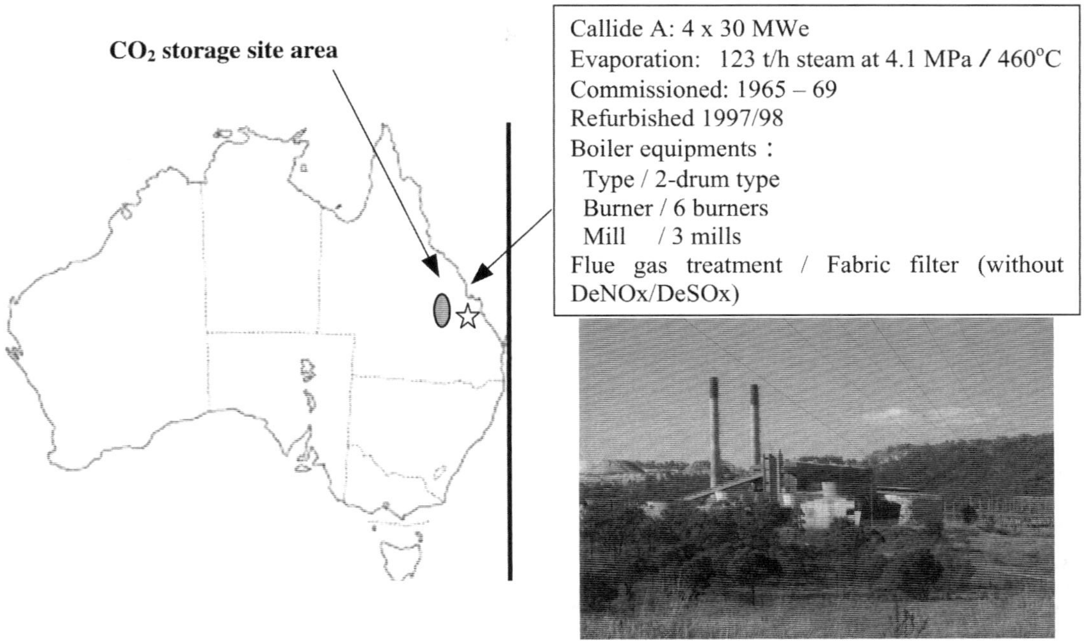

Callide-A Power plant

Fig. 11 Power Plant site and CO_2 storage site

Table 2 Demonstration Project Schedule

Items \ JP-FY	2004	05	06	07	08	09	10	11	12	13	14	15	16
FEED	[bar 2004–06]			**Feasibility**									
Demonstration Project						**Commencement of demonstration**							
Stage 1 CO_2 **capture (ASU, Boiler retrofit, CO_2 compression & purification unit)**			[illegible]	**Oxyfuel retrofit / Commissioning**				**Demonstration**					
Stage 2 CO_2 **storage (Site selection, Drilling Injection)**			**Site selection / Drilling / Design**			**Site work / commissioning**	**CO_2 injection & monitoring**					**Conclusion**	
Stage 3 **Project summary**						**Commencement of CO_2 storage**					[bar 14–15]		

6. Conclusion

These results obtained in this study will be reflected with the design of the demonstration plant for Callide Oxyfuel Project and the mixing simulation like this will also be put into practical use at the commercialization stage of oxy-fuel combustion system.

Oxy-fuel combustion technology has been attracting and increase attention worldwide because it has a potentiality for providing a breakthrough solution that helps reduce CO_2 emissions. In relation to this technology, many countries are conducting research and development towards the realization of this technology which is one candidate of the CO_2 capture from the power plant.

We would like to have much knowledge and information through the studies including above and the demonstration operation at Callide-A, and contribute to reduce CO_2 from the pulverized-coal fired power plant using oxy-fuel technology.

7. Acknowledgments

This paper and the studies for the demonstration project were greatly supported by NEDO (New Energy and Industrial Technology Development Organization), Coal division of METI (Ministry of Economy, Trade and Industry) as well as by Australian organizations such as ACA (Australian Coal Association), University of Newcastle and many others in Australia and Japan. The authors would like to express gratitude to them for their help and support.

6. References

1. K. Kimura et al., JSME-ASME Int. Conf. On Power Eng.-93, Sept. 1993, Tokyo
2. T. Yamada et al., JSME Int. J. Ser. B, vol 41, No. 4, 1998
3. H. Shimoda et al., Proc. 6th Int. Conf. on Technologies and Combustion for a Clean Environment, July, 2001, Portugal
4. Website, *www.callideoxyfuel.com*
5. C. Spero, Proc. Clean Coal Day in Japan 2004, Advanced Clean Coal Tech. Int. Symp. 2004, Sept. 2004, Tokyo
6. NEDO, "Report on the Clean Coal Technology Promotion Project 2004: Study on the Application of Oxygen Combustion Technology to an Existent Pulverized Coal-Fired Power Plant"
7. NEDO, "Report on the Clean Coal Technology Promotion Project 2005: Study on the Application of Oxygen Combustion Technology to an Existent Pulverized Coal-Fired Power Plant"

Oxy-Coal Combustion Using Oxygen Transport Membranes

Joseph Adams, Department of Chemical Engineering, University of Utah, Salt Lake City, UT

Eric G. Eddings, Dept. of Chemical Engineering, University of Utah, Salt Lake City, UT

Praxair, Inc. has developed an Oxygen Transport Membrane (OTM) technology that facilitates separation of oxygen from air through use of an oxygen partial pressure difference across the membrane. This membrane has been placed on a ceramic tube support for the purpose of in-situ oxygen generation in combustion devices. The efficacy of the OTM approach has been previously demonstrated for the firing of natural gas and CO/H2 mixtures by Praxair, Inc., and Praxair has teamed with the University of Utah to determine the viability of OTM operation with coal.

The University of Utah (UofU) has completed construction and shakedown of a bench-scale OTM coal reactor. This presentation will discuss the development of this reactor and present the results of initial testing with both gaseous and solid fuels.

See more of this Session: Oxycombustion of Coal - Needs, Opportunities, and Challenges - I
See more of this Group/Topical: Energy and Transport Processes

An Experimental Investigation On Coal Particles Combustion of Subbituminous Coals in Different Oxygen Concentration Conditions

Yong-Gyun Kim, Mechanical Engineering, Pusan National University, Busan, South Korea

Jae-Dong Kim, Mechanical Engineering, Pusan National University, Busan, South Korea

Gyu-Bo Kim, Pusan Clean Coal Center, Pusan National University, Busan, South Korea

Ju-Hun Song, School of Engineering/Pusan Clean Coal Center, Pusan National University, Busan, South Korea

Young-June Chang, School of Engineering / RIMT, Pusan National University, Busan, South Korea

Chung-Hwan Jeon[†], School of Engineering/Pusan Clean Coal Center, Pusan National University, Busan, South Korea

Oxy-coal combustion is a promising alternative to achieve carbon dioxide capture and clean coal technology which could be directly applicable to the conventional pulverized coal fired utility power plants. In line with this, an 100MWe oxy-combustion power station project was launched in Korea 2007. For successful installation and operation of such facility, it is essential to understand the coal combustion mechanism under unconventional conditions in which the characteristic of pulverized coal combustion under oxy-coal combustion is quite different from that of air-blowing atmosphere. In this study, the characteristic of coal combustion mainly on devolatilization and char combustion, were investigated through flame lengths. The effects of oxygen mole fraction in N_2 and CO_2 diluent gases respectively, particle sizes for sub-bituminous coals were studied in a laminar flow entrained reactor that produces similar environment of utility furnace in terms of high heating rate and temperature. O_2 mole fraction was varied 0~50%. The flame length becomes shorter as increase of O_2 concentration. It means that the reactivity in both devolatilization and char oxidation process is increased in higher O_2 concentration environment. The char particles' burnout time with increase of O_2 fraction becomes shorter but there is not much difference between those of O_2/N_2 and O_2/CO_2 environments.

See more of this Session: Oxycombustion of Coal - Needs, Opportunities, and Challenges - II
See more of this Group/Topical: Energy and Transport Processes

Ignition of a group of coal particles in oxyfuel combustion with CO_2 recirculation

Alejandro Molina*[,1], Ethan S. Hecht[2], Christopher R. Shaddix[2]

[1] *Grupo de Bioprocesos y Flujos Reactivos, Facultad de Minas, Universidad Nacional de Colombia, Sede Medellín, Colombia*
[2] *Combustion Research Facility, Sandia National Laboratories, Livermore, CA USA*

Recent studies for coal ignition in oxyfuel combustion have been conducted as single-particle experiments or in pilot-scale burners. While single-particle experiments have the advantage of well-controlled atmospheres in which it is possible to infer a good understanding of the phenomena related to pulverized-coal ignition, they raise the question of the effect that burning coal in a multi-particle mode, such as exists in industrial burners, may have on particle ignition. On the other hand, pilot-scale experiments are often difficult to control and may not allow a simple understanding of the ignition phenomenon. This paper presents experimental results obtained under group-combustion regime in a laminar flow environment using coal flows larger than those observed in single-particle experiments but smaller than those of industrial burners. We expect the results exhibit some of the synergistic effects observed in industrial-size coal boiler flames. Furthermore, combustion of groups of coal particles can be used to reasonably infer characteristics of flame stability – something difficult to achieve in single-particle combustion experiments. The experiments were conducted with a U.S. high-volatile bituminous coal under oxygen concentrations ranging from 12% to 48%, with N_2 or CO_2 as balance gas, at two gas temperatures (1130 K and 1650 K). The standoff distance from the coal flame to the burner was used as a metric of ignition delay and the variation in time of the flame location as an indication of flame stability. The results showed that oxygen concentration has a strong effect on ignition delay at 1130 K. The presence of CO_2 as balance gas significantly changed how the standoff distance responded to changes in oxygen concentration, particularly at 1130 K. The theory developed for drop group combustion was reviewed for coal particles combustion. The analysis shows that traditional understanding of group combustion based on the G number needs to be revised to take into account a balance between the fast devolatilization process and the laminar flame speed that control group combustion.

---

1. Introduction

During coal combustion in oxygen-enriched environments, a CO_2-rich stream is recycled into the furnace to decrease combustion temperature as a suitable strategy to retrofit existing boilers, keeping low cost and generating a nitrogen-free, CO_2-rich stream than can be used for sequestration, enhanced-oil recovery, and ultimately, reduction of greenhouse emissions.

Although the transition from coal combustion in air (~ 21% N_2) to combustion in a CO_2 stream seems as straightforward as recycling a fraction of the exhaust gas to the burner, the different physical and chemical properties of CO_2 when compared to N_2, particularly the higher molar heat capacity of CO_2, cause differences in flame behavior that motivate changes in burner design and boiler operation, some of them motivated by differences in ignition characteristics of the coal particles [1-4].

Single-particle studies have been the first step towards understanding the complex phenomena occurring during coal combustion. Particularly, in oxygen-enhanced coal combustion with CO_2 recirculation, different studies [5-8] showed that higher concentrations of CO_2 decrease the temperature of the volatile cloud flame surrounding the particle and increase ignition delay.

The required step for extension of single particle experiments to boilers would be to characterize coal flames in industrial-size burners. However, industrial-size burners are difficult to control and limit the possible understanding gain from the experiments.

An alternative to single particle and industrial-size experiments are experiments in where multiple particles burn in such a way that interaction among particles occur.

In this research ignition of groups of particles of a U.S. high-volatile bituminous coal is studied in environments containing 12% to 48% O_2, with N_2 or CO_2 as balance gas, at two gas temperatures (1130 K and 1650 K). The results show that the trends observed for single particles are also evident when a group of particles is ignited. The paper also evaluates the theoretical developments on interactive combustion and suggests new theoretical modifications that may be required for oxyfuel combustion.

2. Theoretical considerations

Although at first hand the difference between a single particle-mode combustion and group combustion seems straightforward, several authors have identified different regimes to characterize combustion of more than one particle. While Law [9] considers four different regimes for drop combustion, Annamalai and Puri [10] describe seven different regimes for drops combustion that can be easily extended to the combustion of a group of particles.

a. Isolated-particle combustion (ISPC). Particles burn completely independent one from another.

b. Individual flame combustion (IFC). The number of particles is high enough to decrease the oxygen content, but the distance between particles is large enough that each particle burns independently.

c. Incipient group combustion (IGC). As the number of particles increases, a point is reached where the oxygen concentration within the clouds decreases to a point that the flame radii around the particle is large enough that flames from different particles touch each other. At this point IGC starts.

d. Partial group combustion (PGC). A further increase in the number of particles causes that the inner particles devolatilize without ignition, because of the low oxygen content, and a group flame is established inside the cloud. Particles far from the center keep burning in the IFC mode.

e. Group combustion (GC). A higher particle load increases volatile concentration inside the cloud to a point where oxygen does not penetrate anymore and a sole exterior flame forms outside the cloud. The onset of this regime is known as Critical Group Combustion (CGC)

f. Sheath combustion (SC). For a very large number of particles, oxygen concentration becomes very low and the cloud behaves as a whole that resembles single-particle combustion.

To characterize the extent of group combustion, different authors (see discussion in [9]) defined a G number as:

$$G = Nd_{p0}/\ell \qquad \text{E 1}$$

where N is given by:

$$N = n\ell^3 \qquad \text{E 2}$$

and n is the particle number density, ℓ a characteristic dimension of the cloud of particles and d_{p0} the initial particle diameter.

Although it is clear that group combustion is favored as for large G, the definition of G is strictly geometrical.

3. Experimental methods

The experiments presented in this section are similar to those already described in [11]. The analysis of results is, however, different as it is based on the theoretical considerations presented above.

3.1 Reactor and gas composition and temperature

The optical entrained flow reactor at Sandia National Laboratories was used in the experiments. The system, extensively described in previous papers (see e.g. [6]) was modified so that a higher coal flow (of the order of 0.4 g min^{-1}) than normally used in the single-particle experiments could be introduced into the reactor. Great care was taken to maintain a constant coal flow. To this aim a modified coal feeding system, similar to that described by Graham [12], was developed. The total gas flow into the reactor was also decreased when compares with previous studies [5] to increase the volumetric load of particles into the system and to increase the available residence time for ignition to occur.

Table 1 shows molar composition, calculated adiabatic temperature and total gas flow of the mixtures used in the experiments. Two different characteristic furnace temperatures were investigated: one in which particle ignition is relatively difficult (~ 1130 K) and one in which all but the most unreactive coals readily ignite (~ 1650 K). The 1130 K furnace environment is produced by N_2-diluent mixtures with an adiabatic temperature of 1200 K and by CO_2-diluent mixtures with an adiabatic temperature of 1240 K. For this low temperature furnace environment, the flat-flame burner was operated so that the oxygen concentration downstream from the burner was 24%, 36%, or 48% O_2. The high temperature furnace environment was produced using gas mixtures with an adiabatic temperature of 1800 K for N_2 mixtures and 1830 K for CO_2 mixtures. In this high temperature environment, experiments were conducted at 12%, 24% and 36% O_2 with a N_2 diluent and at 24% O_2 with a CO_2 diluent.

Table 1. Gas composition used in the experiments

Adiabatic temp.		1200 K			1240 K		
O_2	%v	24.0	36.0	48.0	24.0	36.0	48.0
CO_2	%v		1.1		66.1	54.0	42.0
N_2	%v	64.9	52.8	40.7		0.0	
H_2O	%v	10.0	10.1	10.2	10.0	10.0	10.0
Total gas flow	slpm*		40.0			40.0	

Adiabatic temp.		1800 K			1830 K
O_2	%v	12.0	24.0	36.0	24.0
CO_2	%v		3.3		60.0
N_2	%v	68.7	56.6	44.5	0.0
H_2O	%v	16.0	16.1	16.2	16.0
Total gas flow	slpm*		40.0		40.0

*slpm: standard (1 atm, 298 K) liters per minute

3.2 Fuel

A pulverized sample of high-volatile eastern U.S. bituminous coal (Pittsburgh seam, Bailey mine) was sieved to 75-106 μm size fraction. Sieving was conducted for 30-minute periods, several times, until the mass passing the 75 μm sieve was negligible. This procedure was carried out to ensure that there were no fines in the coal, as flame stability and ignition in a group of particles is very sensitive to the particle size, as has been shown in the literature [13] and as was evident in the flow reactor experiments. This coal sample has been analyzed to have 1% moisture, 7% ash, and 35% volatiles (see [11] for a complete analysis).

3.3 Measurement of standoff distance

In the experiments, the standoff distance from the flat flame burner to the coal flame, caused by the simultaneous combustion of multiple coal particles, was used as a metric of ignition delay and the variation in time of the flame location as an indication of flame stability. To measure the standoff distance, the standard setup [14] used in the entrained-particle flow reactor was modified by decreasing the gain in the photomultipliers (PMs) to a minimum value to prevent saturation. The intensity of the PM outputs at a specific distance from the burner were measured during a 10-second time interval. Time-average radiation intensity detected by the PM, as well as the signal standard deviation, were recorded as the optical probe volume height was increased. The experiment was repeated three times to improve the statistical analysis. This protocol was repeated for all the gas compositions described in Table 1.

Because ignition strongly depends on the heat balance surrounding the particles, care was taken to guarantee stable conditions in the experiments before measurements were collected by allowing the combustion gases to heat the quartz chimney for 10 minutes and then having stable coal injection for 5 more minutes before acquiring any data.

4. Results

4.1 Qualitative Analysis

Figure 1 shows digital pictures of coal particle combustion for the optical furnace operating in N_2 atmospheres, at a nominal gas temperature of 1130 K and at four different oxygen concentrations: a. 12%, b. 24%, c. 36%, and d. 48%. For the 12% O_2

case, there was no visible flame production. However, when the oxygen concentration was 24% (Figure 1b), a volatile flame formed at a certain distance from the burner. Based on the description presented above, while most particles burn in the PGC mode, a region at the bottom of the figure clearly burns under a GC. The presence of the volatile flame at the bottom can be deduced from the strong visible light emission associated with soot oxidation (we are currently working on implementing a means of directly detecting the volatiles flame location and size). As oxygen concentration increases, from left to right in Figure 1b, a qualitative observation suggests that the number of particles burning in PCG mode decreases as group combustion becomes more evident. Additionally, the standoff distance of the volatile flame from the burner decreases as the oxygen concentration increases, as expected. Similar pictures were obtained for a CO_2 atmosphere.

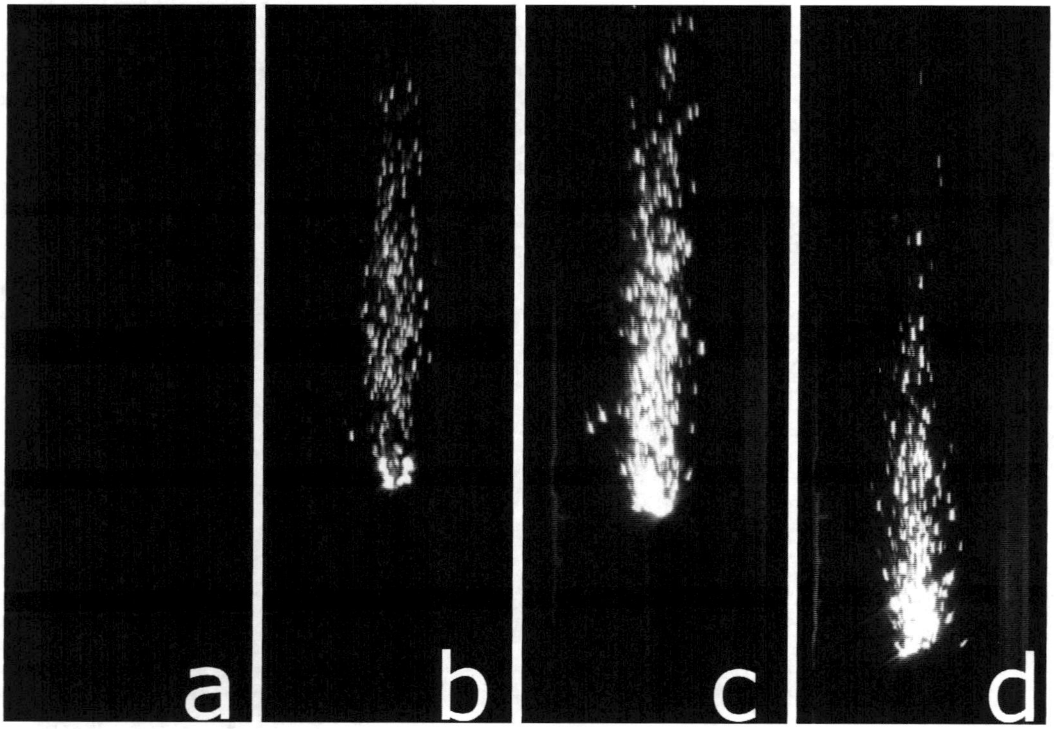

Figure 1. Examples of group of coal particles burning in N_2 at a nominal gas temperature of 1130 K at different oxygen concentrations: a. 12% O_2, 24% O_2,, 36% O_2 and 48% O_2. Particles flow upward from the center of a flat flame burner located at the bottom of the figures.

4.2 Quantitative analysis based on PM data

Figure 2 presents the average, standard deviation, and coefficient of variation (CV defined as standard deviation over average) of the PM signal at different heights (h) above the burner for different oxygen concentrations, for N_2 and CO_2 atmospheres. Figure 2a shows how, as the oxygen concentration increases in a N_2 atmosphere, the distance from the burner where the PM signal becomes detectable, i.e. the standoff distance, decreases. Figure 2b shows the same results for CO_2. While the standoff distances are clearly greater in the CO_2 atmospheres, the effect of the oxygen concentration is not as large as for N_2. It is also clear that the PM signal, an indication of the amount of thermal radiation leaving the flame, is much lower when the background gas is CO_2.

Figure 2c and d show the standard deviation of the PM signal with time. A high value of standard deviation can be associated with an unstable flame that frequently changes with time. However, as the absolute value of standard deviation strongly depends on the magnitude of the PM signal, it might make morasce sense to study the coefficient of variation, which normalizes the standard deviation by the average value. As Figure 2e and f show, the value of CV is higher in the presence of CO_2 than in N_2, and it decreases as the oxygen concentration increases.

Figure 3 presents results equivalent to those of Figure 2 but when the nominal gas temperature was 1650 K and only for a N_2 atmosphere. Figure 3a shows, consistent with the lower temperature results, that as the oxygen concentration increases the standoff distance decreases. However, the difference in standoff distance is minimal and the standoff distances are much shorter than were apparent at 1130 K. The small variation in standoff distance may reflect the stronger effect of particle heat-up (rather than subsequent gas-phase chemical induction time) on ignition at these high temperatures. Perhaps the most interesting result in Figure 3 is the fact that, contrary to what happens at 1130 K, at 1650 K the value of CV decreases as the oxygen concentration increases. This implies that at 1650 K the flame becomes less stable as the oxygen concentration increases, though it should be noted that the CV values are about an order of magnitude lower at 1650 K than at 1130 K.

Figure 4 compares the PM signals for a group of coal particles burning at a nominal gas temperature of 1650 K under a 24% O_2 concentration in N_2 and CO_2. The standoff distance is shorter in N_2 than in CO_2; however, the difference is much smaller than that observed at 1130 K. The lower value of CV for the N_2 environment suggests as well that the particles burn in a more stable flame when in N_2.

The results in Figure 2 to Figure 4 can be explained by results from previous studies [5-8] for single-particle ignition. Higher O_2 concentration decreases the ignition delay time, and, therefore, the standoff distance decreases. This effect is more evident at 1130 K where ignition takes longer, than at 1650 K, where ignition occurs faster. The higher molar heat capacity of CO_2 causes a lower flame temperature for the same oxygen concentration. This is evident from the lower values of PM intensity whenever CO_2 and N_2 atmospheres are compared. The lower flame temperature and, probably, the lower diffusivity of oxygen in CO_2 contribute to longer standoff distances in CO_2. The effect of oxygen concentration on the standoff distance is less pronounced in CO_2 than in N_2, something that needs further analysis to be explained, but that may be related to differences in diffusivity.

4.3 Considerations about G number

For the experiments described above, the G number, as calculated based on E 1, is ~ 3900. This calculation considers a number density of the order of 4.3×10^{7} particles m^{-3} and a characteristic cloud length (ℓ) of 2.54×10^{-2} m, defined as half the length of the squared cross sectional area of the chimney where combustion occurs based on Figure 1. This is clearly an oversimplification, as the value of ℓ in the vertical direction may be higher.

This value of G, as computed from E 1 is constant for all oxygen concentrations, as it will only depend on geometrical factors. Actually, ℓ may change with oxygen concentration. However, this change would be minor and difficult to determine solely based on true color images as those in Figure 1. In fact, Figure 1 suggests that ℓ would decrease as oxygen concentration increases, this would render a lower G, which would indicate a lower propensity for group combustion. The fact that G is basically insensitive to oxygen or background composition suggests the need for further theoretical development to understand interactive combustion during oxyfuel combustion.

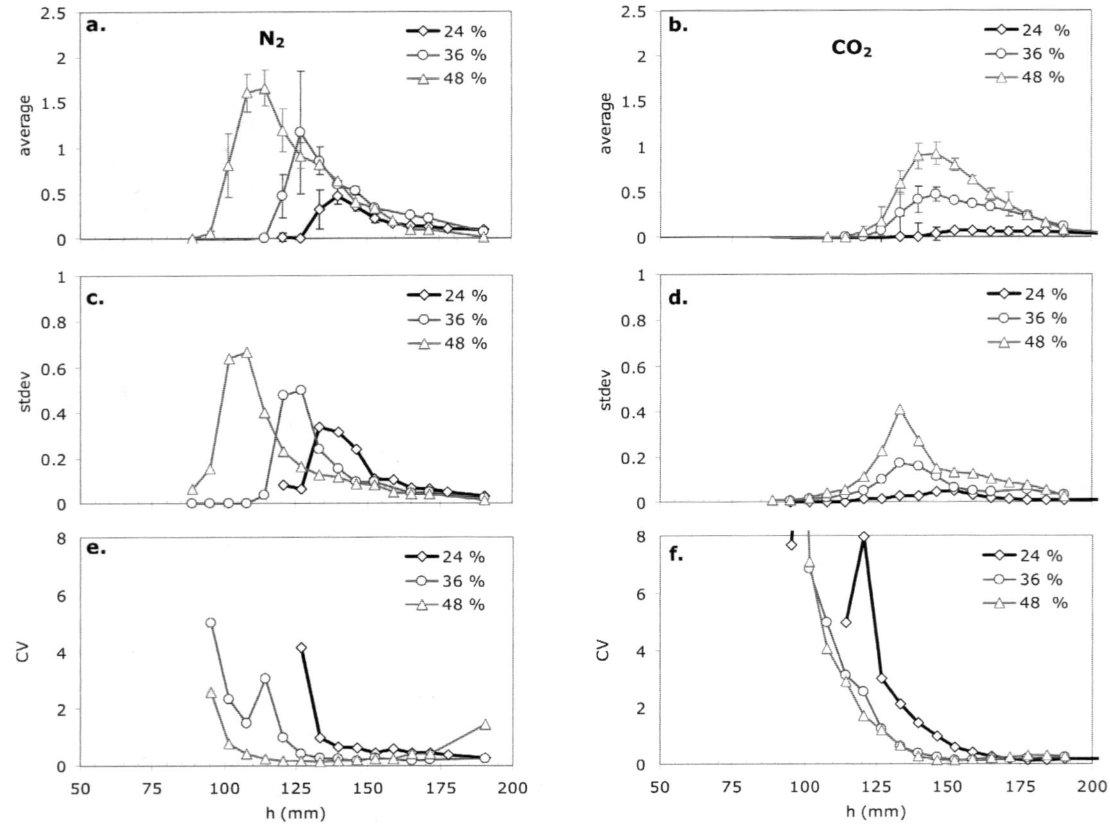

Figure 2. Variation with distance from the burner (h) of the PM signal (a., b.): average, (c., d.) standard deviation and (e., f.) coefficient of variation for different oxygen concentrations for a nominal gas temperature of 1130 K for coal burning in N_2 (a., c., and e.) and CO_2 (b., d and f.) atmospheres . Vertical bars in a. and d. represent the standard deviation from the three tests. Standard deviation in b. and e refers to the standard deviation of the PM signal as it changes in time.

Drop combustion theory [9, 10] suggests that as oxygen concentration increases, particles tend to burn in indifividual flames as a diffusion flame stabilizes given the high amount of oxygen surrounding the particle. Although this concept is valid for drop combustion, its extension to oxy-fuel coal combustion may be limited by the fact that, contrary to drop evaporation, coal devolatilization is an extremely fast process, particularly at high temperatures. Previous studies [6] have shown that as oxygen concentration increases, a volatile flames forms close to the particle surface, for single particle combustion.

For low oxygen concentration, volatiles escape from the particle and will ignite downstream from the particle, at a distance to far from injection at a point where particle density would be too short to grant group combustion. As oxygen concentration increases, volatile ignition occurs sooner in the chimney, when particle density is high enough and group combustion can be sustained.

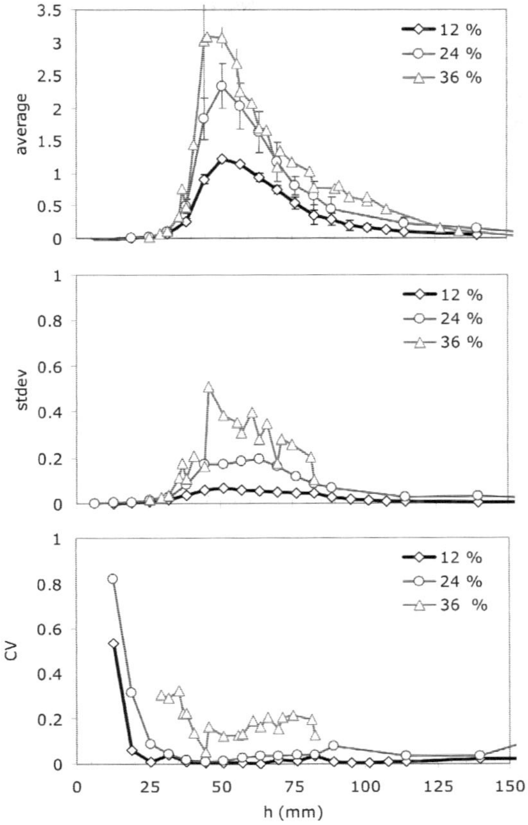

Figure 3. Variation with distance from the burner (h) of the PM signal average, standard deviation, and coefficient of variation for different oxygen concentrations for a nominal gas temperature of 1650 K for coal burning in N_2 atmospheres.

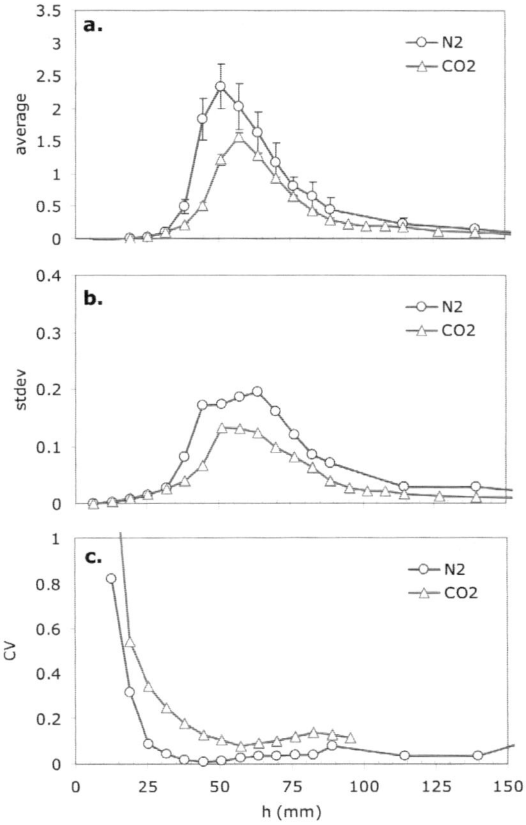

Figure 4. Variation with distance from the burner (h) of the PM signal (a.): average, (b.) standard deviation, and (c.) coefficient of variation for a nominal gas temperature of 1650 K for coal burning under N_2 and CO_2 atmospheres at 24% oxygen concentration. Vertical bars in a. represent the standard deviation from the three tests. Standard deviation in b. refers to the standard deviation of the PM signal as it changes in time.

A thought experiment would suggest that at even higher oxygen concentration (higher than 48% O_2) particles would burn as a single cloud as a volatile flame would form surrounding the cloud. For a higher oxygen concentration to produce single particle combustion, the devolatilization rate should be lower than the laminar flame combustion of the volatiles surrounding the particle.

A simple analysis shows that the laminar flame velocity (S_L) can be approximated as [15]:

$$S_L = \left[-2\alpha(v+1)\frac{\overline{\dot{m}'''_F}}{\rho_{un}} \right]^{\frac{1}{2}} \qquad \text{E 3}$$

where α is the gas thermal diffusivity, v a stoichiometric coefficient, $\overline{\dot{m}'''_F}$ a term that reflects an average combustion rate and ρ_{un} the density of the unburned gases. Combustion with CO_2 recirculation causes a lower laminar flame velocity as the thermal diffusivity is lower than in nitrogen. This analysis suggests that for a constant oxygen concentration, oxyfuel combustion with CO_2 would favor single particle combustion and make the occurrence of group combustion more difficult.

4. Summary

Experiments carried out to determine the effect of gas conditions on the ignition of a group of coal particles in oxyfuel combustion with CO_2 recirculation show that most of the trends can be explained from previous studies using single-particle ignition. In general, the standoff distance from the burner to a volatile flame formed by the group of particles decreases as the oxygen concentration increases and is higher for CO_2 environments. These effects are more noticeable at 1130 K than at 1650 K, where ignition occurs very rapidly.

The group particle experiments give new insight into flame stability. The results show that at a gas temperature of 1130 K, higher oxygen concentration produce more stable flames. An opposite trend was observed when the gas temperature was 1650 K, a surprising result that requires verification with further experiments. At both temperatures, flame stability decreases in CO_2 environments.

At 1130 K and 24% O_2 the particles burn under partial group combustion, however, as the oxygen concentration increases, the system moves to conditions close to total group combustion. The theoretical analysis based on group number theory (G) does not predict clearly the variation of group combustion mode with oxygen concentration or background gas. New theory developments should take into account a balance between the fast coal devolatilization process and laminar flame velocity, which decreases in the presence of CO_2.

Acknowledgment

This research was sponsored by the U.S. Department of Energy through the National Energy Technology Laboratory's Power Systems Advanced Research Program, man-aged by Dr. Robert Romanosky. Susan Maley serves as the NETL technical monitor for this project. Sandia is a multi-program laboratory operated by Sandia Corporation, a Lockheed Martin Company, for the United States Department of Energy's National Nuclear Security Administration under Contract DE-AC04-94AL85000.

References

1. Buhre, B.J.P., Elliott, L.K., Sheng, C.D., Gupta, R.P., and Wall, T.F., *Oxy-Fuel Combustion Technology for Coal-Fired Power Generation.* Prog. Energy Combust. Sci., 2005. **31**(4): p. 283-307.

2. Tan, R., Corragio, G., and Santos, S., *Oxy-Coal Combustion with Flue Gas Recycle for the Power Generation Industry. A Literature Review.* 2005, International Flame Research Foundation: Velsen Noord, Netherlands. p. 1-52.

3. Wall, T.F., *Combustion Processes for Carbon Capture.* Proc. Comb. Inst., 2007. **31**(1): p. 31-47.

4. Wall, T., Liu, Y., Spero, C., Elliott, L., Khare, S., Rathnam, R., Zeenathal, F., Moghtaderi, B., Buhre, B., Sheng, C., Gupta, R., Yamada, T., Makino, K., and Yu, J., *An Overview on Oxyfuel Coal Combustion--State of the Art Research and Technology Development.* Chem. Eng. Res. Des. **In Press, Corrected Proof**.

5. Molina, A. and Shaddix, C.R., *Ignition and Devolatilization of Pulverized Bituminous Coal Particles During Oxygen/Carbon Dioxide Coal Combustion.* Proc. Comb. Inst., 2007. **31**(2): p. 1905-1912.

6. Shaddix, C.R. and Molina, A., *Particle Imaging of Ignition and Devolatilization of Pulverized Coal During Oxy-Fuel Combustion.* Proc. Comb. Inst., 2009. **32**(2): p. 2091-2098.

7. Molina, A. and Shaddix, C.R. *Effect of O_2/Co_2-Firing on Coal Particle Ignition.* in *22nd International Pittsburgh Coal Conference.* 2005. Pittsburgh, PA, USA.

8. Molina, A., Shaddix, C.R., *Efecto Del Co2 En La Velocidad De Combustión De Semicoques De Carbón En Aplicaciones De Oxi-Combustión.* Energética, 2007. **38**: p. 101-106.

9. Law, C.K., *Combustion Physics.* 2006, New York: Cambridge University Press.

10. Annamalai, K. and Puri K., I., *Combustion Science and Engineering.* Crc Series in Computational Mechanics and Applied Analysis, ed. CRC. 2006, Boca Raton, U.S.A.: CRC Press. 1121.

11. Molina, A., Hecht, E.S., and Shaddix, C.R., *Ignition of a Group of Coal Particles in Oxyfuel Combustion with Co2 Recirculation*, in *34th International Technical Conference on Coal Utilization & Fuel Systems.* 2009: Clearwater, Florida, U.S.A.

12. Graham, K., *Submicron Ash Formation and Interaction with Sulfur Oxides During Pulverized Coal Combustion*, 1991, Ph. D. Dissertation. Massachusetts Institute of Technology, Chemcial Engineering: Cambridge.

13. Essenhigh, R.H., Misra, M.K., and Shaw, D.W., *Ignition of Coal Particles: A Review.* Combust. Flame, 1989. **77**(1): p. 3-30.

14. Murphy, J.J. and Shaddix, C.R., *Combustion Kinetics of Coal Chars in Oxygen-Enriched Environments.* Combust. Flame, 2006. **144**(4): p. 710-729.

15. Turns, S.R., *An Introduction to Combustion Concepts and Applications.* 1st ed. 2002: McGraw-Hill, Inc. 565.

Application of Photo-Imaging Techniques to Quantify Flame STABILITY Characteristics DURING Oxy-Coal Combustion

Jingwei Zhang, Department of Chemical Engineering & Institute for Clean and Secure Energy, University of Utah, Salt Lake City, UT

Kerry Kelly, Department of Chemical Engineering & Institute for Clean and Secure Energy, University of Utah, Salt Lake City, UT

Eric G. Eddings, Department of Chemical Engineering & Institute for Clean and Secure Energy, University of Utah, Salt Lake City, UT

Jost O.L. Wendt, Department of Chemical Engineering & Institute for Clean and Secure Energy, University of Utah, Salt Lake City, UT

The purpose of this research is to better understand and to scale the effect of partial pressures of O2 and CO2, two additional degrees of freedom for oxy-coal combustion, on coal jet ignition and flame stability in the near burner zone. A novel methodology is developed to quantify the flame stability and flame length by introducing an image processing technique. Ultimately the experiment is designed to provide data for simulation validation studies that can be used to predict how air fired combustors may be retrofitted to oxy-coal.

Specifically the objective is to explore effects of the partial pressure of O2 and CO2 on coal jet flame stability, which is here quantified by stand-off distance, the distance between burner tip and the base of the detached flame. The experiments are carried on in a 100 kW pilot-scale laboratory furnace outfitted with a coaxial burner (primary jet in the center pipe and secondary stream in the annular sleeve, no swirl, Type 0 turbulent flames) and arrays of electrically-heated panels in the burner zone in order to control variations in near-burner heat loss. Currently fresh, one-through CO2 (instead of flue gas recycle) and O2 streams are introduced to the burner. The furnace also has visual access to the burner zone through quartz windows, which allows for optical measurements. A special CMOS sensor based camera, which is more sensitive to the near infrared wavelength (responsivity: 1.4 V/lux-sec (550nm)), is applied to capture Type 0 axial turbulent diffusion flame shapes for the use of statistical studies of stand-off distance under the influence of changes in various operational parameters, including systematic variations of partial oxygen pressure in both transport and secondary oxidant stream. Statistical analysis is used to scale the effect of oxygen partial pressure in both transport and secondary stream on coal jet ignition and flame stability. Ultimately the statistics from the photo images help to generate probability density functions (PDF's), which can be used to evaluate the experimental accuracy, reproducibility and allow more precise model validation.

Keywords: oxy-coal combustion, flame stability, oxygen partial pressure (PO2), stand-off distance, near burner phenomena, coal jet, type 0 axial turbulent flames, probability density function (PDF), image processing, statistical analysis

See more of this Session: Oxycombustion of Coal - Needs, Opportunities, and Challenges - II
See more of this Group/Topical: Energy and Transport Processes

Utilization of Reburn Reactions for NO_x Control in Oxy-Fuel Combustion

*Fredrik Normann**, *Klas Andersson, Bo Leckner, Filip Johnsson*
Chalmers University of Technology, Göteborg, Sweden

Introduction

Oxy-fuel combustion implies several new options for NO_x control, but for optimization purposes most of these options require further exploration. The present work focuses on reburning of NO_x. Typically, experiments show that recirculation of NO_x into the flame in oxy-combustion lowers the emission with around 70% per unit of fuel supplied, compared to the emission from air-firing under otherwise similar conditions, see review by Normann et al. [1].

Reburning or fuel staging was proposed by Wendt et al. [2] in 1973 and is thoroughly investigated for air-fired applications (see for example Ref. [3-5]). The objective of reburning is to create hydrocarbon radicals (CH_i), which can break up the NO formed and, then, form N_2 instead. A reburning process consists of a reburn zone and a burnout zone. NO is supplied to the reburn zone. In the air-fired application of reburning, the primary combustion is the source of NO and a secondary fuel stream is supplied to create the reburn zone, while in the oxy-fuel application, discussed in the present work, NO is supplied by the recycle and the primary combustion acts as a reburn zone. The reburn zone should be oxygen lean to promote formation of radicals. In the burnout-zone O_2 is added in sufficient amount to complete the combustion.

The influence of high CO_2 concentration on combustion and gas-phase nitrogen chemistry has recently been investigated in a series of studies by Glarborg and Bentzen [6], and Mendiara and Glarborg [7]. Their work is based on a combination of experiments in a lab-scale flow reactor and detailed chemical kinetic modeling. When substituting N_2 with CO_2 in the oxidizer, they detected an increased formation and similar reduction of NO under oxygen-lean conditions and a decreased formation and higher reduction of NO during stoichiometric and oxygen-rich conditions. A similar investigation was performed by Mackrory [8] who used a combination of experiments with a pulverized coal-fired reactor and a model based on established sub-models. Mackrory concludes that the reduction of NO_x under oxy-fuel conditions is less sensitive to an increased stoichiometric ratio than under air-fired conditions. This was mainly explained by the onset of thermal NO_x formation in air combustion when more oxygen becomes available in the primary combustion zone, which does not occur to the same extent under oxy-fuel conditions. Furthermore, Mackrory concluded that the reduction during oxy-fuel conditions is favored by higher stoichiometric ratio, which causes more NO_x to pass through the primary zone due to the increased recycle of flue gas.

Although the reduction of NO_x by reburning reactions has been proven in several studies, the utilization and optimization of this technique is poorly investigated within the framework of oxy-fuel combustion. The present paper is the first step of a comprehensive work on the utilization of the reburning chemistry in oxy-fuel combustion. The results in this paper present the critical reaction paths in reburning chemistry and the main focus is on key differences between air and oxy-combustion. Differences in radical pool composition between O_2/CO_2 and O_2/N_2 atmospheres are also high-lighted due to its direct coupling to the

* Corresponding author. E-mail address: fredrik.normann@chalmers.se

reburning chemistry. An established gas-phase model is applied in the reaction path analysis. Future work will be extended towards experimental studies of the reburning capabilities in oxy-combustion and optimal settings for NO_x control.

Method

The reburning chemistry is investigated with a detailed chemical kinetic mechanism together with the associated thermodynamic library. The mechanism is compiled and validated for oxy-fuel conditions by Mendiara and Glarborg [7]. It includes reactions of importance for the combustion of light hydrocarbons and for the nitrogen chemistry active in reburning of NO_x. The mechanism is implemented in the Chemkin-Pro software [9].

To achieve a similar basis for a comparison between air and oxy-fuel combustion, the replacement of N_2 with CO_2 is the only parameter change, and the inlet oxygen concentration in the oxidizer is set to $21\%_{vol}$ in both cases. In practice, the inlet oxygen concentration in oxy-fuel combustion will probably be higher, and due to the higher heat capacity of CO_2 a similar temperature profile to air combustion is achieved with around $30\%_{vol}$ of O_2. The present simulations are isothermal with the same temperature in both cases to eliminate the influence of temperature. Furthermore, the reburning is simulated as performed during oxy-fuel combustion, i.e. an initial concentration of NO is supplied with the oxidizer also in the air-fired case.

Combustion chemistry

The nitrogen chemistry is a small part of the combustion process compared to the chemistry involved in the CH-oxidation, and small changes to the combustion process can have a large impact on the nitrogen chemistry. Hydrocarbon and nitrogen chemistry are mainly connected through the important radical pool of atomic oxygen (O), hydrogen (H) and the hydroxyl radical (OH). The single most important reaction for a combustion process is the chain branching reaction,

$$H + O_2 \rightarrow OH + O \qquad (1)$$

which consumes one (H) and produces two radicals (O and OH) [10]. Consequently, any combustion reaction increasing the production of H-atoms promotes combustion and any reaction consuming H-atoms inhibits combustion. All hydrocarbon combustion involves H_2 and CO oxidation, Reaction 2, which produces most of the CO_2 in combustion, is also central for the combustion processes [10].

$$CO + OH \leftrightarrow CO_2 + H \qquad (2)$$

The oxidation of CO by Reaction 2 produces H-atoms and is, therefore, of direct importance for Reaction 1.

Figure 1 illustrates the stages of a reburning process. The reburn zone (oxygen lean) is divided in two stages, Stage I and Stage II, and the burnout zone (oxygen rich) constitutes Stage III. In Stage I the fuel is decomposed and oxidized under oxygen-lean conditions and a high level of hydrocarbon radicals is produced. The reaction of hydrocarbon radicals with both H and OH is generally faster than both Reactions 1 and 2 [10]. This leads to a build-up of CO, which starts to oxidize after the oxidation of hydrocarbons is completed. It has been concluded in previous studies [11] that in some parts of an oxy-fuel flame, the elevated concentration of CO_2 may drive Reaction 2 in the reverse direction. The influence of CO_2 on the chain carrying radicals is shown in Figure 2. The reverse Reaction 2 can even compete with the reactions between hydrocarbon and H-radicals and may have an impact on the flame velocity. This is seen in Figure 1 by the slower oxidation of the fuel during oxy-fuel combustion than during air combustion. In Stage II when CO starts to oxidize to CO_2 (the forward Reaction 2) the higher CO concentration in oxy-fuel combustion reduces the relative rate of radical consumption and, eventually, the concentration of radicals in oxy-fuel combustion could exceed that of air combustion, which is seen in the centre diagram of Figure 2. The relative order of the quantity of radicals is changed from H>OH>O in air combustion to OH>O>H in oxy-fuel combustion. In Stage III more oxygen is added to complete the oxidation of the fuel and a similar pattern as in the reburn zone is detected: initially, the concentration of radicals is higher during air-firing but the decrease is slower during oxy-fuel combustion due to the buffering effect by CO_2 in Reaction 2.

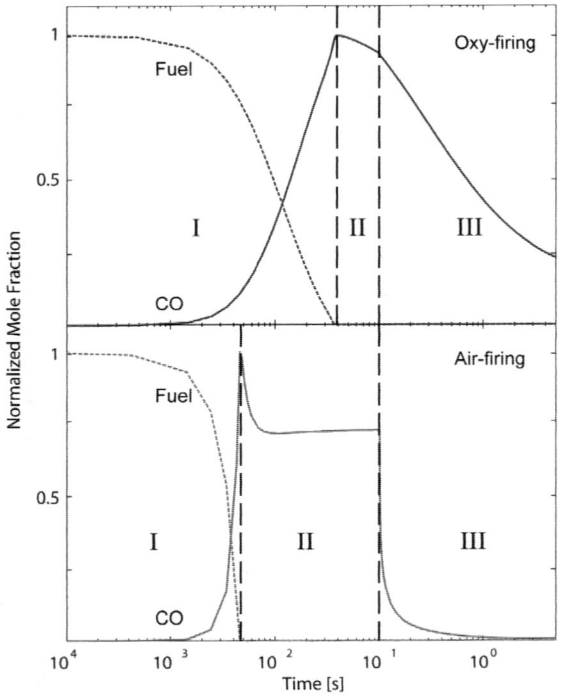

Figure 1. Comparison of the three stages of the reburning process, expressing the normalized mole fractions of fuel and carbon monoxide as functions of resedence time. Dashed verticle lines indicate the break between stages.

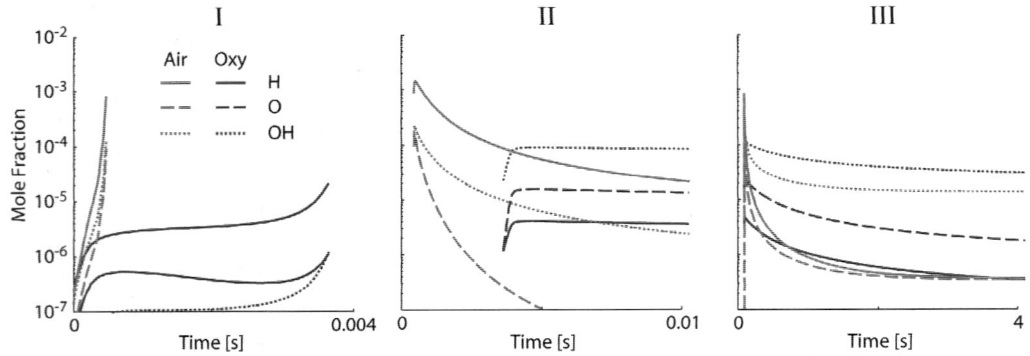

Figure 2. The concentration profile of chain-carrying radicals in the different stages of a reburning process during air and oxy-fuel firing.

Nitrogen chemistry

Figure 3 shows the concentration profiles of NO, HCN and CH_3 in the reburn zone (Stages I and II), which give an indication on how the reduction of NO_x proceeds. The maximal reduction rate is lower during oxy-fuel combustion, due to the lower peak concentrations of important reactants. This is, however, compensated by a longer residence time in the reducing zone (Stage I), due to the slower oxidation of the fuel, which open ups for alternative reaction routes. It can also be seen in Figure 3 that even though there are large differences in combustion conditions and in reduction behavior between air and oxy-fuel combustion in the investigated cases, the overall reduction of NO_x is similar.

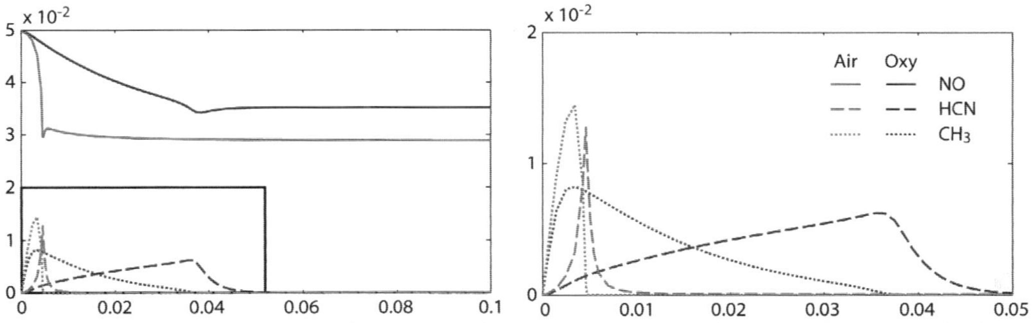

Figure 3. The concentration profile of NO, HCN and CH_3 during air and oxy-fuel firing. The graph on the right hand side is an enlargement of the black square in the figure on the left hand side.

The reaction routes of interest for NO_x reduction in the defined stages of reburning is discussed by means of Figures 4 to 6. Figure 4 shows the governing reaction routes for nitrogen compounds in Stages I and II, while Figure 5 and 6 show the relative importance of each step of these routes during air and oxy-fuel conditions. In Stage I the reduction of NO to HCN dominates nitrogen chemistry. In oxy-fuel combustion less HCN is formed directly from NO, mainly due to lower concentrations of the small hydrocarbon radicals (CH and CH_2). This is, however, compensated by a larger formation of HCN via H_2CN due to higher CH_3 concentration. In oxy-fuel combustion more of the HCN formed reacts to CH_3CN and further to NCO. This is explained by the higher concentration of CH_3 and lower concentration of chain-carrying radicals. In air-firing NCO is also formed directly from the reaction between HCN and O radicals. In Stage II, HCN and CN react to mainly form NCO. In air-firing almost all NCO reacts with an H-radical to form NH and further to N (this route is active already in Stage I). The N atom reacts to form either NO or N_2. In oxy-fuel combustion the concentration of H-radicals are lower, and NCO reacts through alternative paths to form N_2O, NO and N_2. The split between NO and N_2 are dependent on similar reactants in both cases (NO gives N_2, and, O_2 or OH gives NO). The reaction through N_2O to N_2 is more important during oxy-fuel combustion. In summary, the lowered concentration of H-radicals affects the nitrogen chemistry both directly and by changes in the rate of the oxidation of the fuel.

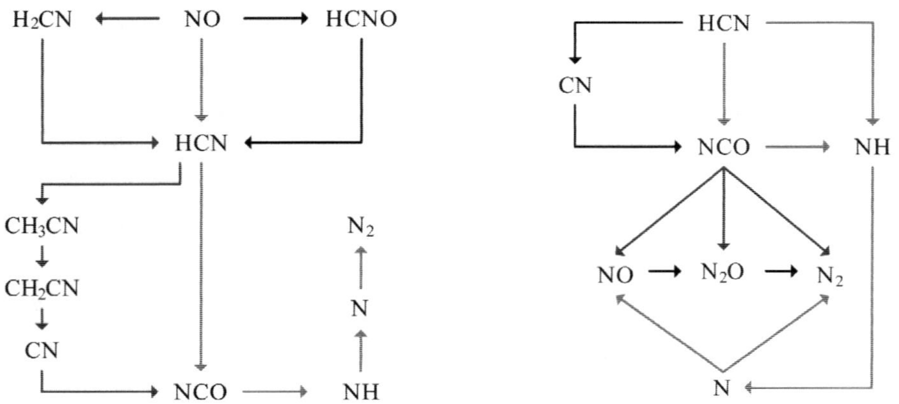

Figure 4. The most important reaction routes in the reburning zone. a) Stage I. b) Stage II. Blue arrows mark routes more important for oxy-fuel combustion; red arrows are the same for air-firing and black arrows mark routes with similar importance for both cases.

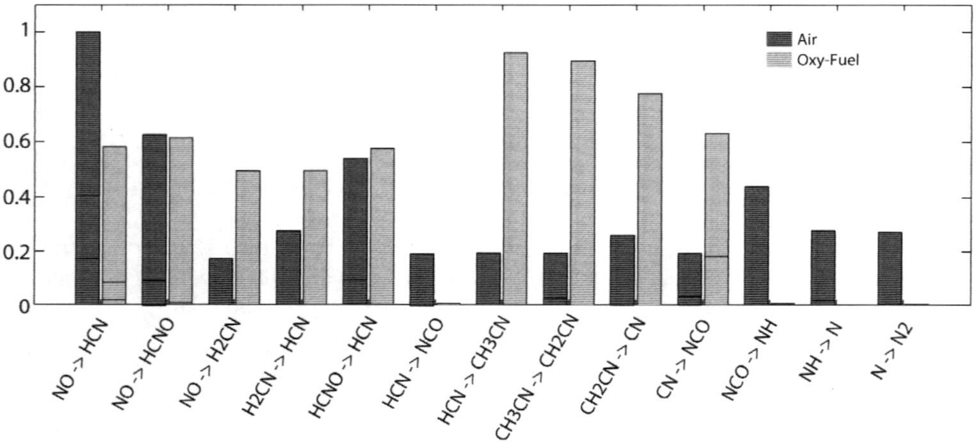

Figure 5. Relative importance of the reaction routes of Stage I presented in Figure 4a.

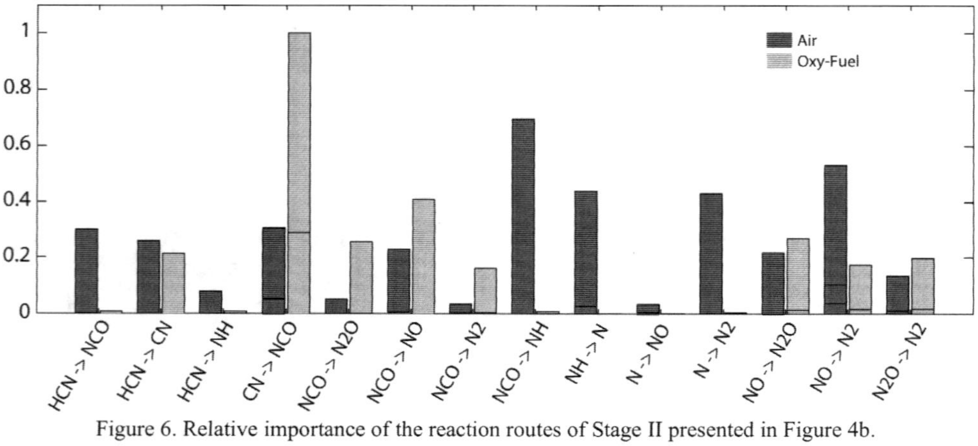

Figure 6. Relative importance of the reaction routes of Stage II presented in Figure 4b.

Conclusions

This work is the first step in an investigation of the reburning process for oxy-fuel conditions. Differences in combustion chemistry and performance of the reburning process during air and oxy-fuel combustion are investigated by means of a detailed reaction mechanism from which it is concluded that there are substantial differences between reaction routes during air and oxy-fuel combustion. The most important factor is the reduced concentration of H-radicals at the elevated concentration of CO_2. The lowered concentration of H-radicals affects the nitrogen chemistry both directly and by changes in the rate of the oxidation of the fuel. Despite the differences in reaction routes the overall reduction of NO_x is similar for the case studied. It is, however, important to further investigate the effect of combustion parameters on the routes of reburning during oxy-fuel conditions to identify eventual differences from the optimizing procedures during air-fired conditions.

Acknowledgements

This work is financed by Vattenfall AB.

References

1. Normann, F., Andersson, K., Leckner, B. and F. Johnsson (2009), "Emission control of nitrogen oxides in the oxy-fuel process", *Progress in Energy and Combustion Science,* 35, pp. 385-397.
2. Wendt, J.O.L, Sternling, C.V. and M.A. Matovich (1973), "Reduction of sulfur trioxide and nitrogen oxides by secondary fuel injection", *Combustion Science and Technology*, 139, pp. 329-363.
3. Kilpinen, P., Glarborg, P. and M. Hupa (1992), "Reburning chemistry: a kinetic modeling study", *Industrial and Engineering Chemistry Research,* 31, pp. 1477-1490.
4. Kristensen, P.G., Glarborg, P. and K. Dam-Johansen (1996), "Nitrogen chemistry during burnout in fuel-staged combustion", *Combustion and Flame*, 107, pp. 211-222.
5. Dagaut, P., Glarborg, P. and M.U. Alzueta (2008), "The oxidation of hydrogen cyanide and related chemistry", *Progress in Energy and Combustion Science,* 34, pp. 1-46.
6. Glarborg, P. and L.L.B. Bentzen (2008), "Chemical effects of a high CO_2 concentration in oxy-fuel combustion of methane", *Energy & Fuels,* 22, pp. 291-296.
7. Mendiara, T. and P. Glarborg (2009) "Reburn Chemistry in oxy-fuel combustion of methane", *Energy & Fuels*, 23, pp. 3565-3572.
8. A.J. Mackrory (2008), "A mechanistic investigation of nitrogen evolution in pulverized coal oxy-fuel combustion", *Dissertation*, Brigham Young University.
9. CHEMKIN-PRO (2008), Reaction design, San Diego.
10. Westbrook, C.K. and F.L. Dryer (1984), "Chemical kinetic modeling of hydrocarbon combustion", *Progress in Energy and Combustion Science*, 10, pp. 1-57.
11. Liu, F., Guo, H., Smallwood, G.J. and Ö.L. Gülder (2001), "The chemical effects of carbon dioxide as an additive in an ethylene diffusion flame: implications for soot and NO_x formation", *Combustion and Flame,* 125, pp. 778-787.

Effects of Gas Residence Time ON NO Reduction IN Oxy-Coal Combustion

Wei-Yin Chen, Chemical Engineering, University of Mississippi, University, MS

Shaolong Wan, Swan@Olemiss.Edu, University of Mississippi, University, MS

We evaluate the potential benefits of long residence time of gas in a recycle system on NO reduction in oxy-coal combustion. In both single-pass and recycle tests, residence time within a range of 0.26 to 1.6 seconds shows only minimal benefits on NO reduction. Thus, the reverse Zeldovich mechanism appears not a governing factor up to 1700 °C, the peak flame temperature in practical boilers. The NO concentrations under equilibrium are much lower than those experimentally observed. Moreover, addition of 500 ppm NO in the feed for simulating the effects of the recycled NO does not change the estimated equilibrium concentrations in burning coal in O2-CO2 environment. Nevertheless, it is not observed in practical operations. Thus, NO reduction in oxy-coal combustion is highly kinetically controlled, not thermodynamically controlled.

See more of this Session: Oxycombustion of Coal - Needs, Opportunities, and Challenges - II
See more of this Group/Topical: Energy and Transport Processes

Cofiring of Coal and Biomass Under Oxycombustion Conditions: NOx Formation and PM Characterization

Benjamin M. Kumfer, Energy, Environmental and Chemical Engineering, Washington University in St. Louis, St. Louis, MO

Scott A. Skeen, Energy, Environmental and Chemical Engineering, Washington University in St. Louis, St. Louis, MO

Richard L. Axelbaum, Energy, Environmental and Chemical Engineering, Washington University in St. Louis, St. Louis, MO

The cofiring of coal with biomass is a promising method under consideration for power generation as a means to reduce net CO_2 emissions and to meet renewable energy standards. Furthermore, if carbon capture and sequestration (CCS) technology is applied to a cofired system, a net removal of atmospheric CO_2 may be realized, since the CO_2 removed by the biomass is not released back into the atmosphere. In this work, the particulate matter and NOx produced from the cofiring of coal and biomass is investigated under both conventional air-fired and oxy-combustion conditions. Oxy-combustion is advantageous since a concentrated stream of CO_2 is inherently produced for CCS. Experiments are performed in a 35 kW, horizontally-fired combustion facility, utilizing pulverized subbituminous coal and various biomass materials. When air is replaced with a mixture of 30% O_2 and 70% CO_2, and no overfire oxygen is utilized, oxy-fuel combustion results in a 20% reduction in NO production. The oxygen concentrations of the primary and secondary streams are systematically varied, while holding constant the total O_2 and CO_2 flow rates, and a minimum emission rate of NO is measured when the primary and secondary O_2 concentrations are both equal to 30% O_2. A further reduction in NO is measured when cofiring with sawdust, primarily due to the reduced nitrogen content of the fuel. The effects of cofiring and oxy-combustion on the fly ash size distribution and the size-fractionated composition are discussed. In addition, the interaction between coal and biomass fly ash in the high-temperature zone and the potential for increased deposition and fouling/slagging is explored.

See more of this Session: Oxycombustion of Coal - Needs, Opportunities, and Challenges - II
See more of this Group/Topical: Energy and Transport Processes

Heat Transfer Fouling: 50 Years after the Kern & Seaton Model

Hans Müller-Steinhagen

Institute of Thermodynamics and Thermal Engineering
Pfaffenwaldring 6, The University of Stuttgart, Germany

Abstract

Fouling of heat exchangers in processing industries is a chronic operating problem. Costs due to over-design, additional fuel consumption and maintenance, loss of production, etc. have been estimated as 0.25% of the GNP of industrialised countries. In addition to the appropriate selection of operating conditions and exchanger geometry, there are numerous chemical and mechanical methods to mitigate fouling and to remove deposits from the heat transfer surfaces. However, all methods to reduce fouling require some understanding of the mechanisms of the deposition process and of the structure and adhesion of deposits on the heat transfer surfaces.

For a long time, fouling was treated as an unavoidable and incomprehensible curse of any heat exchanger operation. However, almost exactly 50 years ago, D.Q. Kern and his co-author R.E. Seaton published a paper, attempting to describe the growth of fouling deposits in terms of an unsteady state heat and mass balance on the heat transfer surface. More or less at the same time, the TEMA fouling factors (now more appropriately named "fouling resistances") have been published based on operational and anecdotal experience of fouling for a range of heat exchanger applications. For the next 20 years, these two approaches formed the basis for most heat transfer fouling models and practical heat exchanger designs. As a result, heat exchangers have typically been over-designed by 10-200%, with an average around 35%

This unsatisfactory procedure would probably have continued if J. Taborek from HTRI would not have reminded the heat transfer community in 1972 that fouling is the major unresolved problem. This important paper triggered a range of empirical investigations, the findings of which have been summarized in 1985 by N. Epstein in his famous 5x5 matrix. Here, for the first time, the different mechanisms of fouling and the different steps in the net deposition process have been brought together and analysed. This has led to a much for systematic and focussed approach to the investigation and mitigation of heat transfer fouling, for both practical and fundamental problems.

Increasing costs of energy, raw material and production downtime have contributed to the growing interest in heat transfer fouling. More recently, environmental legislation has put additional pressure on fouling-related CO_2 emissions and disposal of cleaning chemicals.

In 1995, a bi-annual series of internal conferences on Heat Exchanger Fouling and Cleaning has been initiated which continues to-date and enjoys a continuously increasing number of participants from industry and research institutions.

Significant progress has since been made by following a number of approaches in parallel, such as:

empirical development of mechanical and chemical on-line fouling mitigation techniques

- development of heat exchanger types with reduced fouling rates
- development of advanced guidelines for heat exchanger design
- compilation of best practice guides for heat exchanger operation
- analytical models of the fouling process to optimise design and operation of heat exchangers, as well as complete industrial processes
- numerical models to understand the effects of various operational parameters on the formation of deposits
- investigation of surface-deposit interaction forces and the development of surface modifications to reduce adhesion.

In particular, cooperation of academic researchers and industrial practitioners has been a recipe for success. More recently, fouling research has started to have a closer look at fundamental chemical and physical phenomena taking place at the heat transfer surfaces, in the deposit and in the fluid bulk.

Despite these efforts, fouling of heat exchangers is still far from been understood in its whole complexity. However, design margins are now considerably less than they used to by some 10 years ago and substantial information is available to design and plant engineers to reduce fouling during operation.

The 2009 D.Q. Award Lecture will present some selected aspects of fouling research to-date and identify areas where significant research and development activities are still required. Examples are taken both from industrial and laboratory investigations. Considering the tremendous costs associated with fouling, this will remain an exciting research topic for years to come.

EFFECT OF CARBON DIOXIDE ON SOOT DISTRIBUTION FROM OXY-COAL COMBUSTION

Gyu-Bo Kim, Pusan Clean Coal Center, Pusan National University, Busan, Korea
Do-Wook Choi, Pusan National University, Busan, Korea
Jae-Dong Kim, Pusan National University, Busan, Korea
Ju-Hun Song, Pusan Clean Coal Center, Pusan National University, Busan, Korea
Young-June Chang, RIMT, Pusan National University, Busan, Korea
Chung-Hwan Jeon, Pusan Clean Coal Center, RIMT, Pusan National University, Busan, Korea

Introduction

As an interest in carbon dioxide(CO_2) emission regulation due to climate change and limited energy resource increase, carbon capture sequestration (CCS) technology is being investigated widely. Especially, an oxy-fuel combustion combined with carbon capture sequestration (CCS) technology is known to be one of emerging clean coal technologies for the CO_2 emission reduction. Soot is useful due to an enhanced radiative heat transfer in a coal-fired boiler furnace, while its formation from tar and char should be suppressed to avoid the loss of carbon content in the coal. In this study, soot distribution measurement was carried out with a laser induced incandescence technique in an oxy-fuel diffusion flame burner. A methane-fed diffusion flame was used for ignition of a coal which is independently introduced in the central tube of the burner. The coal is one of coals which are imported to South Korea being used at its coal-fired power plant. An inert gas such as CO_2 was additionally supplied to the oxidant to simulate the N_2-free oxidation condition and to further investigate the effects of CO_2 on coal-derived soot formation and mechanisms. The flame length made solely from coal burning was also observed to provide the oxidation rate of coal as a function of oxygen fractions in a flue gas.

Experimental apparatus and methodology

Laser induced incandescence (LII) is the most powerful method because it can measure soot particle size, soot volume fraction and mass faction etc. simultaneously. It is also preferred in many combustion systems because it is capable of in-situ measurement and high temporal, spatial resolution. Laser-induced incandescence technique is based on the fact that soot particles absorb the laser light and reach, during the laser pulse, well above vaporization temperature of soot.[1] The radiation of the heated particles is then time resolved detected during the cooling phase.

In this experiment, the burner was made for oxy-fuel combustion reactor of coal. The burner consists of three parts: a central tube to supply coal and carrier gas, a middle tube to supply methane and to ignite coal, and an outer tube to supply oxygen and carbon dioxide for forming surrounding gas. A schematic diagram of the burner is shown in Fig. 1. To improve flame stability, sintered metal form is used. Methane was used as the fuel supply for ignition of coal. CO_2 was used as the inert gas and carrier gas. The coal was used one of coals which are imported to South Korea being used at its coal-fired power plant. The optical measurement system for LII is as follows: pulsed Nd:YAG laser (λ_{2nd} = 532 nm, 9 ns FWHM, 10 Hz) as heating source of soot particulate matters, photo multiplier tube for multiply LII signal, 450nm narrow bandpath filter with FWHM=40nm for collecting LII signal, oscilloscope(500Mhz, 2Gs/s) for analyzing LII signal and data acquisition system. A schematic diagram of the system is shown in Fig. 2. Experimental conditions including fuel and oxidant flow rates were listed in the Table 1.

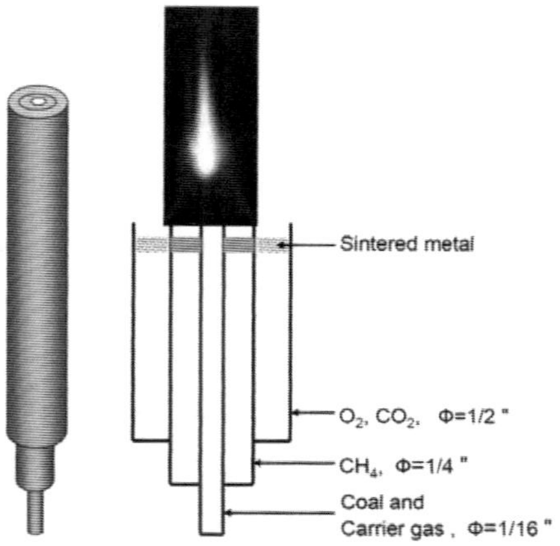

Fig. 1 Schematic diagram of the burner for oxy-fuel combustion

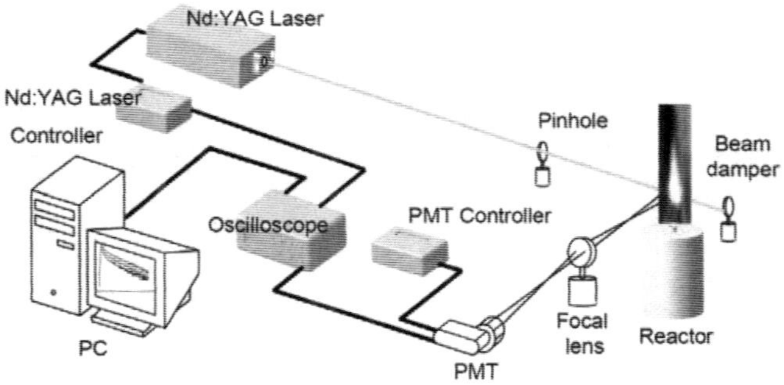

Fig. 2 The optical measurement system for laser induced incandescence

Table 1. Experimental conditions

Item	**Unit**	**Flow rate**
Methane, CH_4	cc/min	36
Oxygen, O_2	cc/min	144
Carrier gas, CO_2	cc/min	63
Inert gas, CO_2	cc/min	210, 350, 490
CO_2 fraction in oxidant stream of O_2-CO_2 mixture	%	65, 75, 80

Soot volume fraction is decided by maximum LII signal.[2] The signal is called prompt LII signal that equal to the LII signal at the maximum temperature of heated particle. Prompt LII signal is proportional to:

$$\text{Prompt LII} \propto N_p\, d_p^{\,x}\,,\ x = 3 + 154\ \text{nm}/\lambda_{det},$$

where d_p is the particle diameter and λ_{det} is the detection wavelength.

In the VIS region, this exponent is about 3. Therefore, the prompt LII signal is proportional to the soot volume fraction.

Experimental results

Fig. 3 shows natural luminosity images of diffusion flame for surrounding gas with air and O_2-CO_2. The flame at far left side image is methane-air diffusion flame. Other images are methane-oxy diffusion flame with carrier gas in the central tube. Flame height of oxy-fuel flame is shorter than one of methane-air flame and the flame shape is symmetrically divided by the effect of carrier gas (CO_2) in the center. It is apparent that there is no indication of soot emitted from all of flame conditions, which would provide a clear background to measure soot amount produced from coal burning under the above oxy-fuel condition.

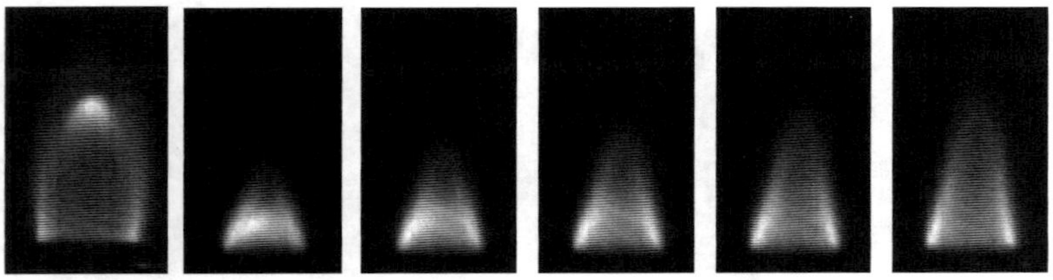

(a) (b) (c) (d) (e) (f)

Fig. 3 Methane diffusion flames in the conditions of air (a) and O_2-CO_2 ; (b) Y_{CO2}=20%, (c) 35%, (d) 45%, (e) 50%, (f) 55%

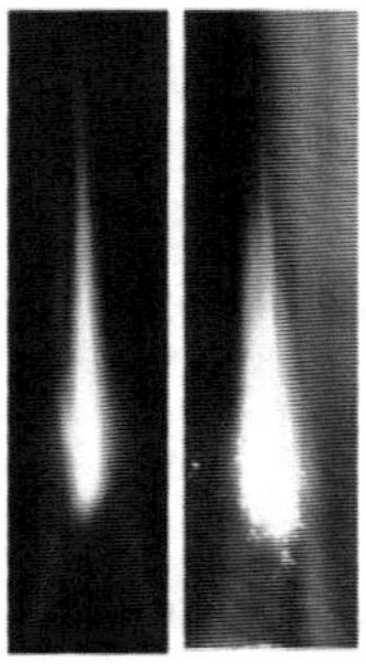

Fig. 4 Comparison of images of oxy-coal firing with 75% CO_2 in the oxidants, left image : without laser, right image : with laser

Fig. 4 shows images of oxy-coal firing and induced by the laser. In the flame image induced by the laser, it will be able to confirm that the scattering light of the laser and incandescence are being mixed. Through the results of the image, it is able to expect that LII signals will include serious noise.

Fig. 5 shows temporal profile of LII signal from methane-air-coal firing as a function of elevation from the flame tip. The signal intensities are decreased as the elevation of coal flame increases. This indicates that soot is formed at early stage in coal combustion which may occur from either the surpression of soot formation due to reduced temperature downstream or a completeness of volatilization process at the height of 10 mm. Fig. 6 shows the comparison with LII signal of methane-air-coal condition and oxy-coal firing. The signal intensity of oxy-coal firing is lower than one of air condition. In the case of oxy-fuel combustion, the soot oxidation is more activated by the effect of the more available oxygen than soot inception and growth is.

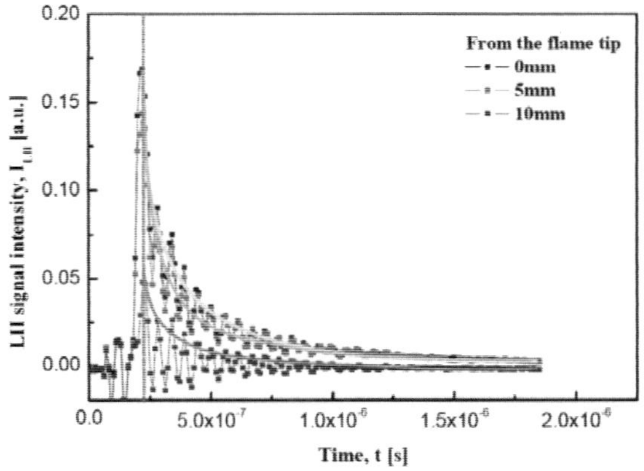

Fig. 5 Temporal profile of LII signal as a function of elevation from the flame tip

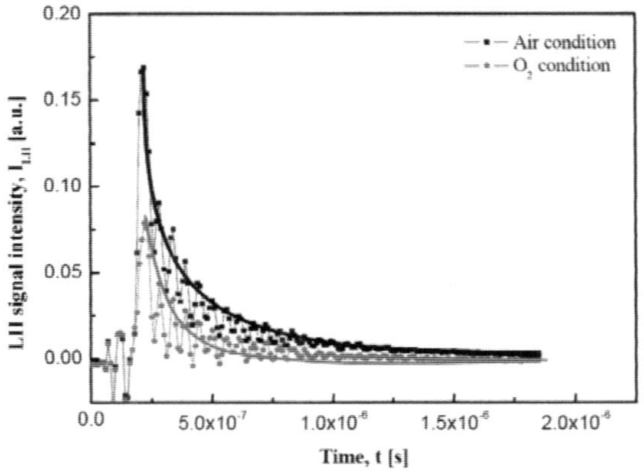

Fig. 6 The comparison of LII signal of methane-air-coal condition with that of oxy-coal firing (Y_{CO2}=75%)

Fig. 7 presents the characteristics of prompt LII signal as CO_2 fraction in oxidants varies over the range between 65 and 80 %. The result shows little dependence of soot formation measured in terms of volume fraction on CO_2 presence in these particular ranges of CO_2 fraction.

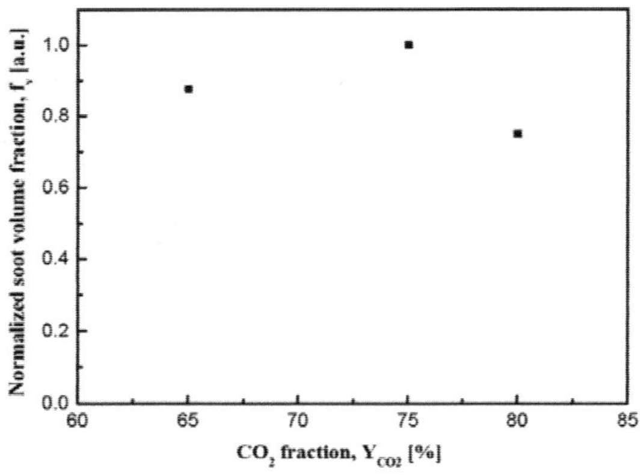

Fig. 7 The characteristics of prompt LII signal as a function of CO_2 fraction during the oxy-fuel combustion

Conclusion

From the results of temporal profile of LII signal obtained from an experiment under oxy-coal firing conditions, the followings are concluded:

1. Soot is observed to be formed at early stage of coal combustion process.
2. Soot volume fraction of oxy-coal firing is lower than one of methane-air-coal condition probably to combined effect of surpressed soot inception and increased oxidation.
3. The effects of CO_2 fraction were not revealed on the soot formation during oxy-coal firing process.

References

1. Bladh H. (2004), "Characteristics of Laser-Induced Incandescence from Soot in Studies of a Time-Dependent Heat and Mass Transfer," *Applied Physics B* 78 pp. 241-248.

2 Speaker, Zizak G. (2000), "Laser Induced Incandescence (LII) of soot," in *the ICS Training course on Diagnostics of Combustion Processes*, Zizak G. NILES, University of Cairo, Egypt

Influence of Biomass On Fine-Particle and Mercury Emissions in Oxy-Coal Combustion

S. Michael Daukoru, Energy, Environmental and Chemical Engineering, Washington University in St. Louis, St. Louis, MO

Achariya Suriyawong, Energy, Environmental and Chemical Engineering, Washington University in St. Louis, St. Louis, MO

Scott A. Skeen, Energy, Environmental and Chemical Engineering, Washington University in St. Louis, St. Louis, MO

Richard L. Axelbaum, Energy, Environmental and Chemical Engineering, Washington University in St. Louis, St. Louis, MO

Pratim Biswas, Energy, Environmental and Chemical Engineering, Washington University in St. Louis, St. Louis, MO

Oxy-coal combustion elicits continued study as a means for obtaining high-concentration carbon-dioxide (CO_2) flue gas that can be readily purified and sequestered. As with most other modalities for addressing the role of CO_2 in global climate change, oxy-coal combustion must contend with the prospect of wide deployment and site-specific access to and preferences for fuels. Biomass co-combustion with coal is one such preference that has the added benefit of net CO_2 reduction and has been widely studied and implemented for existing, conventional coal combustion modalities, in spite of the many technical challenges presented by the introduction of biomass. While the impact of oxy-coal combustion on other emissions such as fine particles and mercury has been examined (Suriyawong et al., 2006), operational demands may result in biomass co-firing in oxy-coal combustion environments.

Therefore, we will compare the effect of sawdust (biomass) on fine particle formation and mercury emissions in oxy-coal combustion to its effect in conventional coal-air combustion for a laboratory-scale self-sustaining combustor. These results will also be compared with those obtained from earlier controlled studies of oxy-coal combustion in drop-tube and self-sustaining combustion environments. It is expected that different temperature-time profiles would result in different vaporization rates, particularly for sawdust, for oxy-coal combustion compared with coal-air combustion. And, increased formation of alkali salts due to the addition of biomass to oxy-coal combustion may increase the potential for mercury oxidation. Practical implications for fine-particle and mercury capture will be suggested.

--

Suriyawong, A., Gamble M., Lee M.H., Axelbaum R. and Biswas P., "Submicrometer particle formation and mercury speciation under O_2 - CO_2 Coal Combustion", Energy and Fuels, 20 (6), 2357-2363, 2006.

See more of this Session: Oxy-Combustion of Coal - Needs, Opportunities, Challenges - III
See more of this Group/Topical: Energy and Transport Processes

Role of Recycle On Fine-Particle Formation, Mercury Speciation, and Electrostatic Capture During Oxy-Coal Combustion

S. Michael Daukoru, Energy, Environmental and Chemical Engineering, Washington University in St. Louis, St. Louis, MO

Pratim Biswas, Energy, Environmental and Chemical Engineering, Washington University in St. Louis, St. Louis, MO

Oxy-coal combustion is being touted as a potential means for increasing concentration of carbon dioxide in exhaust gases to help in potential remediation and sequestration. The impact of this modality of combustion on fine-particle and trace metal formation and mercury speciation has been examined in both drop-tube and self-sustaining coal combustors (Suriyawong et al., 2006; Suriyawong et al., 2008). Most of these studies have been conducted by combustion of the coal in a mixture of oxygen and carbon dioxide. However, in a full-scale operating scenario, the flue gases will be recycled with only the necessary amount of oxygen added at the inlet with the coal.

In this paper, we will compare the effect of flue gas recycle on fine particle formation and other emissions, to that of the controlled studies performed earlier. Due to the difference in the temperature-time histories and the composition of the flue gases, there are expected to be differences in the combustion characteristics and performance of particle capture devices such as electrostatic precipitators (ESPs). First, the sub-oxide formation pathways are expected to result in different vaporization rates due to the different time-temperature relationships. It is expected that the more volatile elements such as mercury will not show any differences, however, the chemistry in the gas phase maybe different due to the potential for oxidation to increase. The performance of added sorbents - both micrometer and nanostructured, will be different - and the results will be reported. Flue gas recycle is also expected to affect particle penetration and corona inception for the ESP due to the variation of flue gas compositions. Thus, strategies for effective modalities for oxy-coal combustion to minimize emissions of fine particles and mercury will be proposed.

--

Suriyawong, A., Gamble M., Lee M.H., Axelbaum R. and Biswas P., "Submicrometer particle formation and mercury speciation under O_2 - CO_2 Coal Combustion", Energy and Fuels, 20 (6), 2357-2363, 2006.

Suriyawong, A., Hogan C.J., Jiang J., and Biswas P., "Charged fraction and electrostatic collection of ultrafine and submicrometer particles formed during O_2-CO_2 coal combustion", Fuel, 87 (6), 673-682, 2008.

See more of this Session: Oxy-Combustion of Coal - Needs, Opportunities, Challenges - III
See more of this Group/Topical: Energy and Transport Processes

The Behavior of Mercury and Sulfur in An Oxy-Fuel and Integrated Pollutant Removal System

Stephen J. Gerdemann, National Energy Technology Laboratory, U.S. Department of Energy, Albany, OR

Casey Carney, National Energy Technology Laboratory, U.S. Department of Energy, Albany, OR

Thomas Ochs, National Energy Technology Laboratory, U.S. Department of Energy, Albany, OR

Danylo Oryshchyn, National Energy Technology Laboratory, U.S. Department of Energy, Albany, OR

Cathy Summers, National Energy Technology Laboratory, U.S. Department of Energy, Albany, OR

Oxy-fuel combustion uses conventional technology and denitrified air to combust fossil fuels resulting in a flue gas which is primarily CO2. Integrated Pollutant Removal (IPR™) compresses the gas to pipeline pressure and uses the cooling at the IPR inlet, as well as intercooling between compression steps, to remove soluble and entrainable material with the precipitated water. Latent and sensible heat given up by the processed gas at the cooling steps are partially recovered and used to preheat the boiler feed water. Because most of the nitrogen has been removed from the oxidizing gas, the relative concentration of the remaining gasses, such as SO2 and Hg, in the resulting flue gas is increased. The increased concentrations have implications for removal and also corrosion. This paper follows sulfur and Hg from an oxy-fuel fired system through the compression and cooling steps of IPR. A model of the system is compared with actual data from a coal-fired burner test facility.

See more of this Session: Oxy-Combustion of Coal - Needs, Opportunities, Challenges - III
See more of this Group/Topical: Energy and Transport Processes

Formation and Impact of Gases Sulfur Components in an Oxyfuel Process

J. Maier, L. Al-Makhadmeh, G. Scheffknecht
Institute of Process Engineering and Power Plant Technology (IVD)
Pfaffenwaldring 23, University of Stuttgart, 70569 Stuttgart, Germany
Corresponding Author: maier@ivd.uni-stuttgart.de +49 711 685 63396

Abstract

The higher concentrations of sulfur components in the furnace during an oxy-fuel process, compared to conventional combustion, increase the interest of studying sulfur components formation and destruction mechanisms under oxy-fuel conditions and its impacts in power stations. Sulfur components have effects on boiler corrosion, ash deposition, ash precipitation, and acid dew point.

In this investigation the formation and destruction mechanisms of sulfur components under oxy-fuel conditions in the boiler were analyzed during different combustion conditions (staged, un-staged). Experimental investigations were conducted in a once-through 20 kW furnace and in a semi technical 500 kW test facility with flue-gas recirculation. Different coal qualities such as lignites and bituminous coals were used. Besides SO_2, H_2S has been measured in the sub-stoichiometric combustion zone. To study effects on destruction and formation of sulfurous species, the SO_2 concentration in the artificial combustion gas mixture was varied between 500ppm and 6000ppm and detailed inflame measurements of gas components (SO_2, H_2S, NO, CO etc) have been performed. During staged combustion, the effect of burner oxygen ratio on sulfur components speciation has been performed at 0.75, 0.85, and 0.95. In addition, the impact of a wet condition was studied by injecting H_2O steam through the burner. Moreover, the conversion of SO_2 to SO_3 and the behavior of the acid dew point have been investigated. On the other hand, SO_2 capture capacity of coals in the convective section of the furnace has been studied.

Keywords: Sulfur components (SO_2, SO_3, and H_2S); Oxy-fuel combustion; Coal; SO_2 capture

1. Introduction

The fate of sulfur during oxy-coal fired power plant with CO_2 capture is an important issue since it could impact the various aspects of the power plant in terms of the process design, plant construction principles and operations. A general flow sheet of oxy-fuel technology is given in Figure 1, which indicates the locations of potential sulfur impacts.

Santos [1] presented an overview in the understanding of fate of sulfur in coal during oxy-fuel combustion, demonstrating the different factors that could impact the level of SO_2 and SO_3 emissions and the efficiency in the capture of sulfur by the ash. Similar to any air fired furnace, the level of SO_2 emission during oxy-coal fired power plant is proportional to the coal's sulfur content. The in-furnace SO_2 concentration has been reported to increase by a factor of 2-3 during O_2/recycled flue gas (RFG) combustion [2, 3]. The presence of other sulfur components i.e. H_2S, SO_2 and SO_3 is associated with high temperature corrosion at the furnace walls and in the super-heater sections and low temperature corrosion through the condensation of sulfuric acid in the economizer or air heater. In terms of the total sulfur mass output (i.e. specific mass per energy input), the SO_2 emissions in oxy-coal fired cases are about 20-40% lower than in the air-fired cases. The lower specific emissions are primarily attributed to the significant retention of the sulfur in the ash [1] and, therefore, depend on fuel and ash properties. This paper is a summary of IVD activities on the impact of combustion media and recycled SO_2 on the formation of gaseous sulfur components in the flue gas (H_2S, SO_2 and SO_3). In addition, the SO_2 capture capacity of coals (ash) in the convective section of a boiler during different combustion conditions when firing different coal qualities such as lignites and bituminous coals are reported.

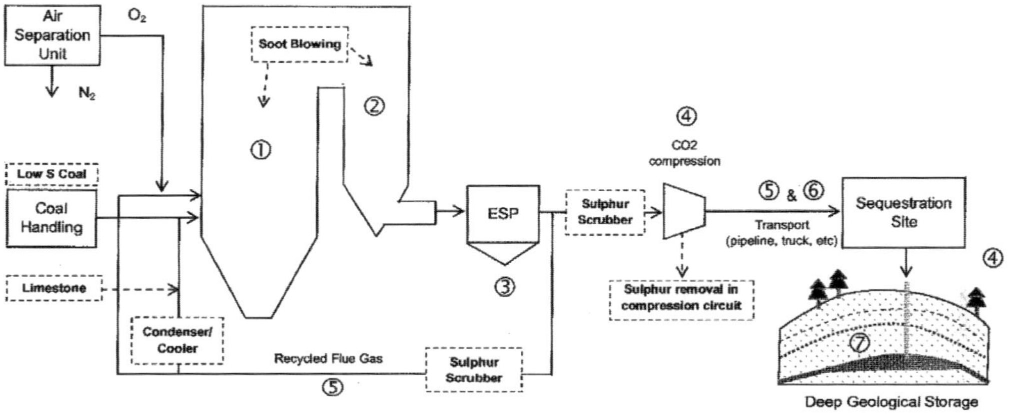

Figure 1: Oxy-fuel flow sheet with locations of the impacts of sulfur and possible mitigation options [4]

2. Experimental Facilities and Approach of investigations

2.1. Coals used

Four coals (Table 1) were used: Klein Kopje (KK), a medium volatile bituminous coal from South Africa, Lausitz brown coal (LA) from Eastern Germany, Rhenish brown coal (RH) from Western Germany and high volatile bituminous coal Ensdorf (EN) from Saar basin in Germany. The brown coals when received from the mine often contains more than 50% moisture but were pre-dried prior to combustion as pre-drying of high moisture lignites is expected to become state-of-the-art for efficiency improvement. The particle size distribution of the coals used is representative of the pulverized coals used in large-scale furnaces.

Table 1: Proximate and Ultimate analyses of the coals used

	LHV [kJ/kg]	Moisture [%, ar[1]]	Ash [%, d]	Volatile [%, daf]	Fixed C [%, daf]	C [%, daf]	H [%, daf]	N [%, daf]	S [%, daf]	O [by diff]
Klein Kopje KK	24932	3.60	19.29	27.76	72.24	83.93	4.50	1.67	0.72	9.18
Lausitz Coal LA	21412	10.20	5.46	57.36	42.64	66.78	5.26	0.65	0.85	26.45
Rhenish Coal RH	20965	11.50	4.07	54.18	45.82	67.96	7.68	0.73	<0.3	23.63
Ensdorf Coal EN	30955	2.42	7.47	37.21	61.90	74.85	5.05	1.59	0.83	17.68

[1] ar: as recieved
daf: dry ash free
by diff: by difference
H, N, S and O: hydrogen, nitrogen, sulfur and oxygen content

2.2. A Once-Through 20 kW Furnace

Figure 2 shows the once-through 20 kW furnace which was used for the major investigations in this summary. It mainly consists of a ceramic tube of 2500 mm length and 200 mm diameter forming the reaction zone. The facility is electrically heated around the reaction zone enabling constant wall temperature up to 1400 °C. The combustion air or O_2/CO_2 mixture is injected through annular clearances, divided into primary and secondary streams. Axial profile measurements of the gaseous and solid samples are carried out by means of an oil-cooled sampling probe which can be moved vertically along the central axis of the reactor from the burnout section to the burner mouth with high flexibility.

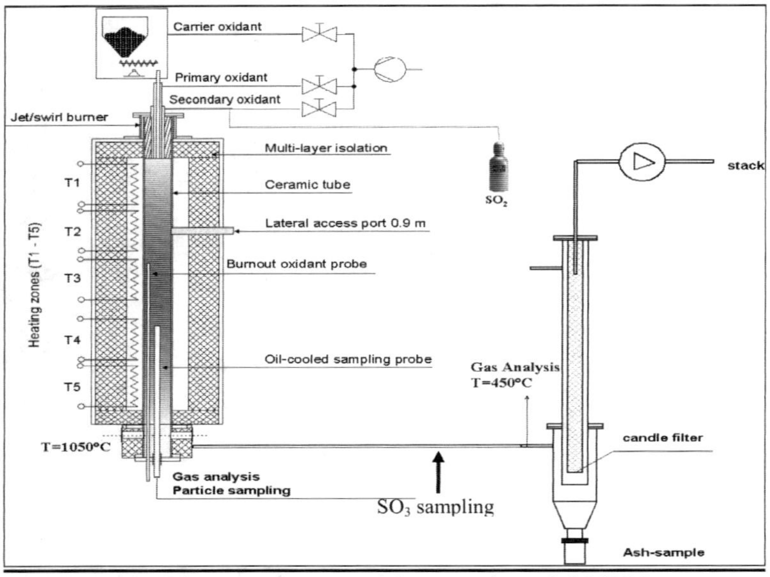

Figure 2: Schematic diagram of the once-through 20 kW furnace

Table 2: Experimental conditions studied in the once-through 20 kW furnace

Staged Combustion in 27%O_2/73%CO_2				
O_2 excess [vol%]		3.52		
Overall O_2 ratio		1.15		
Burner O_2 ratio		0.75	0.85	0.95
Carrier stream	[m³ stp[2]/h]	1.5	1.5	1.5
Primary stream	[m³ stp/h]	2.03	2.03	2.03
Secondary stream	[m³ stp/h]	3.04	3.04	3.04
Staged stream	[m³ stp/h]	3.51	2.32	1.38
SO_2inj	[ppm]	2000, 3000, 6000	2000, 6000	2000, 6000
H_2O	[vol%]	0, 20	0, 20	0

Staged flames for Lausitz coal have been performed in OF27 flame (designating 27%O_2/73%CO_2 flame) at different burner oxygen ratios and to simulate the recycled SO_2 concentration in oxy-fuel process, up to 6000ppm SO_2 was added to the fuel gas through the secondary stream. Moreover, to simulate the recirculated stream in oxy-fuel process where the recycle branch is situated upstream the flue gas condenser (wet recycle), steam injection was also carried out during Lausitz coal combustion. Table 2 summarizes the investigated

[2] stp standard temperature and pressure conditions

experimental parameters for Lausitz coal in OF27. On the other hand, the gaseous concentration measurements were performed at the end of flue gas path and before the candle filter to study SO_2 capture in the convective section of the furnace. For studying the correlation between SO_2 and SO_3 in the 20 kW furnace diluted solution of H_2SO_4 has been injected through the secondary stream to simulate SO_3 concentration in practical oxy-fuel processes. The injection of H_2SO_4 is justified by that the SO_3 in the flue gas forms H_2SO_4 under the temperature window of flue gas path in practical oxy-fuel processes.

2.3. The 500 kW Test Facility

Figure 3 shows the 500 kW test facility. A detailed description can be taken from [3]. As shown in the figure, the SO_3 measurements have been performed upstream the ESP. For SO_3 sampling an adapted controlled condensation procedure, in line with a VDI draft guideline, was applied. In addition information given in the publications [5, 6] where taken into account. Sample gas is extracted from the flue-gas and passed through a condenser helix which is controlled to a temperature of 95 °C. Here, sulfuric acid that has formed in the flue-gas condensates on the helix's inner surface. Afterwards the helix is rinsed with purified water and the sample collected is subjected to analysis by titration. To prevent particles from entering and contaminating the condenser two heated filters with different pore sizes were integrated in the sampling arrangement. The flue-gas temperature during sampling was between 350-380 °C.

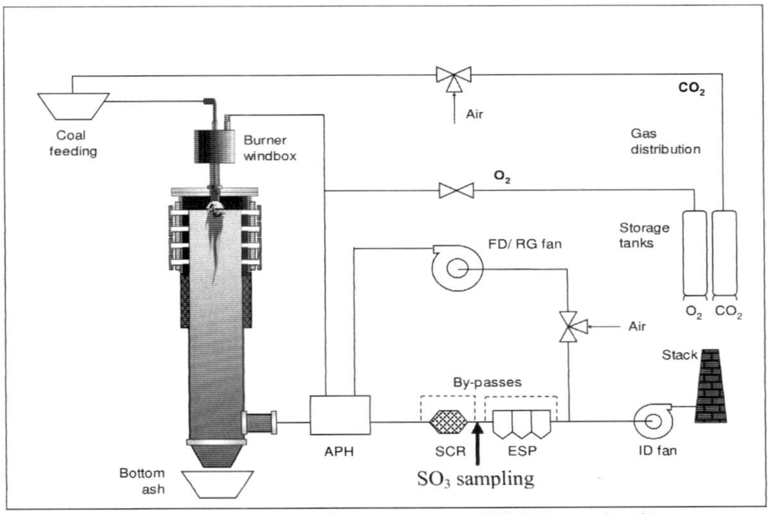

Figure 3: Schematic diagram of the 500 kW test facility

3. Results

3.1. SO_2 and H_2S Formation and Destruction Mechanisms

In conventional combustion, sulfur is only added via the fuel while in oxy-fuel process the recirculated flue gas is another source of sulfur components in the furnace. To study sulfur components speciation in oxy-fuel process, staged flames with SO_2 injection up to 6000ppm in a 20 kW furnace has been performed.

Figure 4 shows the axial concentration profiles of SO_2, H_2S, and O_2 for Lausitz coal combustion in air and OF27 flames with a burner oxygen ratio of 0.75, the SO_2 concentration in the oxidizing agent was set at 3000ppm. For both air-firing and oxy-fuel OF27_3000 cases, it was noticed that the SO_2 concentration increases rapidly up to 0.5 m from the burner followed by a fast reduction to H_2S. The maximum H_2S concentration forms at around 0.9-1.1 m and after the addition of burnout air all sulfur components oxidize to SO_2. On the other hand, the sulfur components speciation is more to the side of SO_2 for oxy-fuel OF27_3000 case, due to the addition of sulfur compound in the form of SO_2 in the oxidizing agent. However the H_2S concentration doubles that of air-firing case due to the reduction of SO_2 to H_2S in the sub-stoichiometric zone. It can be concluded that the H_2S/SO_2 formation and reduction mechanisms in near flame zone are fundamentally similar during air-blown and oxy-coal combustion.

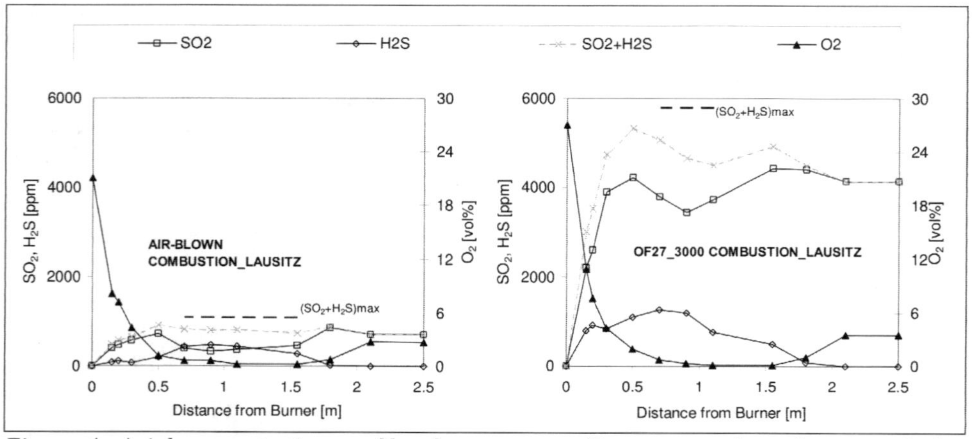

Figure 4: Axial concentration profiles for gaseous sulfur compounds and oxygen during Lausitz coal combustion in air and OF27 with a burner oxygen ratio of 0.75.

Figure 5 shows the absolute and relative H_2S concentration for Lausitz coal combustion between 0.9 and 1.1 m from the burner with different SO_2 concentration in the oxidizing agent, the burner oxygen ratio was 0.75.

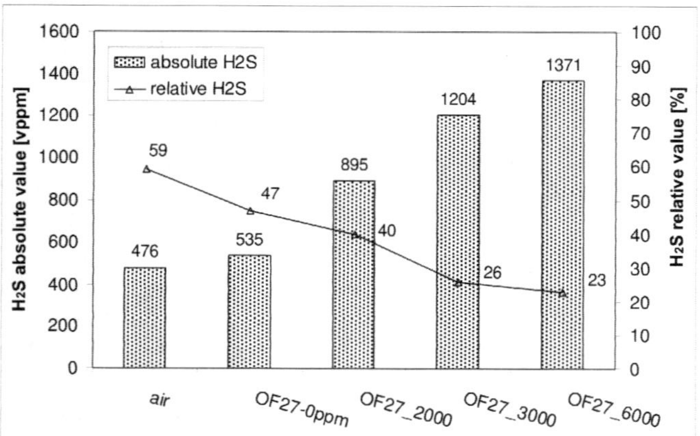

Figure 5: Absolute and relative H_2S concentration between 0.9-1.1 m from the burner during Lausitz coal combustion in air and OF27 combustion

The relative concentration of H_2S (α_{H2S}) or volume fraction at 0.9-1.1 m has been calculated as:

$$\alpha_{H_2S} = \frac{y_{H_2S,d}}{y_{H_2S,d} + y_{SO_2,d}}$$

where $y_{H2S,d}$ and $y_{SO2,d}$ are the volumetric concentration of H_2S and SO_2 in the flue gas (dry), respectively at 0.9-1.1 m.

As can be seen, the fraction of H_2S in the near-flame zone is higher during air-blown combustion compared to oxy-fuel flames, even though the concentration of H_2S is higher and the difference increases as the SO_2 concentration in the oxidizing agent is higher. When the SO_2 concentration in the oxidizing agent was increased from 0ppm to 6000ppm, the H_2S concentration increased from 535ppm to 1371ppm. It can be concluded that the H_2S concentration in the near-flame zone is 2-3 times higher during oxy-coal combustion with SO_2 injection of 2000-6000ppm compared to air-firing case. Therefore, the issue of sulfur induced corrosion in an oxy-coal operated furnace must be considered especially for high sulfur-containing coals. Meanwhile, the relative concentration of H_2S decreases from 47% to 23% since the addition of sulfur compound was in the form of SO_2 in the oxidizing agent as mentioned previously.

Impact of Coal Quality

Besides Lausitz-coal staged flame, staged flames for Klein Kopje and Rhenish coals with burner oxygen ratio of 0.75 have been performed in order to investigate the effect of coal quality on sulfur components speciation. The SO_2 concentration in the oxidizing agent was set at 3000ppm. Generally, it was found that KK and RH coals show the same trend of SO_2 and H_2S concentrations as well as LA coal, even though the chemical speciation of sulfur components differs significantly for the different coals as shown in Figure 6 and Table 3.

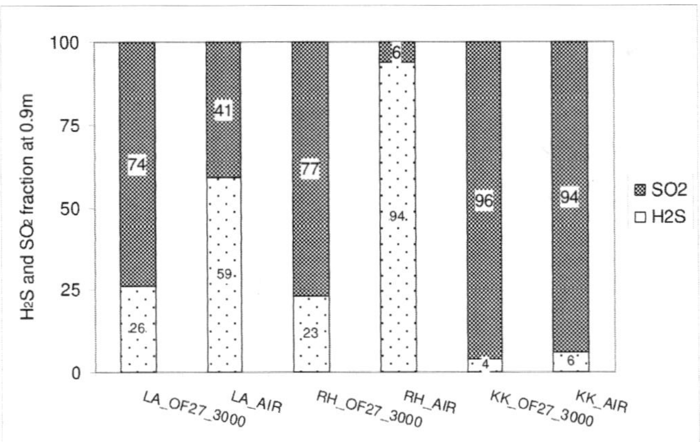

Figure 6: H_2S and SO_2 volumetric fraction for the investigated coals with burner oxygen ratio of 0.75

Table 3: Absolute and relative H_2S values at 0.75 between 0.9-1.1 m for the investigated coals

	LA_OF27_3000	LA_AIR	RH_OF27_3000	RH_AIR	KK_OF27_3000	KK_AIR
H_2S absolute [vppm]	1204	476	910	294	186	37
H_2S relative	26	59	23	94	4	6

[%]							

For example 15% of sulfur (the maximum) was detected as H_2S for Klein Kopje coal while between 25-50% H_2S was detected for Lausitz coal [7]. The quantitative drop in SO_2 level and increase in H_2S level in the sub-stoichiometric region shows a strong relationship to the volatile content of the coal. The high volatile coals, Rhenish and Lausitz lignite showed a strong drop in SO_2 level with high H_2S peak in the sub-stoichiometric region whereas medium volatile coal, Klein Kopje showed only a small drop in SO_2 level and low H_2S peak during air-blown as well as OF27_3000 combustion. On the other hand, from coal pyrolysis investigations in the CO_2 environments at temperature ranges of 700-1150 °C using an entrained-flow reactor, it was found that the S-release for Klein Kopje coal is around 30% while Lausitz coal releases 50% S indicating the importance of sulfur functional groups [8]. It can be concluded that the volatile content and the amount of released sulfur in the sub-stoichiometric zone have a significant effect on sulfur components speciation in the reduction zone.

Impact of Burner Oxygen Ratio

As staging is used increasingly for reduction of nitrogen oxides, the impact of burner oxygen ratio on sulfur components speciation has been studied at 0.75, 0.85, and 0.95 for Lausitz coal. Figure 7 shows the absolute and relative H_2S concentrations at the three investigated burner oxygen ratios without and with SO_2 injection of 2000 and 6000ppm.

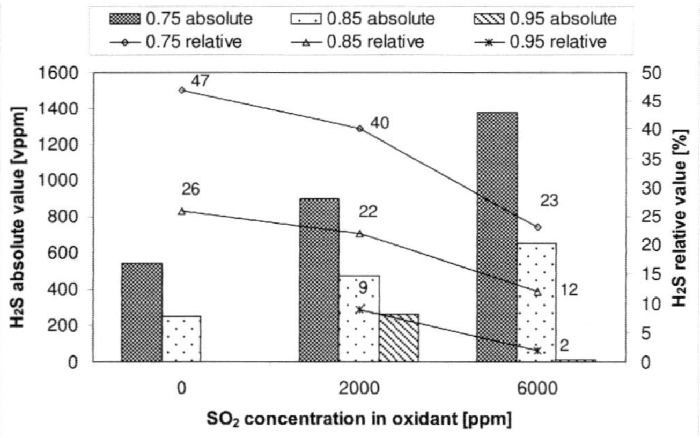

Figure 7: The influence of burner oxygen ratios on the absolute and relative H_2S concentration between 0.9-1.1 m from the burner during OF27 combustion for Lausitz coal

The lower the burner oxygen ratio, the higher the H_2S concentration was found in all flames studied without and with SO_2 injection. It was noticed that the H_2S concentration with burner O_2 ratio of 0.75 doubles that at 0.85 with and without SO_2 injection in the oxidant, e.g. with 6000ppm SO_2 in the oxidant the H_2S concentration increases from 647ppm to 1371ppm as the burner oxygen ratio decreases from 0.85 to 0.75. On the other hand, at burner oxygen ratio of 0.95 the H_2S concentrations are very low. It can be concluded that as the staging is stronger (the burner oxygen ratio decreases) and as the SO_2 concentration in the oxidant increases the reduction of SO_2 and the formation of H_2S increases. In addition, the relative concentration of H_2S decreases as the burner oxygen ratio increases.

Impact of Wet Condition

The effect of the increased water vapor content in the oxidant (27% O_2, 53% CO_2 and 20% H_2O) on the formation of sulfur components has been studied and the results are presented in Figure 8 with burner oxygen ratio of 0.75 and 0.85 and different SO_2 concentrations in the oxidant.

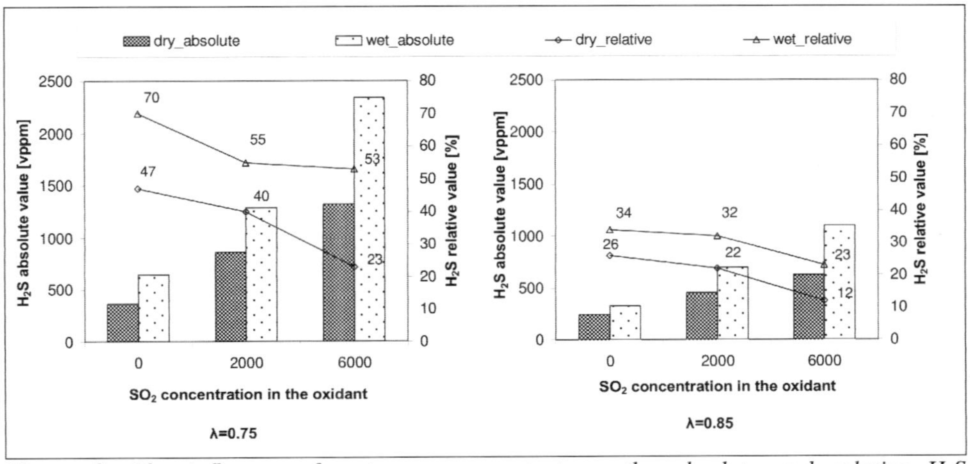

Figure 8: The influence of water vapor content on the absolute and relative H_2S concentrations between 0.9-1.1 m from the burner during OF27 combustion for Lausitz coal with burner oxygen ratio of 0.75 and 0.85(the absolute concentrations on wet basis)

It is clearly seen that the H_2S concentrations under wet condition flames are higher than dry flames at all investigated conditions; the H_2S concentrations are 1.4-1.8 higher compared to dry conditions. The relative H_2S concentration, on the other hand, is 15-30% and 10% higher under wet conditions compared to dry conditions at 0.75 and 0.85, respectively. In addition, as the SO_2 concentration in the oxidant increases with higher water vapor content the higher the H_2S formation will be with both burner oxygen ratios 0.75 and 0.85. The H_2S concentration under wet condition is 2 times higher when SO_2 in the oxidant was 2000ppm with burner oxygen ratios of 0.75 and 0.85 compared to wet condition without SO_2 injection. Moreover, when 6000ppm SO_2 was in the oxidant the H_2S concentration under wet condition was 3-4 times higher than without SO_2 in the oxidant. It can be concluded that the water vapor content intend to increase the H_2S concentration in the reduction zone of the furnace. This can be explained by that even the water vapor retards sulfur conversion to SO_2 in the reduction zone or gives a new hydrogen source which favor H_2S formation. In addition, higher SO_2 concentrations in the oxidant under wet conditions promote higher H_2S formation.

3.2. Reduction of Recycled SO_2 in the Convective Section/Flue Gas Path of the Furnace

Theoretically, the concentration of SO_2 in the furnace in an oxy-fuel process is about 5 times higher than in air firing. Depending on the fuel, however, sulfur is also captured in the ash in the course of flue gas path cooling, so the SO_2 concentration is two-three times greater in oxy-fuel combustion compared to air combustion [2, 3]. In addition, it is reported that the SO_2 concentration did not change significantly with increasing the recycle ratio [9].

For the determination of SO_2 capture potential by the ash in the convective section/flue-gas path, SO_2 measurements were carried out at the end of the flue-gas path just before the candle filter in the 20 kW test facility by gradually increasing the SO_2 injection in the furnace. The flue-gas temperature at the end of radiative section was around 1050 °C and the temperature at the SO_2 measurement point just before candle filter was around 450 °C. Figure 9 shows SO_2 capture in the flue-gas path vs. SO_2 measured at the end of radiative section of the furnace (2.5 m from burner) during air-blown as well as OF27 combustion for all four coals mentioned previously.

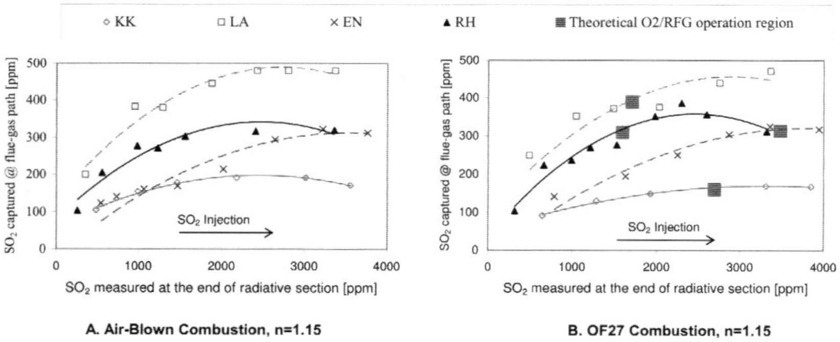

Coal	Ash [wf, mass%]	CaO [wf, mass%]	S in fuel [wf, mass%]	Ca:S molar ratio
Klein Kopje (KK)	19.29	7.8	0.58	2.98
Ensdorf (EN)	7.29	10.50	0.77	1.03
Lausitz (LA)	4.79	23.78	0.35	3.17
Rhenish (RH)	4.07	35.10	0.34	3.96

Figure 9: SO_2 capture in the convective section/flue-gas path during (A) Air-Blown combustion and (B) OF27 combustion (embedded table illustrating coal properties).

As seen from the Figures, SO_2 captured in the flue-gas path increases with increasing SO_2 concentration at the end of the radiative section during air-blown as well as OF27 combustion. However a plateau is reached after a certain concentration of SO_2 at the end of the radiative section for all coals indicating that the SO_2 capture capacity is exhausted and more SO_2 cannot be captured. The capture capacity however shows a weak dependency on the Ca/S molar ratio (see embedded table) as Ca/S molar ratio is highest for Rhenish coal and lowest for Ensdorf coal but SO_2 captured is highest for Lausitz coal and lowest for Klein Kopje coal indicating that the form of calcium present in the ash plays a more important role on SO_2 capture. Klein Kopje char combustion shows that 100% of S is released in the oxidation step while 86-91% is released during Lausitz char combustion indicating again the importance of the form of Ca in the char (ash) [8]. Although the residence time available for interaction between SO_2 in the flue-gas and sulfate forming elements in the ash is not long enough in the flue-gas path, desulfurization efficiency as high as 50% was achieved for Lausitz coal indicating the potential for SO_2 retention in the convective section and flue-gas path if right coal and right temperature gradient is available. The grey symbols indicate the practical SO_2 concentration during combustion for each coal under oxy-fuel condition.

3.3. SO_2 Conversion to SO_3

The conversion from SO_2 to SO_3 has been studied in the 20 kW furnace and the 500 kW test facility (2 campaigns) in air firing and oxy-fuel combustion for Lausitz coal.

Figure 10 shows the correlation between SO_2 and SO_3 concentrations under air firing and oxy-fuel conditions in both test facilities. As can be seen, in oxy-fuel process the SO_2 and SO_3 concentrations are higher than in air combustion. From the 1st campaign results in the 500 kW test facility, the average SO_3 concentration at oxy-fuel conditions has a min/max range of 3.6-12.1ppm while for air conditions the average was 1ppm. In the 2nd campaign, the average SO_3 concentration at oxy-fuel conditions has a min/max range of 2.5-6.0ppm and the conversion rate varies by a factor of 3 to 13 higher than in air firing. Our results are in agreement with others. Wall et al. [9] reported that the SO_3 concentration is approximately two and a half (2.5) to three (3.0) times higher in oxy-fuel combustion than in air combustion. On the other hand, the SO_2 and SO_3 concentrations for Lausitz coal combustion in the 20 kW furnace in air and under oxy-fuel conditions show that as the SO_2 concentration increases the concentration of SO_3 increases.

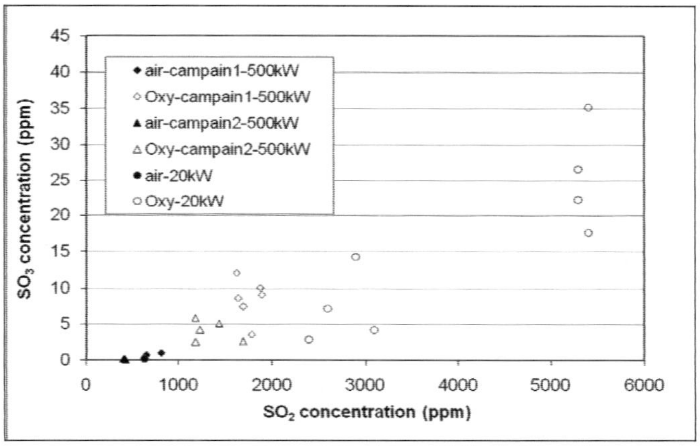

Figure 10: Correlation between SO_2 and SO_3 concentrations results from the 500 kW (1st and 2nd campaign) and 20 kW test facilities

In summary, the trend of higher SO_3 concentration under oxy-fuel conditions and that the SO_3 concentration increases as SO_2 concentration increases have been found from the semi technical facility and the 20 kW furnace despite the variation in the operating conditions for both facilities and the fluctuations during the operation of the 500 kW test facility.

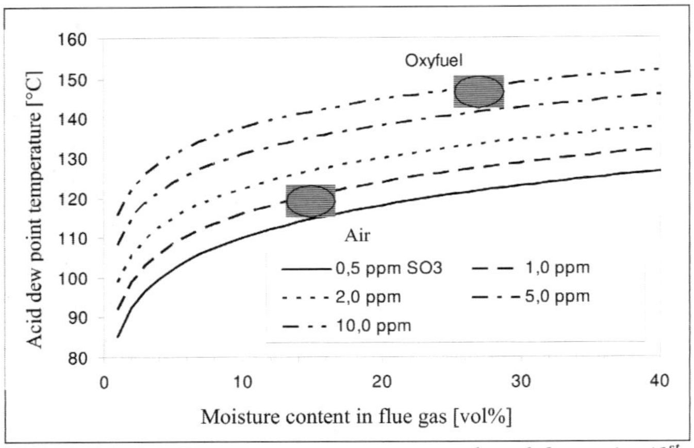

Figure 11: Correlation between SO_3 concentration and acid dew point (1st campaign)

Moreover, the acid dew point temperatures were calculated from the 1st campaign results and presented in Figure 11. It was found that the average acid dew point temperatures are 118 °C for air and 148 °C for oxy-fuel conditions. This difference was to be expected as both the SO_3 concentration and the vapor-phase content of the flue-gas increased significantly at oxy-fuel conditions. The corresponding vapor-phase water contents measured during the SO_3 samplings alternated around the averages by ± 2.3 %-points at air conditions and ± 1.3 %-points at oxy-fuel conditions.

4. Conclusions

Gaseous sulfur components have been investigated in an air and oxy-fuel combustion mode with a 20 kW once-through furnace and 500 kW semi-technical facility with flue-gas recycling. The major conclusions from these results are as follows.

The H_2S/SO_2 formation and reduction mechanisms in near flame zone are fundamentally similar during air-blown and oxy-coal combustion. Although the fraction of H_2S in the near-flame zone or oxygen deficient reduction zone is higher during air-blown combustion, the concentration of H_2S was 2-3 times higher during oxy-coal combustion with SO_2 injection of 2000-6000ppm (representing different O_2/RFG scenarios).

The concentration of H_2S is much higher for high volatile coals compared to medium volatile coals. Therefore, from corrosion point of view, careful attention should be given to coals with high amount of volatiles and sulfur i.e. low rank high sulfur coals.

The burner oxygen ratio, wet condition and the SO_2 concentration in the furnace have a significant effect on H_2S formation and SO_2 reduction mechanisms. Higher H_2S concentration was found when the oxygen ratio was low (< 0.8). In addition, wet condition is a source for higher H_2S concentration in the sub-stoichiometric zone.

Higher SO_3 concentrations and conversions under oxy-fuel conditions compared to air-firing cases have been determined. Considering in addition the increase of the water vapor concentration during wet recycle, the acid dew point could be substantially increased. In addition, good agreement between the results obtained in a semi technical 500 kW test facility and a 20 kW furnace was obtained.

Some of the SO_2 can be reduced in the lower temperature region of the furnace if active sulfate forming elements are available in the coal. Due to high concentration of SO_2 during O_2/RFG combustion, the capture of SO_2 is always higher than in air-blown furnaces by coals containing high shares of sulfate forming elements.

References

[1] Santos, S.:Fate of Sulphur in Coal During Oxyfuel Combustion With Recycled Flue Gas (Review of the Current State of Understanding), CCT2009.

[2] Croiset, E.; Thambimuthu, K.: NO_x and SO_2 emissions from O_2/CO_2 recycle coal combustion, Fuel, 80, pp. 2117-2121, 2001.

[3] Mönckert, P.; Reber, D.; Maier, J.; Scheffknecht, G.: Operation of a Retrofitted 0.5 MW_{th} PF Combustion Facility under Oxyfuel Conditions, 32nd International Technical Conference on Coal Utilization and Fuel System, Florida, 2007.

[4] Wall, T.; Stanger, R.; Maier, J.: Sulfur Impacts in Coal-Fired Oxyfuel Combustion with CCS: Impacts and Control Options. 1st Oxyfuel Combustion Conference.

[5] Fernando, R.: SO_3 issues for Coal_Fired Plant; IEA-COAL, 2003.

[6] Verhoff, Banchero: Predicting Acid Dew Point Temperatures from Flue-Gases; Chem. Eng. Prog., Aug. 1974.

[7] Scheffknecht, G.; and Maier, J.: Firing Issues Related to the Oxyfue Process.VGB PowerTech, 2008. 91-97.

[8] Al-Makhadmeh, L.: Coal Pyroylsis and Char Combustion Under Oxy-Fuel Conditions. Diss. Stuttgart University. Shaker Verlag, 2009.

[9] Wall, T.; Liu, Y.; Spero, C.; Elliott, L.; Khare, S.; Rathnam, R.; Zeenathal, F.; Moghtaderi, B.; Buhre, B.; Sheng, C.; Gupta, R.; Yamada, T.; Makino, K.; Yu, J.: An Overview on Oxyfuel Coal Combustion-State of the Art Research and Technology Development. Chemical Engineering Research and Design, 2009.

Oxy-Coal Combustion in a Pilot-Scale Circulating Fluidized Bed

Lawrence Bool III, Industrial Applications and Healthcare R&D, Praxair Inc, Tonawanda, NY

Eric G. Eddings, Dept. of Chemical Engineering, University of Utah, Salt Lake City, UT

Oxy-coal combustion to facilitate carbon capture and sequestration is the focus of numerous research and development activities. Recently, Praxair collaborated with the University of Utah to study some of the fundamental impacts of oxy-coal combustion in fluidized bed combustion. An existing circulating fluid bed (CFB) combustor was modified to allow firing with oxygen-enriched recirculated flue gas. Experiments with this facility were able to routinely achieve CO2 concentrations greater 85% (dry basis). Bed temperature and pollutant emissions were measured under both air-firing and oxy-coal firing.

INTRODUCTION

Oxy-coal combustion has received a great deal of attention in recent years as a way to facilitate carbon capture and sequestration (CCS). Many groups have demonstrated the feasibility of using oxygen diluted with recirculated flue gas (RFG) as a replacement for air in pulverized coal (PC) fired boilers. Pilot and full-scale demonstrations of oxygen enriched RFG for PC units are planned or ongoing. Others have published limited experimental data and commercial circulating fluid bed (CFB) boiler design concepts based on oxygen enriched RFG. Praxair has teamed with Foster Wheeler, and others, to propose a 50 MW oxy-coal CFB at Jamestown, NY. Even with this work, there are still details that need to be addressed to optimize the use of oxygen in oxy-coal firing. For example, are there mixing strategies to optimize oxygen use and are there trace species that may create problems for the CO2 purification unit (CPU).

Praxair was involved with some of the earliest oxy-coal combustion studies1 and was the first company to inject oxygen into coal-fired utility boilers15. Praxair has teamed with the University of Utah to explore optimized oxygen use for PC and CFB units. Preliminary data has been collected using oxygen-enriched wet RFG in the L1500 furnace, a 1.5 MW pulverized coal combustor, and will be reported elsewhere. Data has also been collected using the laboratory-scale CFB. These data will be discussed in this presentation.

See more of this Session: Oxy-Combustion of Coal - Needs, Opportunities, Challenges - III
See more of this Group/Topical: Energy and Transport Processes

LDL-HDL TRANSPORT IN BLOOD FLOW

Dimitri Gidaspow* and Suttipong Songprawat

Illinois Institute of Technology, Chicago, IL 60616

**(Gidaspow@iit.edu)*

Introduction

Augmented particle transport in shear flow (Zydney and Colton, 1988: Shapely, et al, 2002) has historically been linked to the initiation and progress of artherosclerosis (Ether, 2002). However, transport of nanoparticles, such as LDL, has been modeled using a convective diffusion with a constant diffusivity, often obtained from the Einstein equation (Wada and Karino, 2002). For example, Friedman and Ehrlich (1975) used the Leveque type boundary layer approximate analytical solution to compute wall concentrations in an arterial branch. In the last decade commercial codes, such as FLUENT were used to solve the convective diffusion equation (Kaazempur-Mofrad and Ether, 2001; Hodgson and Tarbell, 2002). Sun, et al (2006) empirically corrected the shear dependent transport coefficients.

The diffusivity in shear flow is often correlated (Zydney and Colton, 1988) by the following equation:

Diffusivity/(square of particle diameter x shear rate) = function of volume fraction (1)

But in channel flow the shear rate is zero at the center of the channel. Hence the correlation gives an unreasonable dependence of the diffusivity on the shear rate.
There is clearly a need for a better theory.
The kinetic theory of granular flow invented by Savage (1983) and others, as reviewed by Gidaspow (1994), shows that shear produces random oscillation of particles whose motion can be measured by an analogue of thermal temperature, called the granular temperature. Hence diffusion of nanoparticles , such as LDL, in blood flow is caused by two mechanisms :

- by the motion of the plasma molecules and
- by shear

Shear enhanced diffusion theory

In 1905 Albert Einstein (1956) showed how to compute particle diffusivity based on the random motion of molecules. The driving force for diffusion was the osmotic pressure, Posmotic, given by the equation

$$\text{Posmotic} = RT\, n/N_{Avagadro} = kBTn \qquad (2)$$

where R is the ideal gas law constant, T the thermal temperature, n the number of particles per unit volume and kB is the Boltzmann constant.

The resistance was the drag on a spherical particle given by Stokes law. At a "dynamic equilibrium " Einstein (1956) obtained the famous expression for the diffusivity, $D_{Einstein}$ given below.

$$D_{Einstein} = \frac{k_B T}{3\pi\mu d_p} \qquad (3)$$

where μ is the fluid viscosity and dp is the particle diameter. The thermal energy of the fluid is kBT . In shear flow there exists an additional energy, $m_i\theta_i$, where mi is the mass of particle I and θ_i is its granular temperature, the random particle energy per unit mass.

Then Einstein's method gives the diffusivity due to shear, D_{shear} given below:

$$D_{shear} = \frac{m_i\theta_i}{3\pi\mu d_p} \qquad (4)$$

Since the energies are additive, the total diffusivity, D is the sum of the Einstein and the shear diffusivities :

D= D Einstein + Dshear (5)

Granular Temperature Equations for Dispersed Phase *i*

For a one granular temperature model the complete multiphase model for blood flow was described by Gidaspow and Huang (2009). The commercial code FLUENT, the publicly available MFIX code (Syamlal, 1998) and the IIT code in the book by Gidaspow and Jiradilok (2009) describe the use of the multiphase Navier-Equations solvers. For developed flow in a channel Gidaspow and Huang (2009) have shown that the Poiseuille flow approximation gives nearly the same granular temperature profiles as the complete model. Here we present a generalization of the kinetic theory model to unequal granular temperatures.

This differential equation for the phase *i* granular temperature is given below in cylindrical coordinates:

$$0=\left(\mu_{i,kinetic}+\mu_{i,coll}\right)\left(\frac{dv_i}{dr}\right)^2+\frac{1}{r}\frac{d}{dr}\left[r\left(k_{i,kinetic}+k_{i,coll}\right)\frac{d\theta_i}{dr}\right]+\sum_{j=1}^{M}\gamma_{ij} \qquad (6)$$

Production of oscillations — Conduction — Dissipation due to inelasticity

i=1,2,3,..., M solid phases

The phase *i* viscosities, conductivities, and the energy dissipation are given in Gidaspow and Jiradilok's book (2009). The above equation is a generalization of Gidaspow and Huang (2009) Equation (23). The contribution of rotation is in Songprawat's Ph.D. thesis (2009).

Boundary conditions

The boundary conditions for granular temperatures at the wall are given by Johnson & Jackson, 1987 type boundary conditions. Flux at the wall equals the dissipation at the wall. Hence,

$$\theta_w=-B_i\left.\frac{\partial\theta_i}{\partial n}\right|_{wall} \qquad \text{where} \qquad B_i=\frac{k_i\theta_i}{\gamma_{ij}} \qquad (7)$$

Developed flow velocities

Since the flow is nearly homogeneous in view of the large drag between the particles, (Gidaspow and Huang, 2009), Poiseuille flow is a good approximation. It has been traditionally used for blood flow (Sherwood, 2004).

$$v_i=2\overline{v}_i(1-r^2/R^2) \qquad (8)$$

Comparison of theory to shear dependent diffusivities

We have compared the above theory to measured shear dependent diffusivities for the nanoparticle BSA flowing in mixtures of red blood cells and serum in the experiment done at IIT. Table 1 summarizes the properties of RBC and BSA, and input restitution coefficients used in the model. Figure 1 shows excellent agreement between the theory and the experiments for the 8 nm diameter particles at several volume fractions of ghost RBC. We have computed the shear dependent diffusivities of HDL and LDL using the information in Table 2. Figures 2 and 3 present typical computed diffusivities of LDL and HDL, respectively. The Einstein diffusivities of LDL and HDL are 4.4 x 10^{-7} cm^2/s and 1.9 x 10^{-7} cm^2/s, respectively. The shear diffusivity made the diffusivities in Figures 2 and 3 more equal. The theory permits us to include particle rotation (Songprawat, 2009).

REFERENCES

Einstein, A., "Investigations on the Theory of Brownian Movement." ed. R. Fürth, translated by A.D. Cowper (1926, reprinted 1956); Einstein, Collected Papers, vol. 2, 170-82, 206-22.

Ethier, C.R., "Computational Modeling of Mass Transfer and Links to Atherosclerosis." *Annals of Biomedical Engineering*. 30 (2002): 461-471.

Friedman, M.H., and Ehrich, L.W., "Effect of Spatial Variations in Shear on Diffusion at the Wall of an Arterial Branch." *Circulation Research*. 37 (1975): 446-454.

Gidaspow, D., "Multiphase Flow and Fluidization." *Academic Press, Boston*. (1994).

Gidaspow, D. and Huang, J., "Kinetic Theory Based Model for Blood Flow and Its Viscosity." *Annals of Biomedical Engineering*. 27(8) (2009): 1534-1545.

Gidaspow, D., and Jiradilok, V., "Computational Techniques: The Multiphase CFD Approach to Fluidization and Green Energy Technologies", Nova Science Publishers (2009)

Hodgson, L., and Tarbell, J.M., "Solute Transport of the Endothelial Intercellular Cleft: The Effect of Wall Shear Stress." *Annals of Biomedical Engineering*. 30 (2002): 936-945.

Kaazempur-Mofrad, M.R., and Ethier, C.R., "Mass Transport in an Anatomically Realistic Human Right Coronary Artery." *Annals of Biomedical Engineering*. 29 (2001): 121-127.

Savage, S.B., "Granular Flows at High Shear Rates." *Theory Dispersed Multiphase Flow. Academic Press, Inc.* (1983): 339-358.

Shapley, N.C., Armstrong, R.C., and Brown, R.A., "Laser Doppler velocimetry measurements of particle velocity fluctuations in a concentrated suspension." *The Society of Rheology, Inc.* 46(1) (2002): 241-272.

Sherwood, L., "Human physcology." Thompson Learning, Belmont, CA. (2004).

Songprawat, S., "Multiphase Kinetic Theory with Rotation." PHD Thesis Book, IIT, Chicago, Illinois. (2009).

Sun, N., Wood, N.G., Hughes, A.D., Thom, S.A.M., and Xu, X.Y., "Fluid-Wall modeling of Mass Transfer in an Axisymmetric Stenosis: Effects of Shear-Dependent Transport Properties." *Annals of Biomedical Engineering*. 34(7) (2006): 1119-1128.

Syamlal, M., "MFIX documentation: numerical technique." *Note DOE/MC31346-5824, NTIS/DE98002029, National Energy Technology Laboratory, Department of Energy*. (1998).

Wada, S., and Karino, T., "Theoretical Prediction of Low-Density Lipoproteins Concentration at the Luminal Surface of an Artery with a Multiple Bend." *Annals of Biomedical Engineering*. 30 (2002): 778-791.

Zydney, A.L., and Colton, C.K., "Augmented solute transport in the shear flow of a concentrated suspension." *PCH PhysicoChemical Hydrodynamics*. 10(1) (1988): Transport of ^{14}C-4-Cholesterol between Serum and Wall in the Perfused Dog Commo77-96.

Table 1 Properties of RBC ghost and Bovine Serum Albumin (BSA)

	RBC ghost	BSA
Density (g/ml)	1.02	2
Normal restitution coefficient, e	0.95	0.998
Wall restitution coefficient, e_w	0.95	0.9995
Diameter (nm)	8000	7.535
O-ring radius (cm)	0.0145	
% RBC Ghost	15%, 30%, or 45%	
Maximum solid volume fraction	0.733	
Shear rate (sec^{-1})	200, 400, 600, 800, or 1000	

Table 2 System and material properties used in the calculations for LDL, HDL, and red blood cells.

Name	Value
Tube diameter	0.19 mm
Tube length	14.4 cm
Plasma density	1020 kg/m^3
Plasma viscosity	0.0012 kg/m's
RBC size	8 μm
RBC density	1092 kg/m^3
Restitution coefficient (e)	0.95
Wall restitution coefficient (e_w)	0.99
LDL diameter	23 nm
LDL density	1020 kg/m^3
LDL level	130 mg/dL
Restitution coefficient of LDL's	0.99
HDL diameter	10 nm
HDL density	1200 kg/m^3
HDL level	30 mg/dL
Restitution coefficient of HDL's	0.99
Mean velocity	10.4 cm/sec

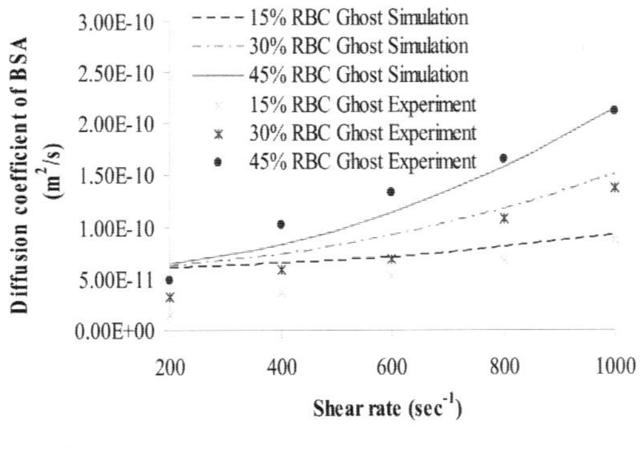

Figure 1 Comparison of computed and measured diffusion coefficient of the protein, BSA at various RBC Ghost concentrations for the Cha and Beissinger (2001) experimental conditions.

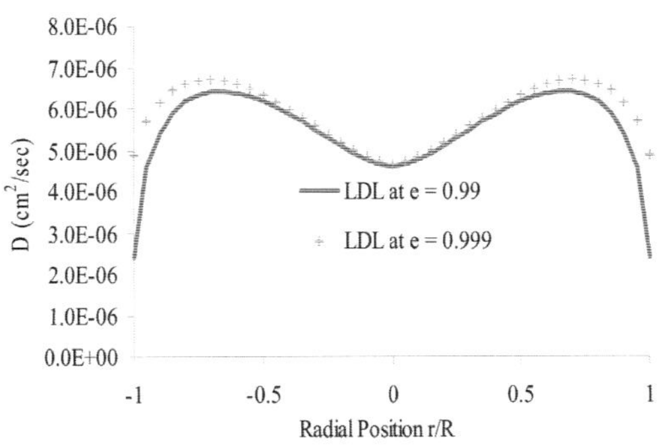

Figure 2. Comparison of computed diffusion coefficients of LDL for two restitution coefficients, e = 0.99 and 0.999 for the conditions in Table 2. Note the small effect of the only empirical parameter in the theory, e, the normal restitution coefficient.

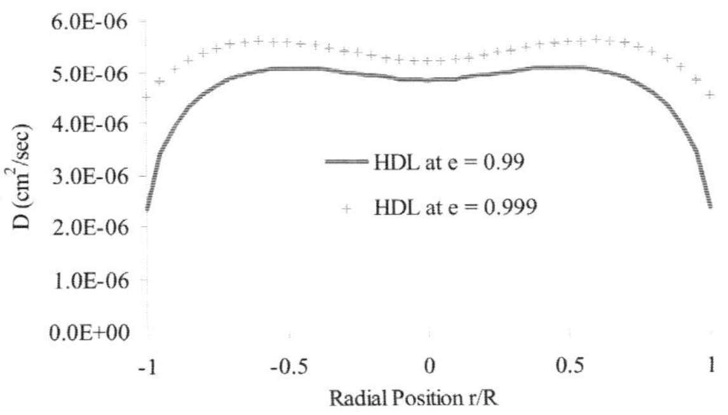

Figure 3. Comparison of computed diffusion coefficients of HDL for two restitution coefficients, e = 0.99 and 0.999 for the conditions in Table 2

Slugging Flow of Water Draining from the Bottom of a Non-Vented Container

By

Charles W. Solbrig
Idaho National Laboratory, Idaho Falls, Idaho

Abstract

Experiments were run to observe the behavior of water exiting through an orifice at the bottom of an non-vented container. Initially, the container is nearly full of water with a small air space on top. Once the orifice was uncovered, the slugging rate and the drain rate of the water leaving the container were measured. Upon initially opening the orifice, water drains out until the air pressure above the water reduces enough that the air pressure drop from inside to outside of the container supports the water column and the water stops flowing. Air then enters the container through the orifice forming a bubble, which grows until it detaches and bubbles through the water to reach the air space. Once the bubble enters, this added air increases the pressure in the air space enough to allow the water to start flowing out again. This cycle of flow out, flow stoppage, air inflow, and bubble breakoff continues over and over until the hole is closed or the container empties. This is referred to as the "slugging cycle."

A mechanism is proposed to describe the slugging cycle which is modeled analytically. This paper presents the description of the experiments, data obtained, the mechanistic model, and comparison of the model to the experimental data. The model predicts outflow rates close to experimental values. Flow rates from non-vented containers are more than 10 to 20 less than vented containers. The bubbles which must enter the container periodically to increase the internal air pressure stop the water flow momentarily so are responsible for this large decrease in flow rate.Swirl induced in the non-vented container causes the flow rates to increase by a factor of two. The flow rate out of a non-vented container is independent of water height which is in direct contrast to a vented container where the flow rate is proportional to the square root of the water height. The constant rate is due to the container pressure. The higher the water level, the lower the air pressure is in the container. This analytical model requires input of the bubble size. The volume recommended is the volume of a cylinder with the base of the orifice area and length of 3.3 cm. Slugging rate varies only a small amount falling in the range to 2 to 4 cycles/sec. Preliminary work with other containers indicates larger containers, larger orifices and nozzle exit shapes produce higher specific flow rates.

The standard multiphase flow equations could not be used to analyze this situation because the two phases are not interpenetrating. Instead one phase must fully stop before the other can flow. Interpenetrating phases allow can pass one another each affecting the other with friction and virtual mass. An interesting observation: The negative air pressure in the container is observable. It equals the water height.

INTRODUCTION

Many instances of emptying a heavier-than-air fluid from a non-vented vessel into an air environment occur in industry and in everyday life. Perhaps the most familiar of these occurs when one obtains a drink from a water cooler. A water cooler consists of an inverted bottle whose contents are prevented from flowing out by a small open container whose water level seals the mouth of the bottle. Before water is extracted from the small container, the air pressure difference between the reservoir surface and surface of the water in the bottle supports the water column in the bottle. As water is retrieved from the reservoir, its water level decreases opening up the bottle. This allows an air bubble to enter the bottle. Buoyancy forces the bubble up through the water until it bursts through the surface. The added air in the bottle increases the air pressure, so the water column is no longer supported by the pressure difference. Water then flows out of the bottle until the air pressure in the bottle decreases enough that the difference between ambient air and the water bottle air pressure supports the water column and the flow stops. This process continues until the water reservoir is again full enough to close off the bottle opening. This is a peculiar physical problem because there is a single opening which has to allow both flow of water out of the bottle and air to enter the bottle. It is inherently an unsteady flow situation where water flows out, then air flows in.

In buoyancy flows such as this (where two fluids are exchanging positions though a single opening), either the flow must oscillate between one fluid leaving the bottle and the other entering through the same opening (oscillatory flow) or one fluid must flow outward through part of the opening and the other flow inward through the remainder of the opening (countercurrent flow). When water is flowing out of the bottom of a container (where the opening is small compared to the diameter of the bottle) the flow oscillates. If a swirl is imparted to the water, the latter can be the case with water swirling out around the edge of the orifice and a center core of air flowing in.

For buoyancy flows through large horizontal or vertical openings, countercurrent flow can occur. For example, buoyancy causes flow through an open door in a building with outside air at a much different temperature than the inside air. This situation has been studied and conclusions drawn that warmer air flows through the top of the door opening and colder air through the bottom part.(Ref. 1). Another example of buoyancy driven flow would occur when a penetration in the floor of a large cell filled with argon is opened to an air atmosphere. The heavier argon gas would exchange positions with the ambient air in a behavior which has similarities to the water draining situation.

In the situation considered here, the flow through the orifice oscillates between liquid flowing out of the container and air flowing in. The force causing the outflow of water is primarily the gravity head of water. However, the inflow of gas represented a difficulty. The analysis of the physics considered here was not amenable to the standard multiphase flow equations (Refs. 2 to 6) because the two phases cannot pass by each other. Instead one phase must fully stop before the other can proceed. Correlations which represent the interaction of two immiscible components or phases represent primarily friction forces and some virtual mass effects. See , for example, Reference 7 for a tabulation of these correlations. These are insufficient to model the inflow of air into the container. The inertia of the liquid column above the orifice and the buoyancy of the bubble are the most significant forces in representing the motion of the gas. This behavior is in direct contrast to a study of solids draining from a non-vented container where countercurrent flow of the solids and air occurs (Ref. 8).

EXPERIMENT DESCRIPTION

The experimental concept is shown in Figure 1 in this figure. Water flows out of the exit at the bottom of the container and air enters. For the experiments described in this paper, an orifice is placed on the exit. The orifice is small relative to the opening so that it may be considered a sharp edged orifice. Short bursts of water exit the container. Each burst is followed by a bubble entering the container. Each out-in cycle occurs within half a second or less. When no orifice is placed on the exit and for larger size orifices, the flow is chaotic. The flow out and in is aided by the shape of the exit. In such flows, it is difficult to distinguish one bubble from the next. When the air space is large enough, the bubbles break through the liquid surface with force throwing liquid high up in the air space. The experiments reported on in this paper were conducted with the smaller size orifices. The bubbles enter more calmly as depicted in Figure 1, and each can be readily distinguished from the other.

In these experiments, measurements of flow rate, water height, and cycle rate were made. Two methods were used to measure the water outflow rate. The first was measured by collecting water in a graduated cylinder for a measured time period. The second was to use a height reference on the container in recording the water level in the

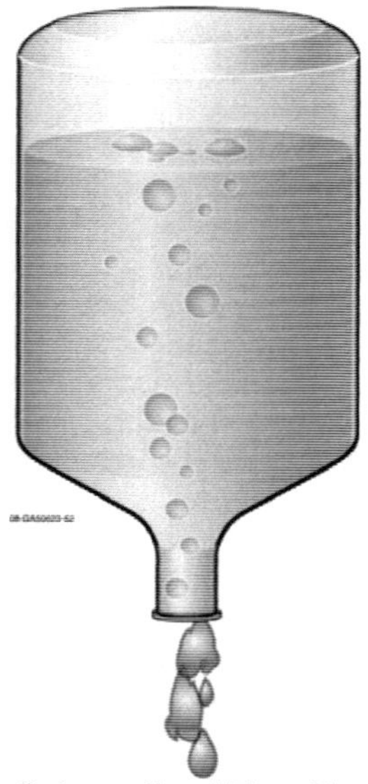

Figure 1 Schematic of Flow Experiment.

container versus time and the water level indicator was used to calibrate the volume versus water height so that the volume drain rate could be determined. The slugging rate was determined by counting the number of slugs over a 5 second period.

EXPERIMENTAL RESULTS

Three experiments are reported on and analyzed in this paper. The container properties are summarized in Table 1 as well as the overall results.

Table 1. Container Properties and Flow Rate Results.

Experiment Number	Inside Diameter (cm)	Hole Diameter (cm)	Distance, Bottom-top (cm)	Container Material	Flow Rate (cc/sec)	Slug Rate (S/sec)
1	13.3	0.794	29.3	Stiff plastic	3.43	3 to 2.2
2	13.3	1.11	29.3	Stiff plastic	9.58	4 to 2.6
3	15.7	1.43	31.3	Glass	16.30	4 to 3

Experiment 1: 13.3–cm-Inner-Diameter Container, 0.794-cm-Hole Diameter. Two tests were run for each experiment to provide confirmation of the flow rates. Although the slugging cycle was evident throughout the tests, the height versus time plots are smooth as evidenced in Figure 2. Data were recorded only every 0.5 to 1 cm. Note that in Test 1, the test was run without stopping, resulting in a curve without jumps. In Test 2, the flow was purposely stopped twice, which resulted in the jumps in the data; however, the slope of the flow data between these jumps is seen to be the same and the same as Test 1. The measured slope of these data is 33.5 sec/cm. The slugging rate ranged from three slugs/sec for the top half of the test and reduced to almost 2 slugs/sec near the bottom. The flow rate is surprisingly constant versus water height, even when the container is less than half full though there is a slight tendency to slow down near the bottom. The container was stiff plastic, but it can deform a small amount during the slugging which affects the flow rate. For example, surface tension prevented flow out of a completely stiff vessel (the glass container of experiment 3) for this size orifice.

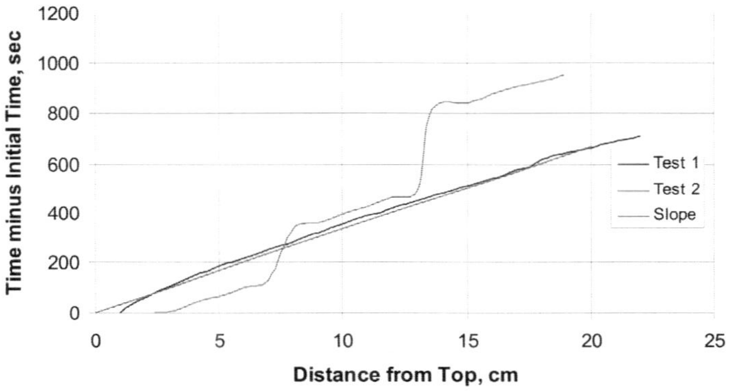

Figure 2 Experiment 1 – Measured Slope 33.5 sec/cm.

The calibration data for the 13.3–cm-diameter container is shown in Figure 3. The slope is 115 cc/cm.

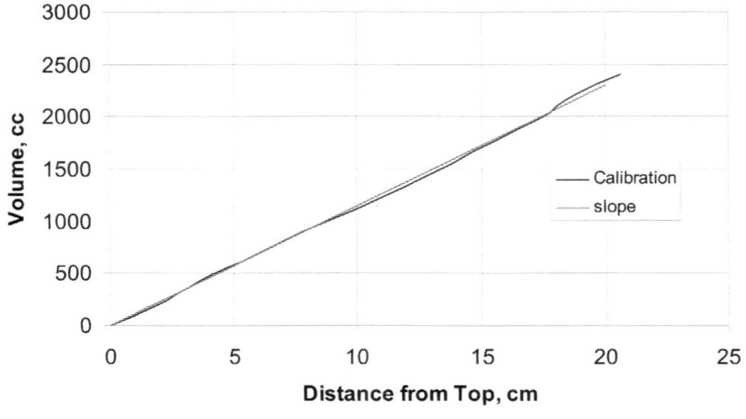

Figure 3. Calibration Data for the 13.3 cm Diameter Container.

The flow rate, calculated as the ratio of these slopes as reported in Table 1 is:

$$Flow\,Rate = \frac{115\frac{cc}{cm}}{33.5\frac{sec}{cm}} = 3.43\frac{cc}{sec}$$

Experiment 2: 13.3-cm–Inner-Diameter Container, 1.11–cm-Hole Diameter. The data from the two tests run are shown Figure 4. The slope is 12.0 sec/cm.

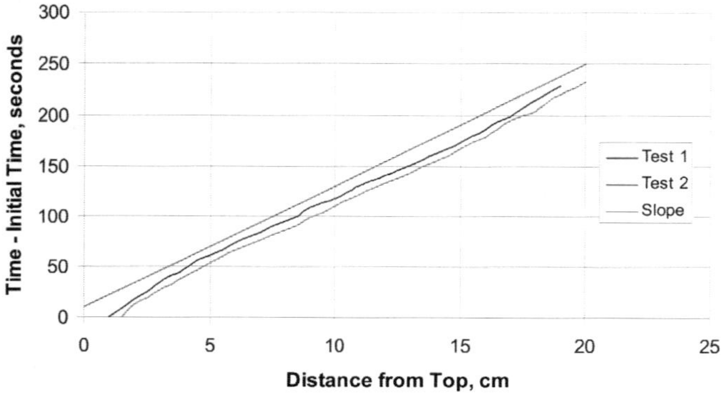

Figure 4 Experiment 2 – Measured Slope 12 sec/cm.

The calibration is the same for this experiment since it is the same container. The flow rate is 115/12=9.58 cc/sec. The slug rate ranged from four slugs/sec for the top half of the test and reduced to 2.6 slugs/sec near the bottom.

Experiment 3: 15.7-cm–Inner-Diameter Container, 1.43–cm-Hole Diameter. (***air water3.xls)*** The data from the two runs of Experiment 3 are shown in Figure 5. The slope in the upper part of the container is 11.5 sec/cm for both tests. Test 1, which was a longer test, shows it is less in the lower section where it starts to neck down at 17.5 cm. The outflow data of Figure 4 were taken in the constant area upper section.

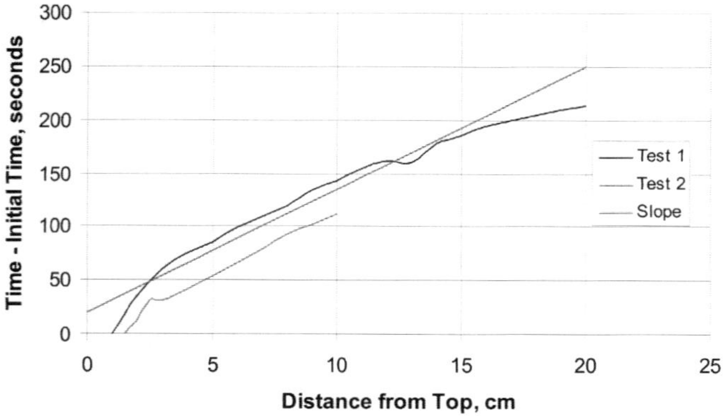

Figure 5 Experiment 3 – Measured Slope 11.5 sec/cm.

The calibration data for this container is shown in Figure 6. The slope of these data is 187.5 cc/cm, resulting in a calculated flow rate of 187.5/11.5=16.30 cc/sec.

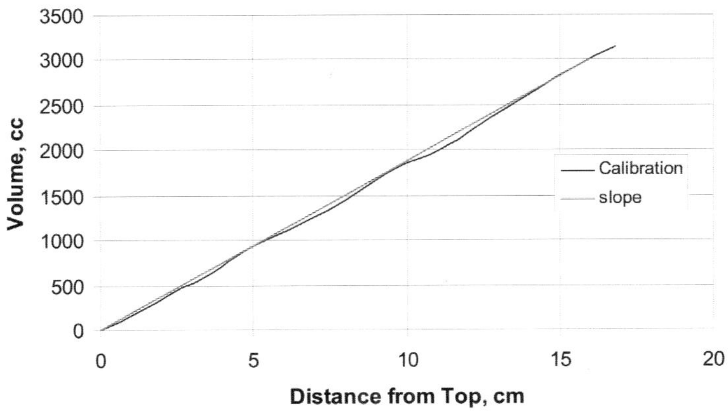

Figure 6 Calibration Data for the 15.7 cm Diameter Container.

It is interesting to note that for hole sizes less than 1.43 cm (9/16 in. and less), the water would not exit the glass container. The fact that the container wall is glass and does not flex during the slugging and the surface tension of the water surface at the hole are probably responsible.

One further measurement was made to measure the flow rate when a swirl was added by moving the container. Above 24 cm, the swirl just damped out. Below this, the swirl turned the flow into a vortex with water swirling out the orifice and air entering a central hole which persisted up to the water surface. The flow rate of water out was 11 cc/sec.

Calibration of Orifice Flow Two runs were made with an open container to provide a basis for determining flow out of an orifice with only a gravity head pressure drop across the orifice. A ¼ inch orifice was used in the first runs (Figure 7) to produce

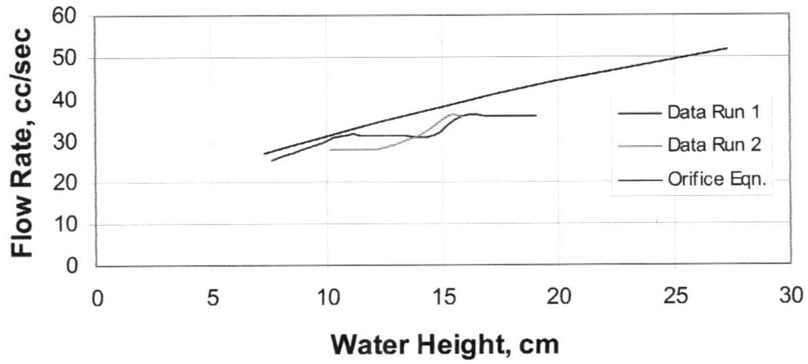

Figure 7 Data and Equation for a ¼ in (0.635 cm) Diameter Orifice

a slow enough flow (250 cc in 7 to 10 sec) to measure with the simple methods used (watch and graduated cylinder). The comparison of the data to the predictions of an orifice equation (described in the next section) show reasonable agreement within the accuracy of the measurements. The second run (5/16 inch orifice) is described later.

ANALYTICAL MODEL

A model is developed here to represent a fluid draining from a vessel in the oscillatory situation described above and in particular for water draining through a hole or orifice in the bottom of an non-vented tank (as occurs when water is taken from a water cooler). The draining of the fluid is made up of a series of cycles. Each cycle consists of water outflow, flow stoppage, air bubble inflow, and bubble breakoff from orifice. Then the cycle starts over again repeating until the container empties or the hole is closed off. This cycle, which is modeled in this paper, is referred to as the "slugging cycle." These repeating cycles are coupled together to form the estimate of the flow rate out of the container. This model considers only flow through a sharp edged orifice. Flow of a non-vented container through a nozzle shaped outlet can be much greater.

The cycle is modeled in two steps, the outflow of water, and the inflow of an air bubble. The water outflow is calculated by the pressure drop across the orifice. Initially, the pressure of the air ,P_A , above the stratified water layer must be less than or equal to the atmospheric pressure, P_{atm}. The pressure of the air on the downside of the orifice at the bottom the container is atmospheric. The pressure just upstream of the orifice is the gravity head plus the air pressure as

$$P_A + (\rho_W - \rho_A) g\, h$$

The velocity, v, at any time is modeled with the pressure drop across the orifice and the orifice equation:

$$\mathrm{v} = C_D \sqrt{\frac{2 * \Delta P}{\rho_W}} \tag{1}$$

where $C_D = \mathit{Orifice\ Discharge\ Coefficient}$

$\Delta P = (P_A(t) + (\rho_W - \rho_A)\, g\, h(t) - P_{atm})$

$P_A(t)$ is the air pressure above the *water* layer at time t

ρ_W is the *water* density, ρ_A is the *air* density

$h(t)$ is the height of the *water column* at time t

The pressure drop, ΔP, is always positive or zero. The discharge coefficient is taken to be an ideal value of 0.707. This equation was used to generate the orifice equation flow rate in Figure 7. Reference 9 recommends a slightly lower discharge coefficient.

The liquid volume flowing out of the breach, *ΔV*, in the time increment Δt is related to both the flow out of the orifice and the decrease in the water height in the container as:

$$\Delta V = A_{orifice} \mathrm{v} \Delta \mathrm{t} = \mathrm{A}_{\mathrm{container}} (h(t + \Delta t) - h(t)) \tag{2}$$

where $A_{orifice}$ = area of the orifice

$A_{container}$ = area of the container

Solving for ΔV from the left side of Equation 2 allows the height as a function of time to be solved for from the right side of Equation 2 as:

$$h(t+\Delta t) = h(t) - \frac{\Delta V}{A_{container}} \qquad (3)$$

The pressure in the air space $P_A(t)$ at any time is calculated by a polytropic expansion as given by:

$$P_A(t) = P_{A,init}\left(\frac{V_{initial}}{V(t)}\right)^k = P_{A,init}\left(\frac{h_T - h_i}{h_T - h(t)}\right)^k \qquad (4)$$

where $P_{A,init} = Container\ air\ pressure\ at\ the\ beginning\ of\ the\ cycle$

$V = Volume\ of\ air\ in\ container$

$h = all\ heights\ are\ relative\ to\ the\ orifice\ height\ of\ zero$

$h_T = \text{the height of the cell}$

$h_i = initial\ height\ of\ the\ water$

$k = effective\ expansion\ exponent\ (\text{taken to be } 1.2)$

The flow of water stops when the air pressure decreases enough to make ΔP=0. This pressure is termed $P_{A,f}$. The final height is h_f which determines the total outflow of water, V_f, of the cycle as:

$$V_f = A_{container}(h_i - h_f) \qquad (5)$$

The time it takes for this outflow, determined by the above when ΔP becomes zero, is defined as t_f. The average volumetric outflow in this time is:

$$\dot{V}_{water} = \frac{V_f}{t_f} \qquad (6)$$

The height, h_f, becomes the new initial height, h_i, and $P_{A,f}$ becomes the new initial pressure, $P_{A,init}$ for the next cycle of outflow. This flow stoppage marks the end of the outflow and the first step of the cycle.

The second part of the cycle begins when an air bubble starts to enter into the opening. During the entire time of air bubble entry, the volume of water in the container remains the same so the effective height of the liquid remains $h = h_f$. The air that enters must fit into the volume that the air occupies in the container at the end of the outflow step because the water volume is not changing so the air is compressed to make room for the additional mass of air entering the container. If the mass of air in the container at the end of the outflow is denoted as m_f and the mass of air that comes in the inflow bubble is denoted by m_B, then the air pressure at the end of the bubble inflow is

$$P = \frac{1}{V}\frac{m_f + m_B}{M_W}RT = \frac{1}{V}\frac{m_f}{M_W}RT + \frac{1}{V}\frac{m_B}{M_W}RT = P_{A'f} + \frac{1}{V}\frac{m_B}{M_W}RT \quad (7)$$

$$where\ V = A_{container} * (h_T - h_f)$$

This compression is assumed isothermal because it occurs slower than the outflow.

The mass of the bubble is determined by the volume of the bubble, V_B, at atmospheric pressure

$$m_B = \frac{P_{atm}V_B M_W}{RT} \quad (8)$$

Substituting this expression into the above equation yields

$$P = P_{A'f} + P_{atm}\frac{V_B}{A_{container} * (h_t - h_f)} \quad (9)$$

The time that it takes for the bubble to enter into the container is calculated by assuming the buoyancy force of the bubble pushes a column of water in the container equal to the area of the hole and the depth of the water. So although the volume of liquid does not change during the bubble ingress, the buoyancy force of the bubble does cause the water to move up. The bubble itself is considered to have the area of the hole as well. The momentum balance for this situation, assuming that the velocity of the bubble entering the hole, v_B, is the same as that of the liquid column of water, is estimated by the equation

$$\begin{pmatrix} mass\ x\ acceleration \\ of\ gas + liquid \end{pmatrix} = \begin{pmatrix} bouyancy \\ of\ air \end{pmatrix} - \begin{pmatrix} gravity \\ of\ liquid \end{pmatrix} - \begin{pmatrix} pressure\ of \\ container\ air \end{pmatrix} + \begin{pmatrix} atmospheric \\ pressure \end{pmatrix} \quad (10)$$

Substituting in yields

$$A_o(\rho_W h_f + \rho_A L_B)\frac{dv_B}{dt} = A_o L_B(\rho_W - \rho_A)g - A_o(\rho_W - \rho_A)gh_f - A_o P_A + A_o P_{atm}) \quad (11)$$

$$where\ v_B = 0\ when\ t = 0$$
$$t = 0\ refers\ to\ the\ start\ of\ in-flow$$
$$A_o = Area\ of\ the\ orifice$$
$$and\ L_B = Length\ of\ the\ bubble$$

The last three terms in this equation are the same as the terms which make up ΔP. The sum of these three terms is zero at the end of the water outflow step so is also zero at the start of bubble ingress. During bubble ingress, ΔP increases because the liquid gravity term and atmospheric pressure remain the same but the air pressure in the container increases. The sum of these three terms is a retarding force causing the bubble to slow down and in some cases, stop.

So at the initiation of bubble ingress, the principal force causing the bubble motion is the buoyancy force of the gas bubble. The majority of mass it must move is the liquid in the column. The mass of the gas is negligible. All other forces which might retard the motion such as friction are extremely small and are neglected. The motion of the bubble may be guided by interphase friction after completely entering the liquid. Models for these forces are reported in Reference 7. But the forces listed in that reference are negligible compared to the liquid inertia. Without the retarding inertia of the liquid column, the gas bubble calculated entry time is negligible compared to that of the liquid egress time and results in a total cycle time much shorter than observed. As will be seen in the results section, with the liquid momentum, the gas bubble ingress time is 3 to 4 times as long as water egress time at high water levels and as low as half at lower levels.

The bubble diameter starts out as that of the orifice but may increase as it moves into the fluid. It is assumed that even though this may occur, the amount of liquid that it moves is still approximately the same as if it were a column of the same diameter as the orifice.

Once the bubble has completely passed through the hole, the gas pressure increase due to that bubble ingress is complete. Even though the bubble continues to rise to the water surface, it does not cause any further pressure increase during its transit. Also the rise time does not add to the total cycle time.since once the bubble has cleared the orifice, the next cycle has begun and fluid is already flowing out of the orifice In fact, during cycles when the fluid level is the top third of the container, several bubbles spaced about 6 cm apart can be rising at the same time.

Equation 11 is solved numerically with the initial velocity at the start of bubble ingress as zero. The unknown time that the bubble takes to pass through the hole, t_B, is determined by the size of the bubble determined in the next section and the integral of the volumetric flow rate from zero to t_B with the relation:

$$V_B = \int_0^{t_B} \mathrm{v}_B A_{hole} dt \tag{12}$$

In some cases, when the water level is high in the container, the bubble velocity may become zero before the value of V_B reached. In those cases, the bubble size and time of entry are taken to be the values at that time.

Bubble Size: Although it would be difficult to measure the bubble size, it can be deduced from the slugging rate by the following reasoning. The volumetric flow rate of the air into the bottle can be determined since it must be equal to the volumetric flow rate of liquid out of the orifice. The slugging rate is a measure of the number of bubbles entering the container per unit time so the bubble size is determined by dividing the volumetric flow rate of liquid out by the measured number of slugs/sec. Using the experimentally determined bubble size, the time that it takes the bubble to enter through the orifice can be calculated with Equations 11 and 12. Table 2 lists the estimated

Table 2 Estimated Bubble Size and Length

Experiment Number	flow rate	Orifice Diameter	Orifice Area	velocity	Estimated	Bubble Volume	Bubble L =Vol/Area
	cc/sec	cm	cm^2	cm/sec	slugs/sec	cc	cm
1.000	3.430	0.794	0.495	6.927	2.200	1.559	3.149
2.000	9.580	1.110	0.968	9.900	3.000	3.193	3.300
3.000	16.300	1.430	1.606	10.149	3.000	5.433	3.383

bubble size for each experiment in column 7. The estimated bubble length obtained by dividing the volume by the hole area is almost constant. The fact that the length of the bubble is almost the same suggests that the gravity head of the water around the bubble determines this length. As will be seen in the results section, the bubble size is smaller than these values when the water level is near the top of the container.

The total cycle time is the sum of the water outflow time plus the bubble inflow time. The average flow rate for the cycle is given as the water outflow divided by the sum of these two times.

$$\dot{V}_{water} = \frac{V_f}{t_f + t_B} \tag{13}$$

The next cycle is started by calculating the increased pressure due to a bubble entering into the container and using this value and the water height at the end of the cycle as the starting height for the next cycle.

Pressure in the air space. It is interesting to note that the gauge air pressure in the container is known when flow stops because at that time, it is equal to the negative of the height of water. That is, since the flow has stopped because the pressure drop across the orifice is zero, the gauge air pressure at that time is

$$\Delta P = P_A + (\rho_W - \rho_A) g\, h(t) - P_{atm} = 0 \; or$$

$$-\frac{P_A - P_{atm}}{(\rho_W - \rho_W) g} = h(t) \tag{14}$$

This is the minimum pressure during the cycle.

COMPARISON OF ANALYSIS TO EXPERIMENT

Experiment 1: The calculated pressure for two seconds is shown in Figure 8

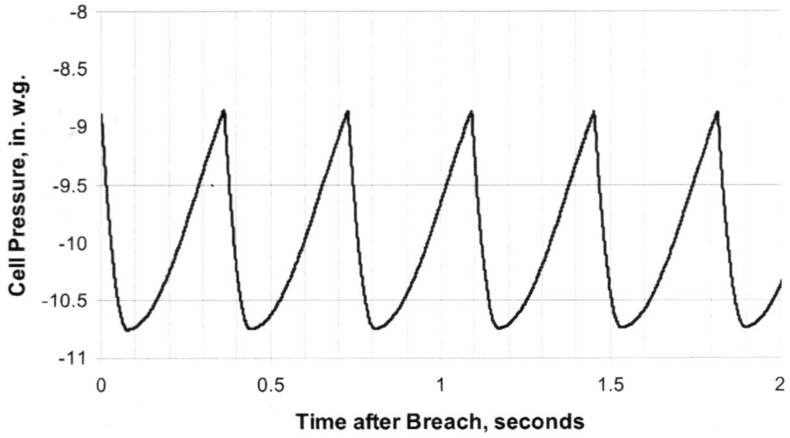

Figure 8 Calculated Pressure Pulsations near the Top (0.794 cm Orifice)

during flow out of the container for the 5/16 inch (0.794 cm) diameter hole with the water level starting near the top of the container. (Water interface is about 27.3 cm above the orifice in a container of height 29.3 cm high.)

The initial pressure was obtained by adding a liquid stoppage pressure (corresponding to 27.3 cm height) to a bubble increase pressure. The bubble stoppage pressure is the pressure P_A calculated in Equation 1 to make ΔP equal to zero and is about -10.75 iwg which is equal to 27.3 cm. The pressure increase caused by a bubble of 1.56 cc (from Table 2) entering into the container is calculated from Equation 9 to be about 1.85 iwg which brings the beginning of cycle pressure up to -8.9 iwg. As the liquid drains out in the first step of the first cycle, the pressure decreases to -10.75 iwg in less than 0.1 seconds. Then the pressure increases again to about -8.90 iwg as the gas bubble flows in and completes the second step of the cycle in about 0.3 seconds. The bubble entry period is about three times as long as the water drain out period. Then the next cycle begins.

The calculated cycle time is about 2.75 cycles/sec. The experimental value included in Table 1 ranges from about 3 near the top to lower cycles/sec in the lower half of the container. The experimental value of 2.2 cycles/sec was used to calculate the bubble size for the calculation. The cycle time is dependent upon the bubble size in the calculation. The initial pressure assumed as the starting point can be changed and the same cycle time will be reached as long as the bubble size is the same and the initial pressure is at or above that chosen in this calculation. So even atmospheric pressure can be chosen and the cycle time and pressure will quickly reach that as shown in Figure 8.

Other experiments have been run which were not reported on in this paper which consisted of flow out of larger containers with nozzle type exits. The flow rates were much faster but the cycle time was always in the range of 2 to 3 cycles/sec.

The minimum pressure is decreasing in Figure 1 although it is difficult to see. This can be seen more clearly in the water level. As mentioned, the minimum pressure is the negative of the water height (27.3 cm = 10.75 cm). The height is shown in Figure 9 out to eight seconds (22 cycles). It shows the water level decreasing during the water outflow step and then remaining constant during the air bubble ingress step.

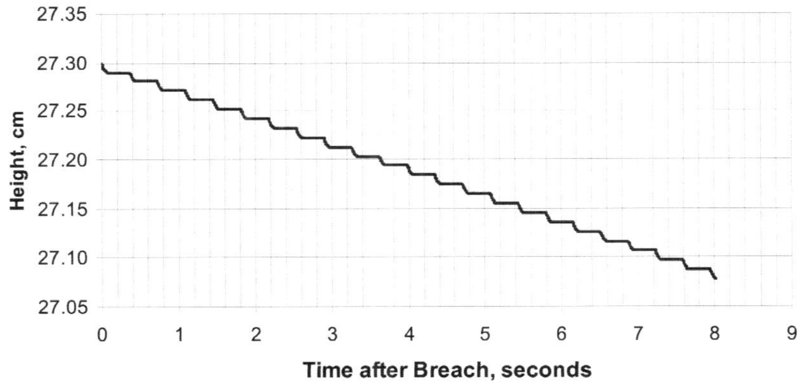

Figure 9 Calculated Water Height near the Top (0.794 cm Orifice)

The velocity of the fluid through the orifice is shown in Figure 10. During the first step of the cycle, the flow starts out high due to the initial gas pressure and it decreases as the liquid leaves the container until it reaches zero at almost 0.1 seconds. At that time, the flow direction reverses and air starts to re-enter the container. The gas velocity increases due to the gas buoyancy effects but eventually starts to decrease as the air pressure in the container increases When the volume of the bubble is calculated to reach the experimentally determined 1.56 cc, the cycle is stopped and a new cycle started so the velocity again switches direction as outflow starts. This abrupt change in the flow direction is probably not realistic and would smooth out if momentum effects were taken into account in the water outflow equations.

It is interesting to note that the flow reversal from bubble inflow does not occur when the bubble velocity becomes zero in this computation. However, for some initial water levels above the 27.3 cm assumed here, the velocity will go to zero. These are discussed in the later two experiments

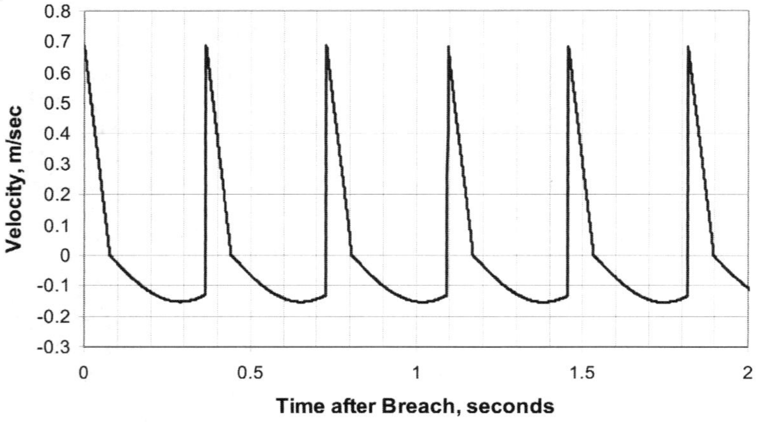

Fig. 10 Velocity Through the orifice (0.794 cm Orifice)

The flow out of the container is shown in Figure 11. Again the outflow increases

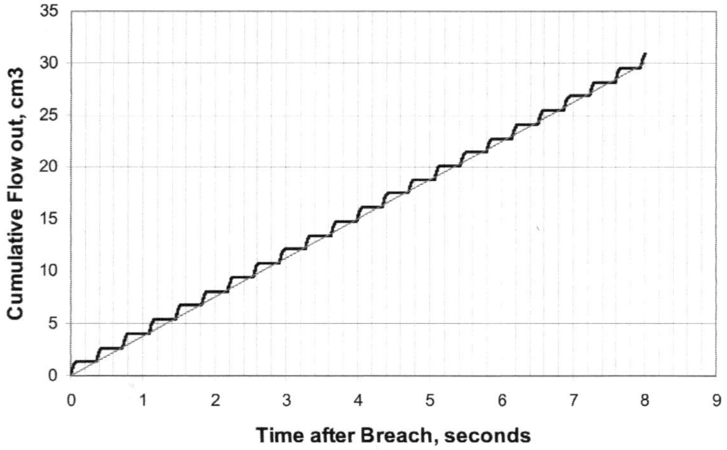

Fig. 11 Outflow From the Container (0.794 cm Orifice)

while the liquid is leaving the container and remains constant during the gas bubble inflow. The flow rate out calculated by the slope of the line in Figure 11 is 3.75 cc/sec which compares to the experimentally measured value of 3.43 cc/sec.

The pressure history predicted for the flow with a water level slightly above midplane at 17.3 cm is shown in Figure 12. The upward movement of the pressure toward atmospheric is a bit more noticeable at this level than the previous. The water outflow takes a bit longer (almost 0.2 seconds) because of the decreased pressure drop across the orifice. The bubble enters quicker (in about 0.2 seconds) because of the reduced back pressure. Combining the two times yields about the same total cycle time.

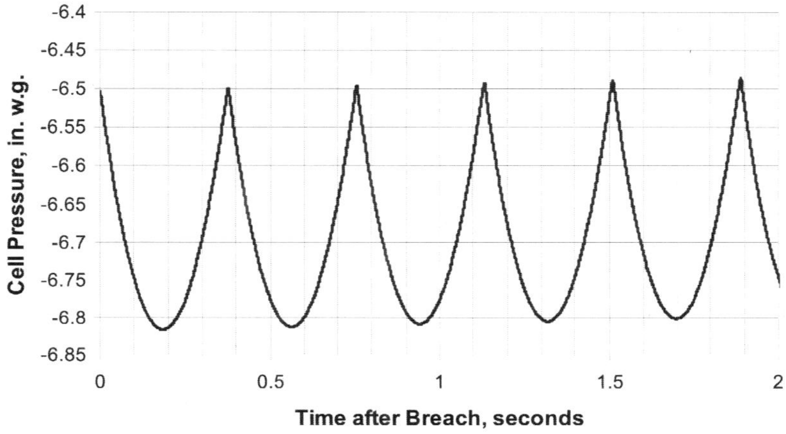

Fig. 12 Pressure History at an Initial Water Height of 17.3 cm (0.794 cm Orifice)

The velocity is shown in Figure 13. The liquid and gas velocity are almost the same slope and there is no discernable slowing down of the gas bubble near the end of travel as was the case for the higher water level.

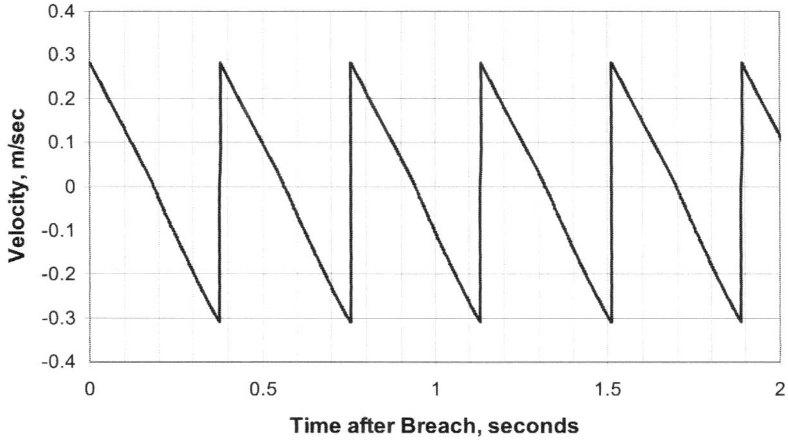

Fig. 13 Velocity History at an Initial Water Height of 17.3 cm (0.794 cm Orifice)

The volumetric flow out is the wavy line shown in Figure 14. The slope of the line parallel to this curve is 3.5 cc/sec which is close to the experimental value of 3.43 cc/sec and is very close to the 3.75 cc/sec for the higher water level.

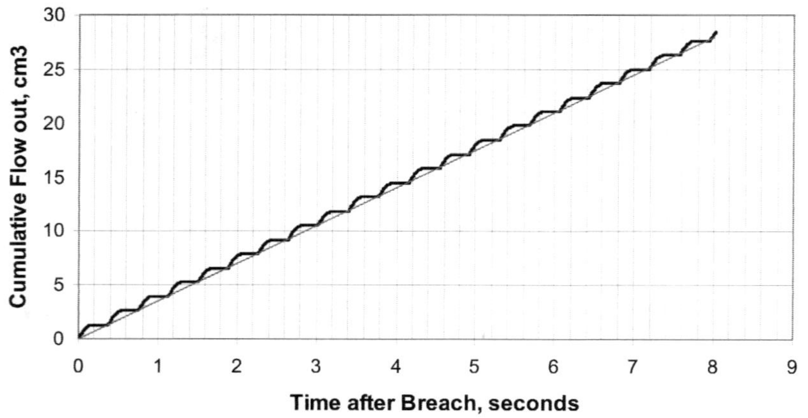

Fig. 14 Volumetric Flow at an Initial Water Height of 17.3 cm (0.794 cm Orifice)

The pressure history for an initial water level of 7.3 cm is shown in Figure 15. The water outflow period has increased, and the bubble inflow time has again decreased so that the total cycle time is again about the same as at the top and middle water levels.

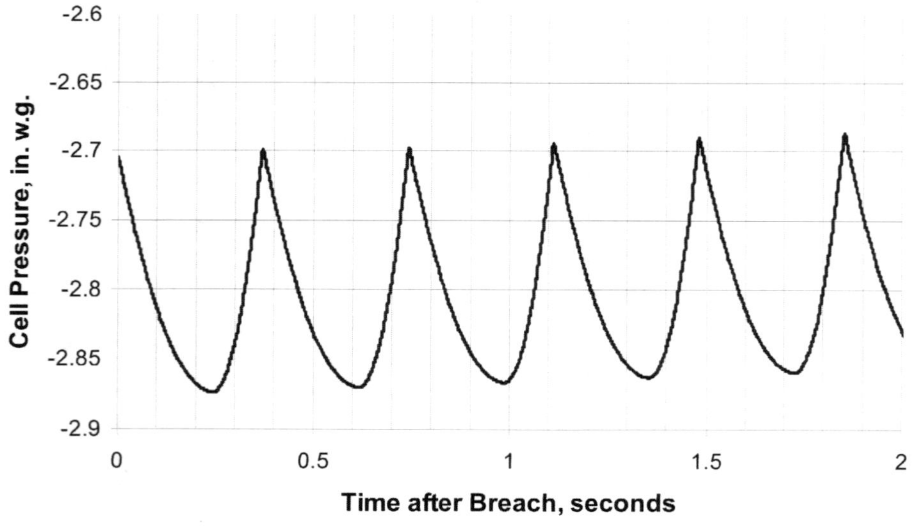

Figure 15. Pressure History at an Initial Water Height of 7.3 cm (0.794-cm Orifice).

The velocity history is shown in Figure 16. The velocity of the gas bubble is seen to have increased considerably in order to get the gas bubble in quick enough to keep the cycle time constant.

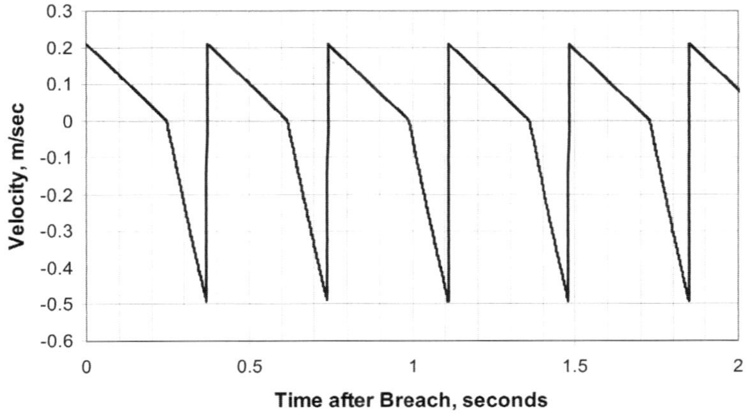

Figure 16. Velocity History at an Initial Water Height of 7.3 cm (0.794 cm Orifice).

The volumetric outflow of water at this lower elevation was again 3.5 cc/sec. Both the middle and bottom regions flow rates were lower than the top but not by much. The flow rates at 27.3 cm, 17.3 cm, and 7.3 cm were respectively 3.75, 3.5, and 3.5 cc/sec. These values correspond to the experimental value of 3.43 cc/sec. So this model shows that the flow rate and cycle time is almost independent of the water level in the container. The experimental data shows a constant flow rate but a slugging rate that decreased slightly from top to bottom (3 to 2.2). The size of the pressure pulses decrease with water level.

Figure 17 shows the magnitude of difference between vented and non-vented containers. The flow from the non-vented container is a small fraction of the open container flow. In addition, the flow rate for the open container increases considerably

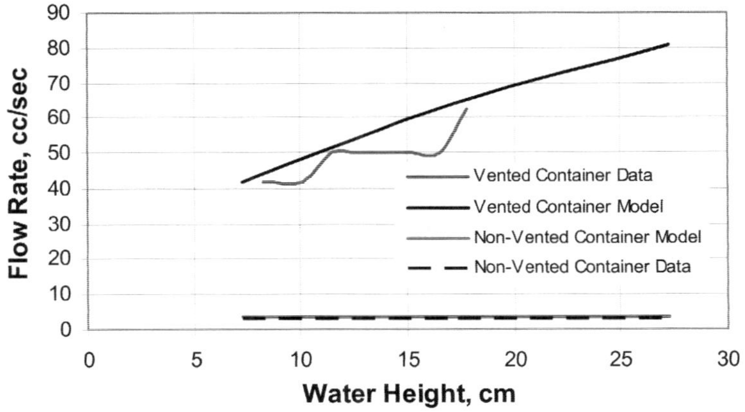

Figure 17. Flow Rates versus Water Height (0.794 cm Orifice).

with water height whereas the flow from the non-vented container is almost constant with height. The reasons for the latter behavior is the container air pressure, which becomes more negative with increasing water height, largely canceling out the effect of an increased gravity head. The reason for the former is that the flow is proportional to the square root of the gravity head.

Experiment 2: The pressure history at an initial water level of 17.3 cm is shown in Figure 18 for outflow of the same container as in Experiment 1 but with a larger orifice diameter of 7/16 in. (1.111 cm). The slugging has now increased to 3 cycles/sec compared to the previous. The bubble size used in the calculations from Table 2 is 3.2 cc. The calculated flow rate is 8.13 cc/sec versus the experimental value of 9.58 cc/sec from Table 1. The pressure oscillations are larger than in Experiment 1 due mainly to the larger sized bubble. Similar behavior was observed from the calculations for an initial water level of 7.3 cm with a slugging rate of 3.26 cycles/sec and an outflow rate of 8.75 cc/sec.

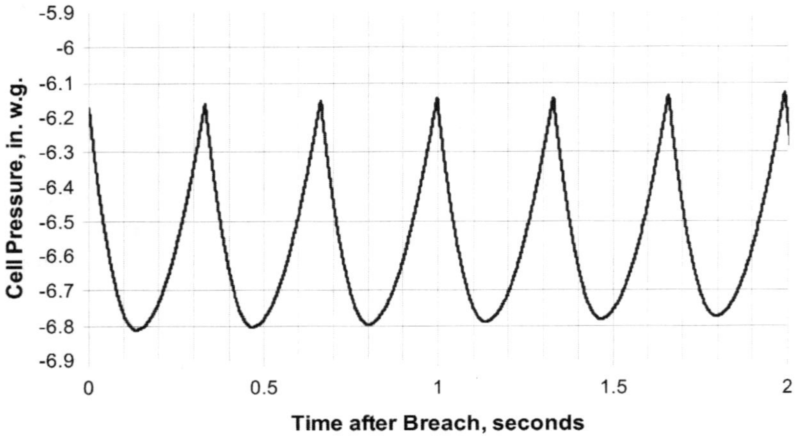

Figure 18. Pressure History at an Initial Water Height of 17.3 cm (1.111-cm Orifice).

Different behavior was observed for the calculation with an initial water level of 27.3 cm. The velocity is shown in Figure 19. In this case, the gas velocity reaches zero

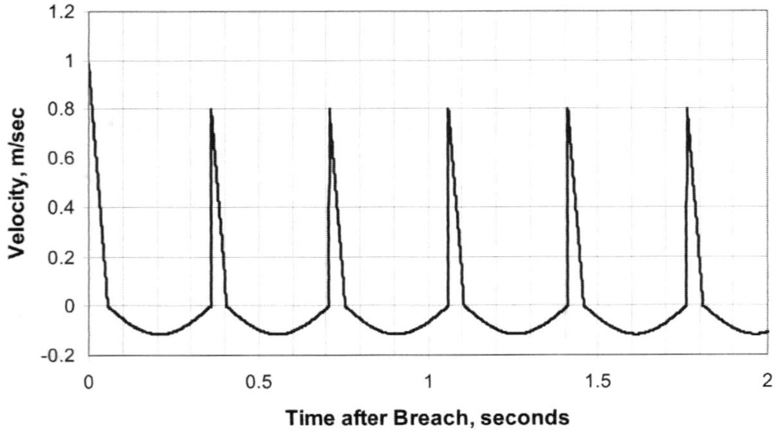

Figure 19. Velocity History at an Initial Water Height of 27.3 cm (1.111-cm Orifice).

before the bubble size of 3.2 cc is reached. The bubble size that was reached at zero gas velocity was 2.14 cc. The flow must reverse at this time, so that determines the bubble size. The slugging rate is smaller at 2.84 cycles/sec. The flow rate is much less at 5.5 cc/sec. So the model indicates that the bubble size and flow rate is lower than average a short time later as explained in more detail in the next section.

Experiment 3: The bubble size listed in Table 2 for Experiment 3 is 5.43 cc. The container is made of glass, and the top is 31.3 cm above the 9/16 in. (1.43 cm) diameter orifice. Calculations made for initial water levels of 19.3 cm and 9.3 cm indicated a flow rate of 13.75 cc/sec and 14.5 cc/sec respectively. The corresponding slugging rates are 2.97 and 3.21 cycles/sec. The experimental flow value from Table 1 is 16.3 cc/sec. The experimental slugging rate is estimated at 3 cycles/sec. Thus, the analysis indicates reasonable agreement with the experimental results although the flow rate is somewhat lower. For this experiment, the following study was made to determine if the analytical model provided a method for determining the bubble size.

Bubble Size Calculations: The calculations made at 29.3 cm produced a smaller bubble size than the 5.43 used in the above two calculations. As mentioned above, the bubble is cut off when the gas velocity reaches zero, which was 3.06 cc in this case. The velocity history looks similar to Figure 19. The pressure history for this case is shown in Figure 20. The initial pressure in this run was chosen to be -6.9 iwg due to the size of the bubble chosen. However, the end of the cycle produces a bubble that only brings the pressure up to -8.9 iwg. So this example shows that even if the initial pressure chosen corresponds to a larger bubble size than the analysis dictates, the code will quickly bring it to an equilibrium oscillation and bubble size.

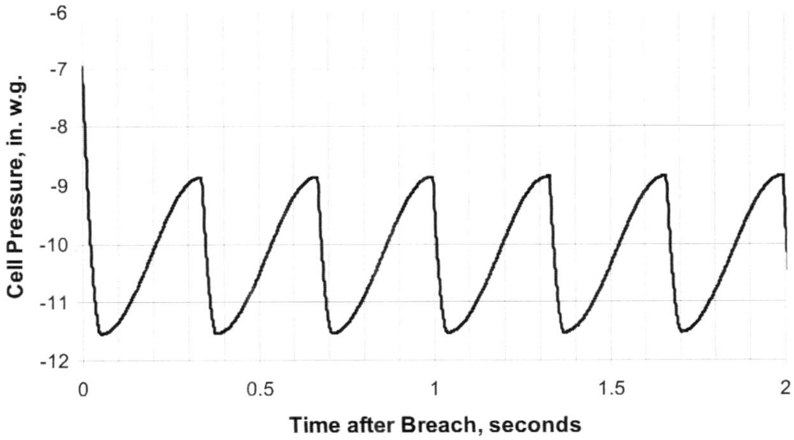

Figure 20. Pressure History at an Initial Water Height of 29.3 cm (1.43 cm).

The calculated flow rate for this case is 8.6 cc/sec, which is about half of the average measured flow rate of 16.3 cc/sec. Since the calculations at 19.3 cm and 9.3 cm produced flow rates comparable to the experimental value, a study was made to determine at what heights the behavior started to deviate from these values. The bubble sizes and slugging rate were determined for several different initial water levels in the upper region. The bubble size results are shown in Figure 21. The points lie on a straight line. The last point at the top of the container (31.3 cm) is estimated. In these calculations, the restriction limiting the maximum bubble size to 5.43 cc was dispensed with. The graph indicates that for water levels lower than 27.5 cm, the bubble size exceeds the measured bubble size (5.43 cm) for this orifice size, so deviation from the from the normal flow rates will occur above 27.5 cm. It was hoped that this analysis would provide analytical justification for the 5.43 cc bubble size. Unfortunately, it does not. Calculations for lower initial water levels without the 5.43 cc limit continue to grow. For example, the bubble size calculated for 9.3 cm initial water level is 33.7 cc with a cycle time of 1.42 cycles/sec. That bubble size lies on the same straight line as in Figure 21. Unfortunately, the slugging rate does not provide a limit on the bubble size

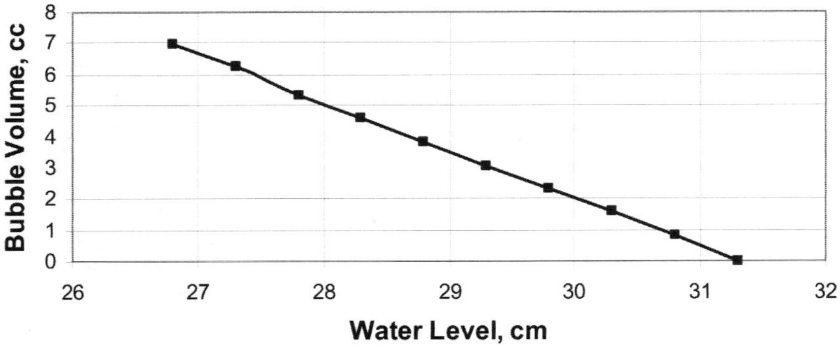

Figure 21. Maximum Bubble Size Possible for Different Initial Water Levels.

either as shown in Figure 22. This rate does not reach a limit at 27.5 cm but cpntinues to decrease with height so it does not suggest a bubble size limit.

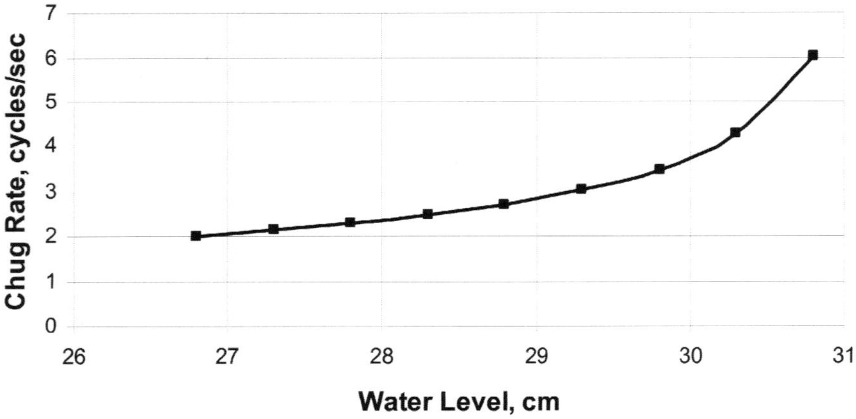

Figure 22. Cycle times for Different Initial Water Levels.

This investigation has not shown a clear indication of how to determine bubble size other than to use the experimentally determined value or to assume the bubble volume is the area of the orifice times a bubble length of 3.3 cm, as indicated by Table 2. Increasing the bubble size does increase the flow rate in the model and decreases the slugging rate. The model indicates that slower flow rates occur when the water level is near the top of the vessel due to decreased bubble sizes, but the flow rate is difficult to measure to confirm this conclusively. The bubble sizes do appear to be smaller, and the slugging rates appear higher in accordance with this study.

CONCLUSIONS

In this paper, a mechanism is proposed to describe the slugging cycle of water draining from a container. The model predicts outflow rates close to experimental values. Flow rates from non-vented containers were shown to be more than 10 to 20 less than vented containers. The bubbles which must enter the container periodically to increase the internal air pressure stop the water flow momentarily so are responsible for this large decrease in flow rate. Swirl induced in the non-vented container caused the flow rates to increase by a factor of two. The flow rate out of a non-vented container is independent of water height which is in direct contrast to a vented container where the flow rate is proportional to the square root of the water height. The constant rate is due to the container pressure. The higher the water level, the lower the air pressure is in the container. This analytical model requires input of the bubble size. The volume recommended is the volume of a cylinder with the base of the orifice area and length of 3.3 cm. Slugging rate varies only a small amount falling in the range to 2 to 4 cycles/sec.

ACKNOWLEDGMENTS

Appreciation is expressed to V..L. Gertman, A .E. Gertman, and C. R. Solbrig for assistance in experimental measurements. J. Sherman is also gratefully acknowledged.

REFERENCES

(1) G. H. Vatistas, D. Chen, T.F. Chen, and S. Lin, “Prediction of Infiltration Rates through an Automatic Door,” *Applied Thermal Engineering*, Vol 27, 2007, Pp 545-550.

(2) C. W. Solbrig, E. D. Hughes, “Two Phase Flow Equations Which Account for Unequal Phase Velocities and Unequal Phase Temperatures,” Aerojet Nuclear Company manuscript, Oct. 8, 1971.

(3) C. W. Solbrig and E. D. Hughes, “Governing Equations for a Seriated Continuum: An Unequal Velocity Model for Two-Phase Flow. In Two-Phase Transport and Reactor Safety,” Proceedings of the Two-Phase Flow and Heat Transfer Symposium-Workshop, Fort Lauderdale, Florida, 1976, October 18-20, Vol. I; Hemisphere, Washington, D.C., 1978; pp 839-886.

(4) C. W. Solbrig, G. A Mortensen, and R. W. Lyczkowski, “An Unequal Phase Velocity, Unequal Phase Temperature Theory Applied to Two-phase Blowdown from a Horizontal Pipe,” Proceedings. of the 1976 Heat Transfer and Fluid Mechanics Institute, University of California, Davis, California, 1976. June 21-23, Stanford University Press, Stanford; pp 60-76.

(5) R. W. Lyczkowski, “Twenty Five Years of Modeling Multiphase Flow and Heat Transfer,” Proceedings of the 1999 National Heat Transfer Conference, Albuquerque, New Mexico, 1997, August 15-17; p 167.

(6) D. Gidaspow, *Multiphase Flow and Fluidization Continuum and Kinetic Theory Descriptions,* Academic Press: San Diego, CA: 1994.

(7) C. W. Solbrig, J. H. McFadden, R. W. Lyczkowski and E. D. Hughes, "Heat Transfer and Friction Correlations Required to Describe Steam-Water Behavior in Nuclear Safety Studies," AIChE Paper No.21, 15th National Heat Transfer Conf., San Francisco, August 1975, published AIChE Symposium Series, No.174, Vol. 74, 1978, pp. 100-128.

(8) R. W. Lyczkowski, I. K. Gamwo, and D. Gidaspow, "Analysis of Flow in a Non-Aerated Hopper Containing a Square Obstacle," Powder Technology, Vol 112, 2000, pp 57-62.

(9) R. B. Bird, W.E. Stewart, and E. N. Lightfoot, *Transport Phenomena,* Wiley: 1960, P. 226.

The History of Multiphase Computational Fluid Dynamics

Robert W. Lyczkowski, Energy Systems Division, Argonne National Laboratory, Argonne, IL

Charles W. Solbrig, Nuclear Engineering Division, Idaho National Laboratory, Idaho Falls, ID

It all started with Charles "Chuck" W. Solbrig in the remote wilds of Idaho. He was the first PhD student of Dimitri Gidaspow when he was an Adjunct Professor in the Department of Gas Technology at Illinois Institute of Technology. After Chuck finished his Doctorate in 1966, he joined the department as an Adjunct Professor. Dimitri was my Master's Thesis advisor and Chuck was my co-advisor. Sarvajit S. Sareen and I were Dimitri's next two PhD students. Chuck left the Gas Technology Department in 1968 and joined Westinghouse Atomic in Pittsburgh before I earned my Doctorate in 1970. While there he developed this idea of modeling the hypothetical "Loss of Coolant Accident (LOCA)" using a new set of equations he called the "Seriated (series of) continuum" approach. He moved to Idaho to work for Aerojet Nuclear Company (ANC) which was a part of what was then the Nuclear Reactor Testing Station. He and his boss Larry Ybarrondo in George Brockett's Division sold the program to the Atomic Energy Commission in about 1971.

See more of this Session: Festschrift for Professor Dimitri Gidaspow's 75th Birthday
See more of this Group/Topical: Energy and Transport Processes

AICHE Annual, Meeting, 2009
November, 8-13, Nashville, TN, USA

Safety Aspects of Reactive NanoPowders.

Jacques Bouillard[1], Alexis Vignes[1], Olivier Dufaud[2], Laurent Perrin[2] and Dominique Thomas[2], (1) INERIS, Verneuil en Halatte, 60550 France, (2) LSGS-CNRS Rue Grandville, Nancy 54001, France.

Abstract

Emerging nanomanufactured products are being incorporated in a variety of consumer products ranging from closer body contact products (i.e. cosmetics, sunscreens, toothpastes, pharmaceuticals, clothing) to more remote body-contact products (electronics, plastics, tires, automotive and aeronautical), hence posing potential health and environmental risks. The new field of nanosafety has emerged and need to be explored now rather than after problems become so ubiquitous and difficult to treat that their trend become irreversible. Such endeavour necessitates a transdisciplinary approach.

A commonly forgotten and/or misunderstood risk is that of explosion/detonation of nanopowders, due to their high specific active surface areas. Such risk is emphasized and illustrated with the present development of an appropriate risk analysis. For this particular risk, a review of characterization methods and their limitations with regard to nanopowders is presented and illustrated for a few organic and metallic nanopowders.

Keywords: nanodust explosions, MWNT, carbon nanotube, nanopowders, Aluminum, combustion kinetics.

1. Introduction

For the past 10 years, there had been a growing interest in developing nanomaterials in various industrial sectors such as catalysts, in cosmetics, paints and coatings, because of their unique properties. Some of these materials that have traditionally been manufactured are now being developed as new products on the nanoscale. Others show truly new structures, such as carbon nanotubes. Concerning carbon nanotubes (CNTs), their market is positioned to "grow explosively" to reach 540 millions of dollars per year in 2007-2008 and more specifically 340 millions of dollars per year, specifically for multi-walled carbon nanotubes (MWNT) (Frost & Sullivan 2004). CNTs production is going to increase in the coming years and consequently potential risks associated to the handling of CNTs will be proportionally more and more important.

Aluminum powders have long been employed as fuel ingredient, explosive, propellants to enhance energy density and stability of propulsion systems. Aluminum is added to improve the burn rate and enhance blast effects in explosives (Kwok et al. 2002, Huang et al. 2005).

For any powders set in suspension in air, the ease with which they ignite, burn and propagate vary considerably with important factors that have to be characterized so as to be able to control their explosion. One of them is obviously the particle size, which takes nanopowders now under higher scrutiny for OH&S issues, others are the particle concentration, the degree of de-agglomeration, and the degree of turbulence (Eckhoff 2003, Dahoe et al. 1996). So far, literature studies concerning the evaluation of explosion and flammability hazards of powders were essentially carried out on microsized powders, albeit one on nanomaterials explosivity published in 2004 (Pritchard 2004). This work proposes to review recently measured main explosion safety parameters of multi-walled carbon nanotubes, carbon blacks and aluminum nanoparticles performed in course of the EU Nanosafe2 project.

2. Experimental setups and methods:

2.1 Particle geometrical characteristics

In this study, multi-walled carbon nanotubes were provided by an industrial partner of the Nanosafe2 consortium. Some geometrical characteristics were unknown and were determined during this work. The particle-size distribution, determined by using a laser diffraction analyzer (Mastersizer, Malvern Instrument) and performed in isopropanol, was characterized by the d_{10}, d_{50} and d_{90} of the volumetric distribution. d_{10}, d_{50} and d_{90} values respectively 20, 200 and 560 µm.

It has to be noticed that carbon nanotubes seem to form large agglomerates, which is confirmed by the transmission electronic microscope analysis.

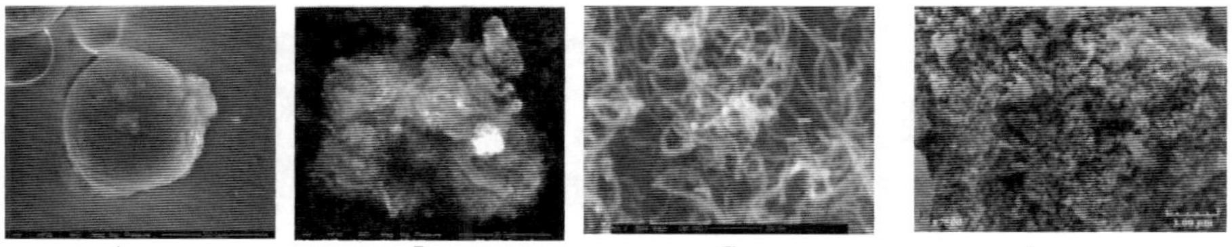

A B C D

Figure 1: TEM images of the MWNTs carbon nanotubes (A-C) and aluminum nanoparticles (D) at different resolutions

TEM observations of CNTs shows particles appearing as large spheres with imbedded nanotubes networks having a diameter ranging between 15 and 20 nm.

Similar Investigations were carried out with three nano-sized aluminium powders supplied by industrial partners of the consortium. Their surface is covered with an alumina layer representing approximately 20-30 % by weight.

BET specific surface areas (S_{BET}) and helium pycnometer density have been measured. By coupling the results obtained by BET and helium pycnometer, an average grain size can be calculated by considering that particles are isolated, cylindrical (or) spherical with a monodisperse size distribution:

$$d_{BET} = \frac{4(or)6}{\rho_{helium} \cdot S_{BET}} \quad \text{(Eq.1)}$$

2.2 Sample preparation for safety parameter characterisation:

A good reproducibility between the different tests was ensured by systematically drying the samples at 50°C under dynamic vacuum of 1 Pa during 15 minutes and then under a static vacuum of 10 Pa during 4 hours before handling. Particular attention was taken to handle the nanopowders. Nitrile gloves, airline hood, non-woven coverall and HEPA/ULPA vacuum cleaner have been used in addition to the more classical safety barriers.

2.3 Explosion sensitivity and severity

Explosion sensitivity is often refered to as a combination of three parameters characterizing the ability of dust to be ignited and to generate dusts explosions. These parameters are the following: the minimum ignition energy (MIE), the minimum ignition temperature (MIT) and the minimum explosive concentration (MEC) (Jacobson, Cooper & Nagy 1965) (Nagy & Verakis 1983).

In our study, MIE was determined using a modified Hartmann tube (Kühner AG). It consists in a 1.2 L glass cylindrical vessel in which the powders are dispersed and ignited by an electric spark. Both spark's energy and dust concentration could be modified to characterize the MIE. Turbulence degree is also variable. This apparatus also allows the determination of the MEC (IEC standard 1241-2-3 1994).

MIT in cloud can be tested by using a Godbert-Greenwald furnace (Chilworth Technology) set in an unconfined environment. This parameter was not systematically measured in our study.

The measurements of dust explosion severity, i.e. P_{max} (maximum overpressure), and $(dP/dt)_{max}$, (maximum rate of pressure rise), were performed in a 20 L spherical vessel in accordance with the ASTM Method E 1226 (See figure 3). Explosion severity is quantified by an explosion index K_{st} which is defined as

$$K_{st} = \left(\frac{dP}{dt}\right)_{max} \cdot V^{1/3} \quad \text{(Eq. 2)}$$

where V is the explosion volume.

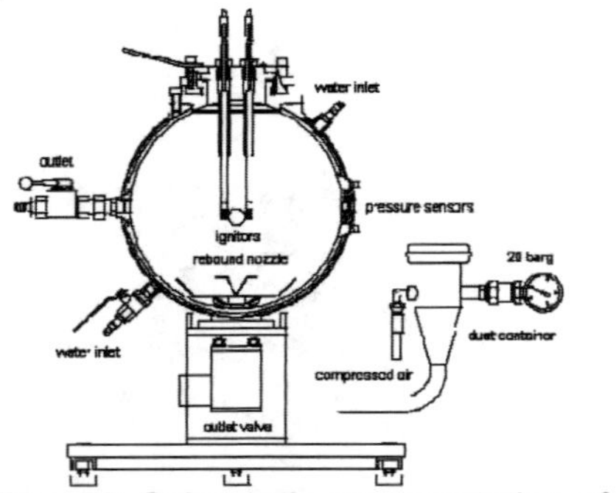

Figure 2. Schematic representation of the 20 L apparatus (Kühner AG).

3. Results and Conclusions

The main results of this study are summarized in the following table 1.

Table 1: Main safety parameters for CNTs, Carbon blacks and Aluminum

Products	CNT	Carbon blacks				Aluminum		
		Corax N115	Corax N550	Thermal Black N990	Printex XE2	Alex	100 nm	200 nm
MIE (mJ)	> 1J	> 1J	> 1J	> 1J	> 1J	<8mJ	< 1mJ	7 mJ
MEC (g/m3)	45	60	60	60	60	60	-	-
Pmax (bars)	6,6	7,7	7,5	6,7	7,2	8.3	8.2	9.5
(dP/dt)max (bars/s)	227	326	503	240	343	933	1340	2480
Kst	62	88	136	65	93	253	365	675
Explosivity Class	St1	St1	St1	St1	St1	St2	St3	St3
d_{agg} (nm)	200 000	90 000	15 000	2000	88000	-80000		
d_{BET} (nm)	10	23	70	330	3	-	100	200
S_{BET} (m^2/g)	195	128	42	9	900	-	23	10.5

From this table, it is shown that although the primary particles estimated by the BET method, the resulting average particle diameter (of the order of 10-200 µm) is much larger then the particle primary size indicating that the particles are agglomerated.

The maximum explosion pressure for the studied carbon blacks is 6.6 bars which is slightly lower than the value for MWNTs. The maximum rate of pressure rise is 227 bar/s and the explosion index K_{st} is 62, lower that that of carbon blacks, ranking the explosivity of these powders in a low hazard range. It is seen, however, that for aluminum nanopowders, the explosion sensitivity is high and that the severity can also be very high (Kst of 675), greater than that of hydrogen which is considered as a detonable gas. Hence, metallic nanopowders can possibly detonate under certain conditions, thus providing grounds for the need to perform a thorough risk assessment for such types of nanopowders

Acknowledgement

We thank the EC and the French Ministry of Ecology and Energy for Sustained Development, for the funding of NANOSAFE2 (project N° 515843-2) and NANORIS projects.

References :

Dahoe, A.E., Zevenbergen, J.F., Lemkowitz, S.M. and Scarlett, B., (1996), Dust explosions in spherical vessels: The role of flame thickness in the validity of the 'cube-root law', *Journal of Loss Prevention in the Process Industry*, 9 (1):33-44.

Eckhoff, R. K., 2003, *Dust explosions in the process industries* (3rd edition). Amsterdam – Boston – Paris: Gulf Professional Publishing.

Frost&Sullivan (2004), *An Assessment on the Future of Carbon Nanotubes - Strategic Analysis of the Market & Potential, Market engineering deliverable*, www.frost.com

Hamor, R.J., Smith, I.W. and Tyler, R. J. (1973), Kinetics of combustion of a pulverized brown coal char between 630 and 2200 K. *Combustion and Flame*, 21:153

Huang Y., Risha G. A., Yang V., and Yetter R. A. (2005). Analysis of Nano-Aluminum Paricle dust Combustion in Different Oxidizer Environments, 43rd Aerospace Sciences Meeting and Exhibit, January 10-13, Reno.

IEC standard 1241-2-3 1994

ISO 6148-1 (1985). *Explosion protection systems – Part 1: Determination of explosion indices of combustible dusts in air.* Geneva: ISO – International Organization for Standardization.

Jacobson, M., Cooper, A. R., & Nagy, J. (1965). *Explosibility of metal powders. Report of investigation 6516,* Bureau of Mines, U.S. Department of the Interior.

Kwok, Q. S. M., Fouchard, R. C., Turcotte, A., Lightfoot, P. D., Bowes, R., & Jones, D. E. G. (2002). Characterization of aluminium nanopowder compositions. *Propellants, Explosives, Pyrotechnics*, 27, 229 – 240.

Kwon, Y., Gromov, A. A., Ilyin, A. P., Popenko, E. M., & Rim, G. (2003). The mechanism of combustion of superfine aluminium powders. *Combustion and Flame,* 133, 385–391.

Laurent, A. (2003). *Sécurité des Procédés Chimiques - connaissances de base et méthodes d'analyse de risques*. Paris: Tec&Doc.

Nagy, J. & Verakis, H.C., (1983), *Development and Control of dust explosions*, New York, Marcel Dekker.

Pritchard, K ., (2004). *Literature Review-explosion hazards associated with nanopowders*, HSL/2004/12.

High Resolution Simulations of a Coal Gasifier

Madhava Syamlal, Office of Research and Development, National Energy Technology Laboratory, U.S. Department of Energy, Morgantown, WV

Chris Guenther, National Energy Technology Laboratory, U.S. Department of Energy, Morgantown, WV

Aytekin Gel, Aeolus Research, Inc., Phoenix, AZ

Sreekanth Pannala, Computer Science and Mathematics Division, Oak Ridge National Laboratories, Oak Ridge, TN

Integrated Gasification Combined Cycle (IGCC) is a promising technology for meeting the growing demand for power using fossil fuel resources, while economically controlling the emission of CO2 and other pollutants. The centerpiece of an IGCC system is the gasifier, which converts coal or other carbonaceous materials such as biomass into syngas, a mixture of CO and H2. The syngas can be used for the production of liquid fuels and chemicals or for power generation. In an IGCC system the syngas is shift converted into a mixture of CO2 and H2, CO2 is captured and stored, and H2 is used for power generation.

A reliable gasifier is critical for the commercial viability of IGCC. The gas-solids multiphase flow such as occurring in the gasifier is known to make the design of commercial-scale units using traditional scale-up methods difficult. Multiphase computational fluid dynamic (CFD) models are being developed at NETL to address that challenge, and this presentation discusses the potential for using high-resolution CFD simulations for gasifier design. A gasifier model based on the code MFIX was ported to several high performance computing (HPC) platforms ranging from Cray XT series to clusters with high bandwidth, low latency interconnects (e.g., Infiniband). MFIX is based on a continuum multiphase flow model that considers gas and solids to form interpenetrating continua. Low resolution simulations of a commercial scale gasifier with a validated MFIX model revealed interesting physical phenomena with implications on the gasifier design, which prompted the study presented here. To be predictive, the simulations need to model the spatiotemporal variations in gas and solids volume fractions, velocities, temperatures with any associated phase change and chemical reactions. These processes occur at various time- and length-scales requiring very high spatial resolution and large number of iterations with small time-steps. We were able to perform perhaps the largest known simulations of gas-solids reacting flows, providing detailed information about the gas-solids flow structure and the pressure, temperature and species distribution in the gasifier. Simulations were conducted with up to 6032 computational cores with a 10 million cells grid resolution. Porting the code to run on 1000's of cores instead of 10's or 100's of cores required changes in the code algorithm, parallelization methods, and the method of writing the output. This presentation will discuss the experiences in porting and running the code on HPC, the results of the high resolution simulations, and the prospects of using such high resolution simulations for scale-up.

See more of this Session: Festschrift for Professor Dimitri Gidaspow's 75th Birthday
See more of this Group/Topical: Energy and Transport Processes

Boiling of Dilute Emulsions: A State-of-the-Art Review

Frank A. Kulacki, Department of Mechanical Engineering, University of Minnesota, Minneapolis, MN

M. L. Roesle, University of Minnesota, Minneapolis, MN

Boiling of dilute emulsions of two components has been long standing problem in two-phase heat transfer which has received modest attention over the past 25 years. Recent research however has identified some surprising and potentially useful characteristics of such systems. This paper reviews the current state of the art for modeling the boiling process and experimental data, and proposes a new modeling approach.

Emulsions are mixtures of two immiscible liquids in which one liquid is suspended in the other in the form of fine droplets. These droplets may range in size but can be as small as 1 μm. We consider a system in which the low-boiling-point component makes up less than five percent of the mixture by volume. One remarkable feature of such emulsions is that a very large degree of superheat is necessary before boiling begins. This delay occurs because most of the droplets never contact a solid surface, so bubble nucleation must occur separately in each droplet in the volume of the emulsion. Currently, the most detailed description of boiling dilute emulsions is based on a simple model of the thermal boundary layer and upon the buoyancy of the vapor bubbles. When applied to experimental data, the model shows that nucleation in the droplets occurs more quickly than spontaneous nucleation would occur in the pure liquid. It is speculated that the rapid nucleation is due to chain-boiling between droplets that results when a boiling droplet forms a shock wave that breaks up floccules of nanoparticles in neighboring droplets. This hypothesis has not been proven experimentally, and is not predictive.

In light of past kinetic and thermodynamic models and experiments, a new model of boiling dilute emulsions is proposed. First, the behavior of a single boiling droplet must be simulated accurately. These data can then be used to model the effects of boiling droplets on the surrounding liquid and nearby droplets, including agitation of the mixture and collisions between droplets and bubbles. Finally, the large scale behavior of the emulsion can be modeled with the theory of interacting continua, using the data of the detailed simulations to define parameters for the large-scale model. The model thus formed is amenable to use in numerical simulations of boiling dilute emulsions.

See more of this Session: Festschrift for Professor Dimitri Gidaspow's 75th Birthday
See more of this Group/Topical: Energy and Transport Processes

Finite domain Complete Trial Functions Method of Moments (FCMOM) for Particulate Flow

Hamid Arastoopour, Department of Chemical and Biological Engineering, Illinois Institute of Technology, Chicago, IL

Matteo Strumendo, Department of Chemical and Biological Engineering, Illinois Institute of Technology, Chicago, IL

To account for continuous variations of the particle size and density distributions due to phenomena such as reaction, agglomeration, nucleation and growth using a low computational time, a new approach to solve PBE (Population Balance Equations), the FCMOM (Finite size domain Complete set of trial functions Method of Moments), was developed. The solution of the PBE is sought, instead of the [0,Å‡] range, in the finite interval between the minimum and maximum particle size; their evolution is tracked imposing moving boundaries conditions. After reformulating the PBE in the standard interval [-1,1], the size distribution function is presented as a series expansion by a complete system of orthonormal functions. Moments evolution equations are developed from the PBE in the interval [-1,1]. The FCMOM is implemented through an efficient algorithm and provides the solution of the PBE both in terms of the moments and in terms of the particle size distribution function. The FCMOM was validated with applications to particle growth, dissolution, nucleation, particle aggregation, and linked to the fluid phase governing equations (CFD) to solve problems in gas/particle systems.

See more of this Session: Festschrift for Professor Dimitri Gidaspow's 75th Birthday
See more of this Group/Topical: Energy and Transport Processes

Carbon Dioxide Injection In Complex Subsurface Formations and Issues In Thermodynamics and Fluid Mechanics

Abbas Firoozabadi, Yale University, New Haven, CT

A recent study by the Energy Information Administration of the US Department of Energy indicates that the share of natural gas, oil, and coal as primary energy sources will increase further by 2025. The same study gives a share of about 87% in 2025. A major concern would be global warming and efforts towards environmental stewardship. Provided carbon dioxide can be captured efficiently, its sequestration and its use in improved oil recovery can be much more efficient than current estimates. There may be a need to re-examine much of the material in the literature with respect to the interface modeling both in thermodynamics and in fluid mechanics. Some of the carbon dioxide injection projects may also need re-examination. Our recent formulation of non-equilibrium phenomena when carbon dioxide is placed above a column of oil or water shows many interesting features. The most important aspect is rapid dissolution. Diffusion aspects of mixing and momentum balance across a moving interface coupled with special behavior of carbon dioxide when dissolved in water and in petroleum fluids all add to features that if properly recognized can give enormous benefits. In order to study various features of carbon dioxide there may be need to develop new numerical models for subsurface flow studies. Results using high-order numerical modeling(combined discontinuous Galerkin and mixed finite element) from carbon dioxide injection in fractured permeable media will be shown to demonstrate that carbon dioxide may not travel fast through the cracks in the subsurface. The mixing from diffusion results in high oil recovery.

See more of this Session: Festschrift for Professor Dimitri Gidaspow's 75th Birthday - II
See more of this Group/Topical: Energy and Transport Processes

A Lattice Boltzmann Method with Adaptive Mesh Refinement for Direct Simulation of Bubble Dynamics

L. -S. Fan, Chemical and Biomolecular Engineering, The Ohio State University, Columbus, OH

The lattice Boltzmann method (LBM) for two-phase flow simulation is often hindered by insufficient resolution at the gas-liquid interface. As a result, the LBM simulation of bubbles is commonly limited to spherical or slightly deformed bubble shapes. In this study, the adaptive mesh refinement (AMR) method for the LBM is developed to overcome such a problem. The approach for this new method is based on the improved interaction potential model, which is able to maintain grid-independent fluid properties in the two fluid phases and at the interface. The LBM-AMR algorithm is described, especially concerning the LBM operation on a non-uniform mesh and the improved interaction potential model. Numerical simulations have been performed to validate the method in both single phase and multiphase flows. The 2D and 3D simulations of the buoyant rise of bubbles are conducted under various conditions. The agreement between the simulated bubble shape and velocity with experiments illustrates the capability of the LBM-AMR approach in predicting bubble dynamics even under the large bubble deformation conditions. Further, the LBM-AMR technique is capable of simulating a complex topology change of the interface. Integration of LBM with AMR can significantly improve the accuracy and reduce the computation cost. The method developed in this study may appreciably enhance the capability of LBM in the simulation of complex multiphase flows under realistic conditions.

See more of this Session: Festschrift for Professor Dimitri Gidaspow's 75th Birthday - II
See more of this Group/Topical: Energy and Transport Processes

Filtered Two-Fluid Models for Fluidized Gas-Particle Suspensions

Sankaran Sundaresan, Chemical Engineering, Princeton University, Princeton, NJ

Ysim Igci, Princeton University, Princeton, NJ

Gas-particle flows exhibit large fluctuations in velocities and local suspension density. In riser flows, these fluctuations are associated with the random motion of the individual particles and with the chaotic motion of particle clusters, which are repeatedly formed and broken apart. These clusters occur over a wide range of length scales and their dynamics span a broad range of time scales. We construct filtered hydrodynamic models for such systems by averaging over fluctuations occurring on short length and time scales, and determine the associated constitutive models by filtering results obtained from highly resolved simulations of a kinetic theory based model for fluidized suspensions. In this presentation, we will describe results obtained by simulating such filtered hydrodynamic models and compare them against simulations of the kinetic theory model which formed the starting point of the filtering analysis. Such a verification study exposes conditions when the filtered model is able to afford grid-size independent solutions at much coarser resolutions than the original model and when the macroscale character predicted by the filtered and kinetic theory models are in agreement.

See more of this Session: Festschrift for Professor Dimitri Gidaspow's 75th Birthday - II
See more of this Group/Topical: Energy and Transport Processes

LDV Measurements of Liquid-Solid Flow in a Vertical Pipe

Jennifer Sinclair Curtis, University of Florida, Gainesville, FL

Mark Pepple, Department of Chemical Engineering, University of Florida, Gainesville, FL

Caner Yurteri, Department of Chemical Technology, Delft University of Technology, Netherlands

Fluid-particle flows are prevalent across a diverse range of industrial and geophysical processes. In many of these applications, particles engage in collisions both with each other and the confining vessel. In these same applications, there are also complex interactions between the fluid and the solid phases that significantly influence particle interactions.

The vast majority of research and our best understanding of fluid-particle flows within the scientific community are for processes operating exclusively in either the inertia-dominated regime - where the influence of the fluid phase on the direct interactions between particles is neglected - or the macro-viscous regime - where the fluid phase plays the significant role in the mechanics of particle momentum transport. These regimes of flow occur in both gas-solid and liquid-solid systems. And, the fundamental predictive models that currently exist do not adequately describe fluid-particle flow in the "transitional" regime, between the viscous and interia-dominated regimes.

One key limitation impeding improved understanding and the development of fundamental models is the lack of detailed, non-intrusive flow measurements for this "transitional" regime. There is also a lack of measurements that bridge the transitional regime with both the inertia-dominated and viscous-dominated regimes. Hence, this talk will present detailed, non-intrusive LDV/PDPA measurements obtained in a unique, pilot-scale, slurry flow loop. By varying the flow velocity, particle concentration, and particle size, the range of regimes of flow behavior are spanned - from the inertia to the viscous-dominated regime. The resulting flow measurements enable significant advancements in the understanding of fluid-particle flows. In addition, these flow measurements will be a benchmark in the scientific literature for model development and validation.

See more of this Session: Festschrift for Professor Dimitri Gidaspow's 75th Birthday - II
See more of this Group/Topical: Energy and Transport Processes

Molecular Dynamics-Driven Transport Coefficients for Granular Flows: A Perspective

Christine M. Hrenya, Chemical and Biological Engineering, University of Colorado at Boulder, Boulder, CO

Over the past decade, the closure of constitutive quantities appearing in the continuum description of high-velocity, particle-laden flows has relied heavily on kinetic-theory-based closures. Such descriptions have found considerable success, though do not come without their limitations. For example, the extension of kinetic theory descriptions to non-spherical particles is non-trivial at best. Molecular dynamic (MD) simulations, on the other hand, can be extended in a relatively straightforward manner to accommodate such complexities. Furthermore, because the location of each particle is tracked as a function of time, constitutive quantities can, in principle, be extracted. In practice, however, the identification of the corresponding transport coefficients does not follow easily. In this talk, several complications will be discussed, including the unknown form of the constitutive relation, multiple transport coefficients, clustering instabilities, and Knudsen effects

See more of this Session: Festschrift for Professor Dimitri Gidaspow's 75th Birthday - II
See more of this Group/Topical: Energy and Transport Processes

Integration of CFD and Experimental-Based Design and Development of Fluidized-Bed-Reactor-Based Processes

L. J. I. Coleman, Center for Energy Technology, RTI International, Research Triangle Park, NC

B. S. Turk, Center for Energy Technology, RTI International, Research Triangle Park, NC

R. P. Gupta, Center for Energy Technology, RTI International, Research Triangle Park, NC

RTI International (RTI) has an active research and development (R&D) program for commercial development of catalyst/sorbents and associated fluidized-bed processes. One of RTI's most successful development projects has been a high temperature transport desulfurization process based on a regenerable ZnO-based sorbent with operating conditions of 370ºC to 550ºC and up to 1,200 psig. The key reasons for using a transport reactor for this application came from the advantages associated with the hydrodynamics, mass transfer, heat transfer, and reaction rate between the gas and solids. The success of this particular project was strongly based on the experimental R&D program completed. However, with the advances in computational fluid dynamics (CFD) modeling, CFD modeling represents a new tool for development of fluidized-bed processes. RTI is rapidly finding opportunities to integrate CFD modeling into its R&D program with the objective of accelerating process development.

Although the transport desulfurization process has been successfully demonstrated in a pilot plant system with syngas from a commercial coal gasifier, potential commercial end users of this technology would like to see a larger scale demonstration of this technology before installing the first commercial system. Test results collected during 3,000 hours of operating the pilot plant system have been used to develop a CFD model for the process. This CFD model is being used to help optimize the design of this larger demonstration unit and the full scale commercial design.

RTI is also developing the Dry Carbonate Process (DCP) to capture CO2 from power plant flue gas. In DCP, an inexpensive, dry, regenerable, solid sorbent is used to effectively capture CO2 from the flue gas in one fluidized-bed reactor and subsequently regenerated in a second fluidized-bed reactor by heating the sorbent. Design of this reactor system is particularly challenging because

- Pressure drop across the reactor must be very low
- Rapid heat transfer to the sorbent is required for both CO2 capture and regeneration
- CO2 capture reaction is exothermic
- Reaction rate for CO2 capture rapidly decreases with increasing temperature
- Regeneration reaction is endothermic
- Removal of large amount of CO2 requires rapid gas-solid mass transfer rates

To address these technical challenges, an integrated R&D effort combining an experimental test program and CFD modeling was initiated. This integrated approach

See more of this Session: Festschrift for Professor Dimitri Gidaspow's 75th Birthday - II
See more of this Group/Topical: Energy and Transport Processes

A Fully Coupled Third-Order Quadrature-Based Moment Method for the Simulation of Gas-Particle Flows

Alberto Passalacqua, Department of Chemical and Biological Engineering, Iowa State University, Ames, IA

Rahul Garg, Department of Mechanical Engineering, Iowa State University, Ames, IA

Shankar Subramaniam, Department of Mechanical Engineering, Iowa State University, Ames, IA

Rodney O. Fox, Department of Chemical and Biological Engineering, Iowa State University, Ames, IA

A third-order quadrature-based moment method (Fox, 2008) for simulating poly-disperse fluid-particle flows has been implemented in a computational fluid dynamics code, accounting for the full coupling between the gas and the dispersed phase. The particle phase is described by solving the moments transport equations obtained from kinetic theory, using a kinetic-based finite-volume technique, which is able to predict phenomena that lead to locally discontinuous velocity fields like the particle trajectory crossing. The moment equations are fully coupled with a fluid solver. The partial elimination algorithm (Spalding, 1980) is adopted to ensure a stable numerical solution in strongly coupled flows (Passalacqua et al., 2009). The algorithm has been extended to deal with locally packed zones, where the particle mean free path has to be limited to prevent the particle phase volume fraction from reaching unphysical values and compromising the stability of the numerical procedure.

The robustness of the implementation is tested by simulating a gas-particle flow in a vertical channel, with typical riser-flow particles (density: 1500 km/m3 , particle diameter 80μm and 250μm, corresponding to a Stokes number of 0.061 and 0.61 respectively). The results are compared to the predictions of the classical two-fluid models (Gidaspow, 1994) in both the considered cases, and with Euler-Lagrange simulations (Garg et al, 2009) in the case of Stokes number equal to 0.61.

The fluid and particle velocities, particle-phase volume fraction and granular temperature are observed to reach a steady state in the case of Stokes number 0.061, while the two-fluid model predictions shows unsteady flow structures.

In the case with the higher Stokes number, instabilities are observed that led to the formation of structures and initiate the particle segregation phenomena, inducing the formation of the typical flow structures experimentally observed in risers. The process that causes the development of the instability predicted by the quadrature-based method is confirmed by Euler-Lagrange simulations, while the two-fluid model provides an inconsistent result. The inconsistencies observed in two-fluid models are explained by considering the local particle Knudsen number, which is significantly higher than the limit of 0.1, above which the hydrodynamic model is unable to properly describe the flow properties, even adopting partial slip boundary conditions, and higher order approximation of the Boltzmann equation are required (Bird, 1994).

References

Bird, G. A. 1994. Molecular Gas Dynamics and the Direct Simulation of Gas Flows. 2° ed. Oxford University Press, USA.

Fox, R.O. 2008. A quadrature-based third-order moment method for dilute gas-particle flows. Journal of Computational Physics 227, no. 12: 6313-6350.

Garg, R., C. Narayanan, and S. Subramaniam. 2009. A numerically convergent Lagrangian-Eulerian simulation method for dispersed two-phase flows. International Journal of Multiphase Flow 35, no. 4: 376-

388.

Gidaspow, D. 1994. Multiphase Flow and Fluidization. 1° ed. Academic Press.

Passalacqua, A., R. Garg, S. Subramaniam, and R. O. Fox. 2009. A fully coupled quadrature-based moment method for dilute to moderately dilute fluid-particle flows. Chemical Engineering Science (Submitted).

Spalding, D.B. 1980. Numerical computation of multi-phase fluid flow and heat transfer. In Recent Advances in Numerical Methods in Fluids, Ed. C. Taylor. Pineridge Press.

See more of this Session: Festschrift for Professor Dimitri Gidaspow's 75th Birthday - II
See more of this Group/Topical: Energy and Transport Processes

Risk Assessment for CO2 Geologic Sequestration

Yan Zhang, Chemical Engineering, Carnegie Mellon University, Pittsburgh, PA

Panagiotis Vouzis, Chemical Engineering, Carnegie Mellon University, Pittsburgh, PA

Nick Sahinidis, Chemical Engineering, Carnegie Mellon University, Pittsburgh, PA

Sequestration of CO2 in existing and abandoned oil reservoirs, deep saline aquifers, and coal seams are currently among techniques intensely studied by the international research community for their potential to reduce greenhouse gases. In the case of sequestration to abandoned oil reservoirs, CO2 is pressurized and stored in these reservoirs, and a cement plug is placed at the bottom of the injection well. This cement plug, combined with an impermeable layer of caprock, creates a seal for the storage system, thus preventing CO2 from escaping.

A main concern for any sequestration system is whether it may leak CO2 over a long-term time horizon consisting of up to many thousand years. The outcome depends on the competition between CO2 sequestration and leakage processes. Leakages after sequestration in an abandoned reservoir may occur from failure of the manmade cement seal, or through naturally existing fractures in the caprock. Additionally, the CO2 source pool may be dissolved into formation water or migrate to water of surrounding formations. On the other hand, chemical reactions, such as mineralization, may prevent leakage. Modeling all these processes from first principles, while possible, requires knowledge of a number of parameters, such as various permeability and porosity values, that are not known with certainty. This motivates the development of simulators that can accurately deal with uncertainty in model parameters.

A number of simulators, including CQUESTRA [1, 2] and CO2-PENS [3], have been developed to assess the long term risk of CO2 sequestration. Monte Carlo (MC) simulations have been used in the context of these simulators to assess the uncertain behavior of sequestration processes. The main challenge in this context is that MC simulations are computationally demanding due to the large number of required iterations to produce a statistical representation of the output. In this work, we use a Graphics Processing Unit to parallelize the application of MC simulation to sequestration process models. We have implemented a sequestration simulator, mostly following [1] and [2], and used nVidia's Compute Unified Device Architecture (CUDA) [4] to develop a parallel MC implementation. Extensive computational results will be presented, illustrating that the proposed methodology provides a reliable and effective tool for performing sequestration studies, including assessing the effect of uncertain parameters and model equations.

References:

1. D. M. LeNeveu, "CQUESTRA, A Risk and Performance Assessment Code for Geological Sequestration of Carbon Dioxide," Energy Conversion and Management, 49(1), 32- 46, 2008.

2. F. B. Walton, J. C. Tait, D.M. LeNeveu, M.I. Sheppard, "Geological Storage of CO2: A Statistical Approach to Assessing Performance and Risk," Proceedings of 7th International Conference on Greenhouse Gas Control Technologies, 2004

3. P. H. Stauffer, H. S. Viswanathan, R. J. Pawar, and G. D. Guthrie, "A System Model for Geologic Sequestration of Carbon Dioxide," Environmental Science and Technology, 43 (3), 565-570, 2009.

4. V. Podlozhnyuk, and M. Harris, "Monte Carlo Option Pricing," nVidia Corporation Tutorial, June 2008.

See more of this Session: Advanced Modeling Methods in Carbon Dioxide Sequestration
See more of this Group/Topical: Energy and Transport Processes

CO_2 Sequestration: Temperature and Gas Compositional Effects On the Kinetics of Mineralogical Reactions

Prashanth N. Mandalaparty, Chemical Engineering, University of Utah, Salt Lake city, UT

Milind Deo, University of Utah, Salt Lake city, UT

Joseph Moore, Energy and GeoSceince Institue, University of Utah, Salt Lake City, UT

It may be possible to lower costs of Carbon Capture and Sequestration by keeping constituents such as sulfur dioxide (SO_2) in the flue gas stream. The reactive behavior of pure CO_2 and CO_2 + SO_2 mixtures within a geologically realistic environment was examined. The experimental apparatus consisted of a series of four high-pressure reactors operated at different conditions and with different feed gas compositions to observe changes in both the rock and water compositions. The rock consisted of equal proportions of quartz, calcite, andesine, dolomite, chlorite and magnesite (constituents in arkose or dirty sand stone). Several long term batch experiments with pure CO_2 were carried out at different temperatures. Each mineral in the mixture showed evidence of participating in the geo-chemical reactions. Layers of calcite were seen growing on the surface of the arkose. Analcime deposits were omnipresent, either occurring as large connected aggregates or as deposits on the surfaces of other minerals (Quartz). Ankerite and calcite depositions were observed as amorphous masses intergrown with the feed. The CO_2 + SO_2 mixture experiments showed growth of euhedral anhydrite crystals and pronounced dissolution patterns over the examined surfaces. The growth of these new phases indicates the complex geo-chemical interactions taking place and the alterations due to the presence of SO_2. The trends in ionic concentration changes in the aqueous phase complemented the changes in the rock chemistry. The rates of these reactions were computed by measuring the changes in the compositions of the initial and the final samples.

See more of this Session: Advanced Modeling Methods in Carbon Dioxide Sequestration
See more of this Group/Topical: Energy and Transport Processes

CO2 Pipeline Integrity Modelling Challenges

Haroun Mahgerefteh, Chemical Engineering, University College London, London, United Kingdom

Pressurised pipelines are widely considered to be the most practical, and in the case of fossil fuel power generation plant, the only option for transporting captured CO2 for subsequent sequestration. Typically, such pipelines may be several hundred kilometres long, transporting enormous amounts of CO2 under supercritical conditions.

Given the large quantities involved and the hazardous nature of CO2 (at concentrations greater than 10% the gas is instantly fatal [1] the understanding of the hazards that may undermine the mechanical integrity of such pipelines is central to the success of CCS as an effective means for combating the effects of global warming.

It is widely acknowledged [2] that existing experience with the use of pressurised CO2 pipelines is simply too limited to draw any meaningful conclusions. Additionally, the widely different hazard profile of CO2 as compared to hydrocarbons means that any confidence that existing experience with operating hydrocarbon pipelines can be wholly extended to CO2 pipelines is dangerously misplaced [3].

This paper presents an overview of some of the most important hazards affecting CO2 pipeline safety that can be mitigated using robust mathematical modelling through informed operating conditions and pipeline materials selection. These include consideration from the molecular level, such as the impact of impurities on thermodynamics and phase behaviour of CO2, corrosion resistance and hydrate formation to macroscopic failures such as fast running ductile or brittle fracture propagation.

References

1.http://www.energyinst.org.uk/content/files/hseinfopaper.pdf

2. Intergovernmental Panel on Climate Change, IPCC, Carbon Capture & Storage ISBN 92-9169-119, 2005.

3. Mahgerefteh, H., Denton, G., Rykov, Y., CO2 Pipeline Rupture, HAZARDS XX Process Safety and Environmental Protection, IChemE Sympsium Series No. 154, 869 -879, 2008.

See more of this Session: Advanced Modeling Methods in Carbon Dioxide Sequestration
See more of this Group/Topical: Energy and Transport Processes

A Generalized Negative Flash Procedure for Phase Equilibrium Computations Related to Carbon Dioxide Sequestration

Alireza Iranshahr
Denis Voskov
Hamdi Tchelepi

Introduction

Predictions of subsurface CO_2 sequestration in depleted oil and gas reservoirs involves solving coupled systems of transport equations. A particular challenge is resolving the complex multiphase thermodynamic equilibrium behaviors associated with CO_2-oil-water mixtures at reservoir pressures and temperatures. Here, we propose a multiphase equilibrium method based on a generalized negative-flash strategy. For this class of problem, the standard strategy based on two-phase equilibrium is inadequate, since three-phase states have been shown to exist.

In standard methods, a stability test is performed to determine the phase-state of a mixture for a given overall composition at fixed pressure and temperature [1]. For multiphase states, a flash procedure is used to determine the phase amounts and compositions. Flash calculations usually include simultaneous solution of the chemical-potential constraints for all the components that may be present in the equilibrium phases [2].

For gas injection processes, Method of Characteristics (MoC) solutions for two-phase multi-component displacements have been derived for ideal systems (one-dimensional dispersion-free), where tie-lines are used to construct the solution path in compositional space [3]. Motivated by this theory, a general tie-line based compositional space parameterization method has been developed in order to improve the computational efficiency of isothermal compositional flow simulation [4, 5]. The mathematical framework for the extension of this methodology to arbitrary numbers of phases has been presented in [6].

Here, we describe a negative-flash method for mixtures that can form an arbitrary number of phases. The overall algorithm combines Successive Substitution Iteration (SSI) methodology with Newton iterations. For faster convergence of the flash iterations, a good initial guess for the phase partitioning coefficients (K-values) is necessary for both SSI and Newton procedures. We propose a general methodology to provide this initial guess for a system with an arbitrary number of phases. The procedure is based on tie-simplex parameterization of the compositional space and uses a negative-flash procedure [6].

During any SSI iteration, the K-values are assumed to be constant, and therefore, a robust constant K-value negative-flash procedure (i.e., multiphase extension of the Rachford-Rice problem) is required to obtain a stable SSI algorithm. A gradient-based methodology to solve the generalized Rachford-Rice problem is not guaranteed to converge. The problem is, however, amenable to a bracketing based scheme, such as bisection. We first show the existence and uniqueness of the solution to the three-phase, constant K-value, negative-flash problem. Then, we provide a nested bisection solution procedure. This procedure is different from a two-dimensional bisection method, which was developed to solve the Rachford-Rice equations for the three-phase flash problem [7]. The proposed method is used for a general negative-flash procedure (when phase fractions are not necessarily bracketed by $[0, 1]$ interval) and is recursively extendible to systems with any number of phases. We demonstrate the robustness of the algorithm using challenging problems, such as multiphase multi-component mixtures for a very wide range of K-values.

Performing multiphase negative-flash with the highest dimension is not sufficient for phase identification. For three-phase systems, tie-lines parameterize the space around the three-phase region, and for the compositions which lie outside the tie-triangle region, additional two-phase negative flashes are required to accurately identify the phase-state. As such, the method provides not only the phase state of the mixture, but also the compositions of the actual phases at equilibrium.

Several examples of CO_2 injection in depleted oil reservoirs with complex multiphase behaviors are used to demonstrate the effectiveness and computational efficiency of the proposed method.

Constant K-Value Multi-Phase Negative Flash

In the following, we assume that $0 \le i < N_c$ and $0 \le j < N_p$. The overall molar fraction of a component, z_i, is given by:

$$z_i = x_{i,0}\nu_0 + \sum_{j=1}^{N_p-1} x_{i,j}\nu_j, \qquad 0 \le i < N_c, \tag{1}$$

where we have arbitrarily chosen a phase as the 'base' phase, and we denoted its index by 0. From,

$$\begin{aligned} \nu_0 &= 1 - \sum_{j=1}^{N_p-1} \nu_j, & (2)\\ x_{i,j} &= K_{i,j}x_{i,0}, \qquad 0 \le i < N_c,\ 1 \le j < N_p, & (3) \end{aligned}$$

one obtains:

$$z_i = x_{i,0}\left(1 + \sum_{j=1}^{N_p-1} \nu_j(K_{i,j} - 1)\right) = x_{i,0}m_i(\boldsymbol{\nu}), \qquad 0 \le i < N_c. \tag{4}$$

The equations to be solved simultaneously for $\boldsymbol{\nu}(\equiv \nu_{1\leq j<N_p})$ are:

$$
\begin{aligned}
f_j(\boldsymbol{\nu}) &= \sum_{i=0}^{N_c-1} (x_{i,0} - x_{i,j}) \\
&= \sum_{i=0}^{N_c-1} \frac{z_i(1-K_{i,j})}{m_i(\boldsymbol{\nu})}, \qquad 1 \leq j < N_p. \qquad (5)
\end{aligned}
$$

The system contains $N_p - 1$ equations to be solved simultaneously for $N_p - 1$ unknowns.

Existence of the Solution

We show the existence of the solution to Eqs. (5) for a three-phase system. For this purpose, consider the domain of the problem (ν_1, ν_2). Once the K-values are fixed, the tie-triangle and its associated plane in compositional space is defined. Clearly, a subset of this plane, which is located inside the compositional space is physical. This physical region is defined by the intersection of the plane with the boundaries of the space. We map this physical plane in compositional space onto the space of unknowns ($\boldsymbol{\nu}$-space). For a physical $\boldsymbol{\nu}$,

$$m_i(\boldsymbol{\nu}) \geq 0, \qquad \forall i, \qquad (6)$$

must hold, which is obtained from Eq. (4). For a three-phase flash, $m_i(\boldsymbol{\nu}) = 0$ is represented by a line in the two-dimensional $\boldsymbol{\nu}$-space. Each line is associated with one component. The corners of the physical domain are determined from the intersection of such lines. Consider the intersection of the pth and q^{th} lines. This intersection is considered to be a corner of the physical plane if it satisfies:

$$m_i(\boldsymbol{\nu}) \geq 0, \qquad \forall i \setminus \{p, q\}. \qquad (7)$$

Once all the corners of the extended plane in compositional space have been determined, the minimum and maximum values of ν_1 and ν_2 can be obtained. For a given ν_1 such that $\nu_{1,min} < \nu_1 < \nu_{1,max}$, one can calculate the range of physical ν_2. Let j and k represent subsets of component indices for which $K_{j,2} > 1$ and $K_{k,2} < 1$, respectively. The physical bounds of ν_2 are obtained from,

$$\max_j \left(\frac{\nu_1(1-K_{j,1}) - 1}{K_{j,2} - 1} \right) \leq \nu_2 \leq \min_k \left(\frac{\nu_1(1-K_{k,1}) - 1}{K_{k,2} - 1} \right). \qquad (8)$$

Similarly, the physical domain for ν_1 can be calculated for a given ν_2 in $\nu_{2,min} < \nu_2 < \nu_{2,max}$. Figure (1) shows the physical boundaries of the two variables (shown by black lines) for a 15-component mixture with random K-values. The blue triangle shows the region of unknowns for which all three phase molar ratios (ν_0, ν_1, ν_2) have values between 0 and 1. One can show that this triangle is always inside the physical domain.

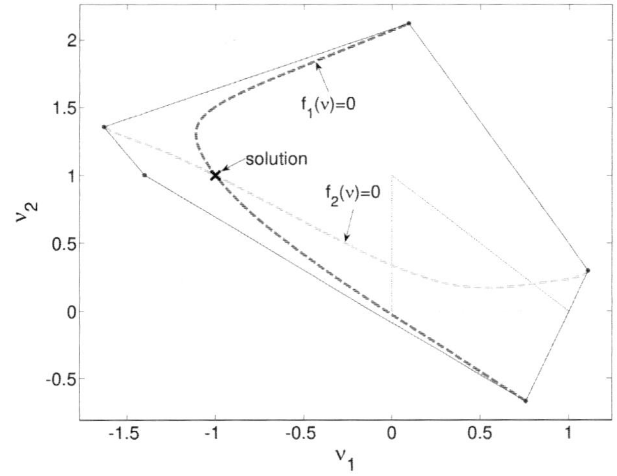

Figure 1: Physical plane of unknowns (ν_1 and ν_2) for a 15-component mixture with random K-values. The boundaries of the plane are represented by black lines. Zero loci of each function are shown by dashed curves. Their intersection is the system's solution, which is outside the tie-triangle.

For a three-phase negative flash, we simultaneously solve $f_1(\boldsymbol{\nu})$ and $f_2(\boldsymbol{\nu})$ for ν_1 and ν_2 (See Eqs. 5). We have:

$$\frac{\partial f_j(\boldsymbol{\nu})}{\partial \nu_j} = \sum_{i=0}^{N_c-1} \frac{z_i(1-K_{i,j})^2}{m_i^2}, \qquad 1 \le j \le 2, \tag{9}$$

$$\frac{\partial f_1(\boldsymbol{\nu})}{\partial \nu_2} = \frac{\partial f_2(\boldsymbol{\nu})}{\partial \nu_1} = \sum_{i=0}^{N_c-1} \frac{z_i(1-K_{i,1})(1-K_{i,2})}{m_i^2}. \tag{10}$$

From Eqs. (9), $\frac{\partial f_1}{\partial \nu_1} > 0$ and $\frac{\partial f_2}{\partial \nu_2} > 0$. Moreover, both f_1 and f_2 are indefinite at the boundaries of the compositional space. As a result, for a given ν_2, there is one and only one ν_1 such that $f_1(\nu_1, \nu_2) = 0$. Similarly, for a given ν_1, there is one and only one ν_2, such that $f_2(\nu_1, \nu_2) = 0$. Therefore, the loci of $f_1 = 0$ and $f_2 = 0$ are represented by two curves extending from $\nu_{2,min}$ to $\nu_{2,max}$, and from $\nu_{1,min}$ to $\nu_{1,max}$, respectively (See Figure 1). These curves intersect in at least one point, which can be a solution to Eqs. (5).

Uniqueness of the Solution

Next, we show that $f_1 = 0$ and $f_2 = 0$ intersect at one and only one point, i.e., Eqs.(5) have a unique solution. For this purpose, we need to show that the directional derivative of either function along the zero loci of another function is strictly positive (or negative). We have,

$$\frac{\mathrm{d}f_1(\nu_1,\nu_2)}{\mathrm{d}\nu_1} = \frac{\partial f_1(\nu_1,\nu_2)}{\partial \nu_1} + \frac{\partial f_1(\nu_1,\nu_2)}{\partial \nu_2}\frac{\partial \nu_2}{\partial \nu_1}, \tag{11}$$

where, $\partial\nu_2/\partial\nu_1$ is defined by the locus of $f_2(\nu_1,\nu_2)=0$. That is,

$$\frac{\mathrm{d}f_2(\nu_1,\nu_2)}{\mathrm{d}\nu_1} = \frac{\partial f_2(\nu_1,\nu_2)}{\partial\nu_1} + \frac{\partial f_2(\nu_1,\nu_2)}{\partial\nu_2}\frac{\partial\nu_2}{\partial\nu_1} = 0, \tag{12}$$

$$\frac{\partial\nu_2}{\partial\nu_1} = -\frac{\frac{\partial f_2(\nu_1,\nu_2)}{\partial\nu_1}}{\frac{\partial f_2(\nu_1,\nu_2)}{\partial\nu_2}}. \tag{13}$$

Substituting Eq. (13) in Eq. (11) yields,

$$\begin{aligned}\frac{\mathrm{d}f_1(\nu_1,\nu_2)}{\mathrm{d}\nu_1} &= \frac{\partial f_1(\nu_1,\nu_2)}{\partial\nu_1} - \frac{\partial f_1(\nu_1,\nu_2)}{\partial\nu_2}\frac{\frac{\partial f_2(\nu_1,\nu_2)}{\partial\nu_1}}{\frac{\partial f_2(\nu_1,\nu_2)}{\partial\nu_2}} \\ &= \frac{\partial f_1(\nu_1,\nu_2)}{\partial\nu_1} - \frac{\left(\frac{\partial f_2(\nu_1,\nu_2)}{\partial\nu_1}\right)^2}{\frac{\partial f_2(\nu_1,\nu_2)}{\partial\nu_2}},\end{aligned} \tag{14}$$

where, the derivatives on the right hand side are given by Eqs. (9) and (10). Consider the following two vectors in N_c-dimensional space:

$$v_i = \frac{\sqrt{z_i}(1-K_{i,1})}{m_i}, \tag{15}$$

$$w_i = \frac{\sqrt{z_i}(1-K_{i,2})}{m_i}. \tag{16}$$

From the Cauchy-Schwarz inequality,

$$\left(\sum_{i=0}^{N_c-1} v_i w_i\right)^2 \le \left(\sum_{i=0}^{N_c-1} v_i^2\right)\left(\sum_{i=0}^{N_c-1} w_i^2\right). \tag{17}$$

Therefore,

$$\begin{aligned}\left(\sum_{i=0}^{N_c-1}\frac{z_i(1-K_{i,1})(1-K_{i,2})}{m_i^2}\right)^2 &\le \left(\sum_{i=0}^{N_c-1}\frac{z_i(1-K_{i,1})^2}{m_i^2}\right)\left(\sum_{i=0}^{N_c-1}\frac{z_i(1-K_{i,2})^2}{m_i^2}\right), \\ \left(\frac{\partial f_2(\nu_1,\nu_2)}{\partial\nu_1}\right)^2 &\le \frac{\partial f_1(\nu_1,\nu_2)}{\partial\nu_1}\frac{\partial f_2(\nu_1,\nu_2)}{\partial\nu_2}.\end{aligned} \tag{18}$$

Moreover, since v_i and w_i are linearly independent, the equality in Eq. (18) does not hold. Therefore, $\mathrm{d}f_1(\nu_1,\nu_2)/\mathrm{d}\nu_1$ (and similarly $\mathrm{d}f_2(\nu_1,\nu_2)/\mathrm{d}\nu_2$) is strictly positive (See Figure 2), and the solution to Eqs. (5) is unique.

A Bisection-Based Solution Methodology

Based on the properties of the problem discussed in the previous section, a bisection based method is guaranteed to converge to the solution. We note that gradient-based algorithms could be faster; however, even with a very good initial guess, they may not converge.

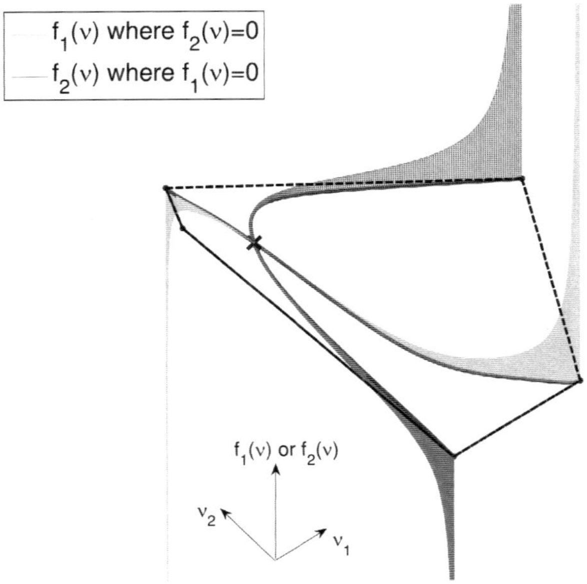

Figure 2: $f_1(\boldsymbol{\nu})$ and $f_2(\boldsymbol{\nu})$ at zero loci of the other function

Here, we use a nested bisection scheme to solve for the two unknowns ν_1, and ν_2. In the outer and inner loops, we solve for ν_2 and ν_1, respectively. This algorithm is summarized as follows:

Step 1: Given the set of K-values, find the boundaries of the physical plane, which yields $\nu_{2,min}$ and $\nu_{2,max}$.
Step 2: Let $\nu_{2,mid} = (\nu_{2,min} + \nu_{2,max})/2$.
Step 3: Solve $f_1(\nu_1, \nu_{2,min}) = 0$, and $f_1(\nu_1, \nu_{2,mid}) = 0$ for $\nu_{1,min}$, and $\nu_{1,mid}$, respectively. A bisection based methodology is suggested.
Step 4: If $f_2(\nu_{1,min}, \nu_{2,min}) \times f_2(\nu_{1,mid}, \nu_{2,mid}) > 0$, $\nu_{2,min} = \nu_{2,mid}$; else $\nu_{2,max} = \nu_{2,mid}$.
Step 5: Stop if $\nu_{2,max} - \nu_{2,min} < \varepsilon$, go to Step 2 otherwise.

Figure (3) illustrates the first iterations of the procedure for a sample system. The SSI method for solving two-phase negative-flash problems can be generalized to three-phase systems by introducing this algorithm as a constant K-value flash solver at each iteration. An initial guess for the K-values is required for such multiphase SSI procedure. For this purpose, we flash the overall composition, which yields the 'largest' tie-triangle at the given temperature and pressure conditions. The resulting tie-triangle is then used as an initial guess to flash the target overall composition. This method of computing initial guess is applicable for systems with arbitrary numbers of phases.

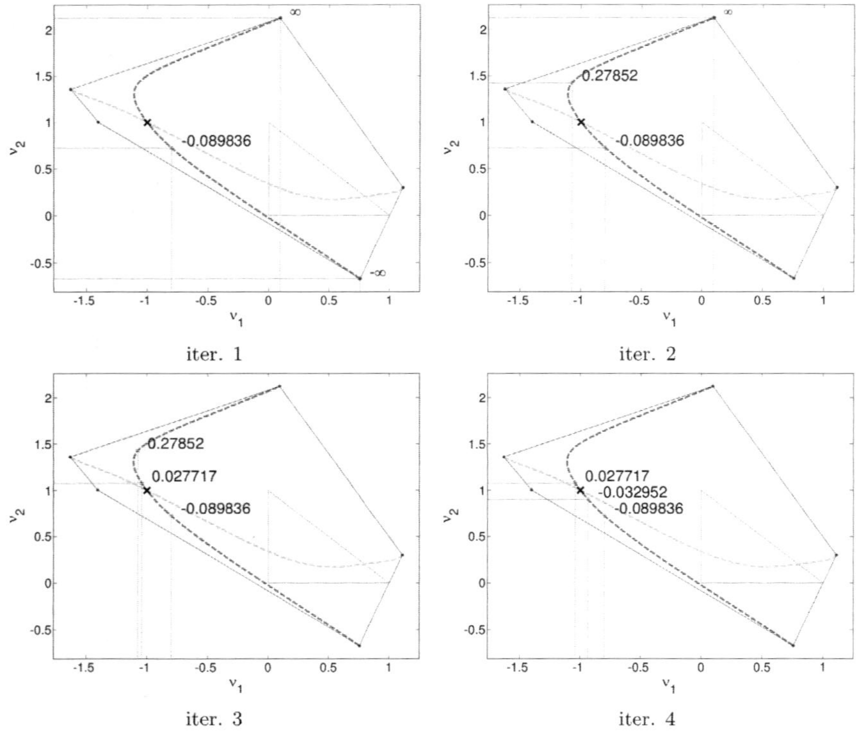

Figure 3: f_2 at $\nu_{2,min}$, $\nu_{2,mid}$, and $\nu_{2,max}$ during the first four iterations of the methodology.

Multi-Stage Three-Phase Negative Flash

The three-phase negative flash introduced in the previous section is not enough for determining the phase state of a given composition. This is because the actual two-phase behavior outside the three-phase region may be different from the three-phase negative-flash results (the straight lines which define the tie-triangle, see Figure 4).

In order to accurately determine the number and compositions of the existing phases for a given overall composition at specified temperature and pressure conditions, we use a multi-stage negative flash algorithm. In the first stage, we perform a three-phase negative flash. The state of the composition is confirmed if it is inside the tie-triangle (three-phase). Otherwise, based on the output of the three-phase negative flash, we perform additional two-phase negative flash(es). In this framework, we assume that all side tie-lines are in the plane of the tie-triangle. While actual systems show deviations from this behavior, several numerical experiments indicate the robustness of the proposed framework. We also note that side compositions of the tie-triangle provide very good initial guess for the two-phase negative flash.

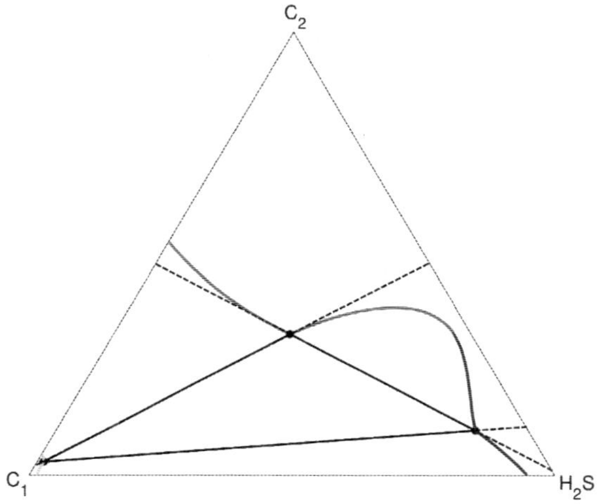

Figure 4: Complete parameterization of $\{C_1, C_2, H_2S\}$ at $T = 180\text{K}, p = 15bar$. The accurate two-phase behavior is shown by the solid curves outside the three-phase region.

Numerical Examples

We present two numerical examples to illustrate the multiphase equilibrium calculations described before (See Figure 5). For each system, we choose two compositions as initial oil and injection gas compositions. We find the state and parameterize the tie-triangle/tie-line for all the intermediate compositions. One tie-triangle and two tie-lines are used to parameterize a single-phase composition. A two-phase composition is parameterized by one tie-triangle and one tie-line and is inside the tie-line region. Similarly, only one tie-triangle is required to determine the state of a three-phase composition. In Figure (5), phase boundaries show where the phase state changes along the connecting path between the injection and initial compositions.

References

[1] Michelsen M.L., The Isothermal Flash Problem. Part I. Stability Analysis. Fluid Phase Equilibria, 9:1-19 (1982a).

[2] Michelsen M.L., The Isothermal Flash Problem. Part II. Phase Split Calculation. Fluid Phase Equilibria, 9:21-40 (1982b).

[3] Orr, F.M., Jr., Theory of Gas Injection Processes, Tie-Line Publications, Copenhagen, Denmark (2007).

[4] Voskov D., Tchelepi H., Compositional Space Parameterization: Theoretical Background and Application for Immiscible Displacements, SPE Journal 14

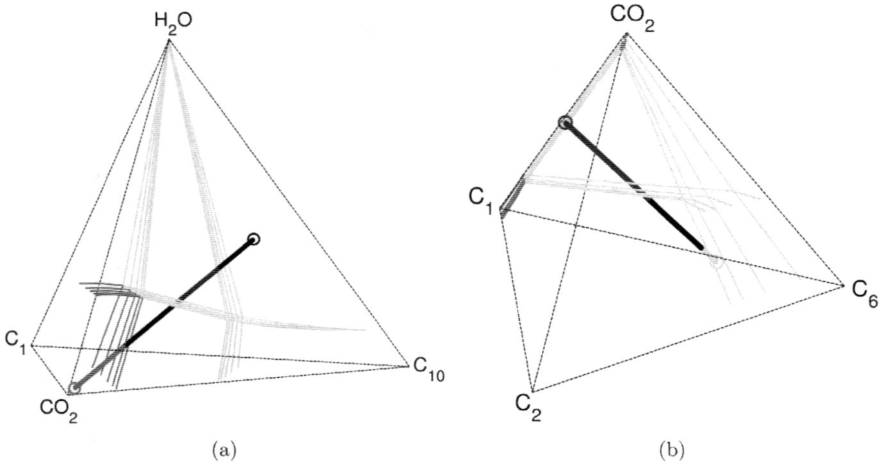

Figure 5: (a) Injection of CO_2 into $\{C_1(0.1), C_{10}(0.4), H_2O(0.4), CO_2(0.1)\}$ at $T = 500$ K and $p = 140$ *bar*, (b) Injection of $\{C_1(0.5), CO_2(0.5)\}$ into $\{C_6(0.5), CO_2(0.2), C_2(0.3)\}$ at $T = 180$ K and $p = 4$ *bar*. Linear path includes compositions in gas-phase (•), oil-phase (•), two-phase (•), and three-phase (•) states.

(3), 2009, (paper SPE 106029 presented at 2007 SPE Reservoir Simulation Symposium, Woodlands, TX, February 26-28).

[5] Voskov D., Tchelepi H., Compositional Space Parameterization: Multi-Contact Miscible Displacements and Extension to Multiple Phases, SPE Journal 14 (3), 2009, (paper SPE 113492 presented at 2008 SPE/DOE Symposium on Improved Oil Recovery, Tulsa, Oklahoma, April 20-23).

[6] Voskov D., Tchelepi H., Tie-Simplex Based Mathematical Framework for Thermodynamical Equilibrium Computation of Mixtures with an Arbitrary Number of Phases, Fluid Phase Equilibria, vol. 283 (1-2) pp. 1-11, 2009.

[7] Haugen K.B., Sun L., Firoozabadi A., Three-Phase Equilibrium Calculations for Compositional Simulation, paper SPE 106045 presented at 2007 SPE Reservoir Simulation Symposium, Woodlands, TX, February 26-28.

NATIONAL ENERGY TECHNOLOGY LABORATORY

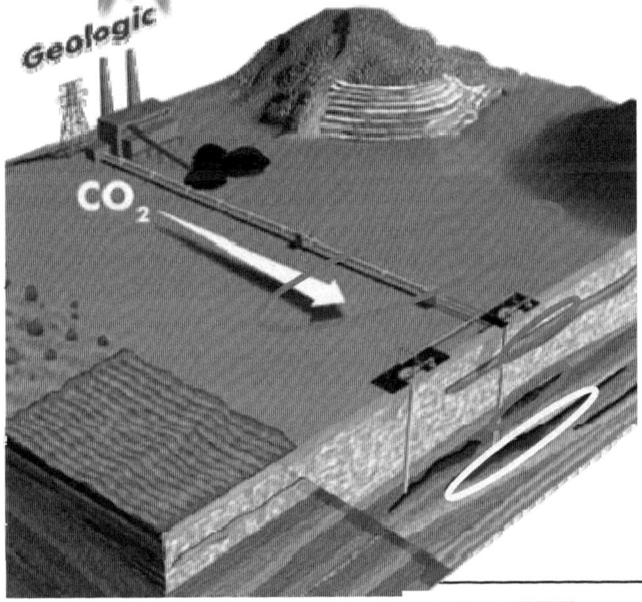

Numerical Estimates for CO_2 Leakage through Geological Seals Considered for Saline Reservoir Sequestration of Carbon Dioxide

AIChE

November 8-13th, 2009

Nashville, TN

Craig Griffith

Greg Lowry (CMU)

David Dzombak (CMU)

Yee Soong (NETL)

Grant Bromhal (NETL)

Carnegie Mellon ENGINEERING

Outline

I. Introduction

II. Saline CO_2 sequestration Field Test Validation overview

III. Study objectives

IV. Data collection methods

V. Results

VI. Implications and summary

NATIONAL ENERGY TECHNOLOGY LABORATORY

DOE proposed field test validation sites for CO_2 sequestration

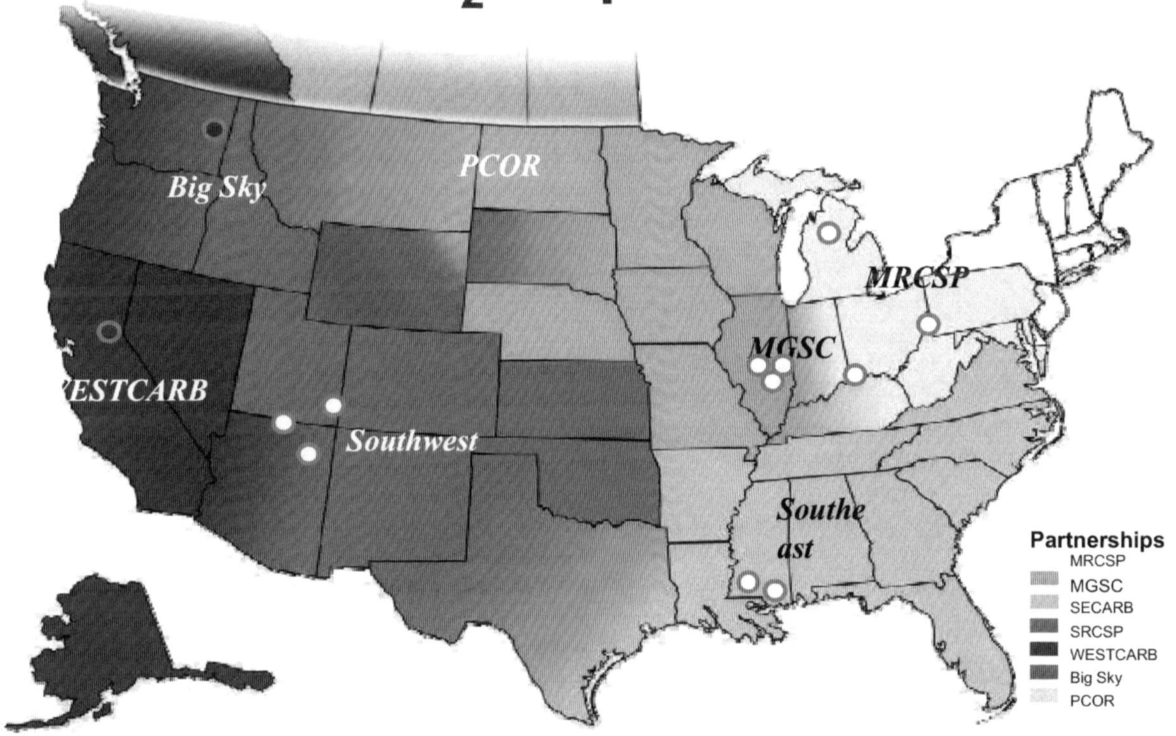

Goal: 99% storage permanence over 100 and 1,000 years

NATIONAL ENERGY TECHNOLOGY LABORATORY

Main objectives considered with CO_2 Risk Assessment

- **Need a robust framework and methodology to uniformly assess CO_2 sequestration sites**
  - Accommodate breadth of sites and adapt to site-specific details for potential storage reservoirs
  - Address complexity, heterogeneity and uncertainty in natural systems
  - Relate risk assessment to fundamental physical and chemical phenomena

(Reference: National Plan for Assessing the Risks of Engineered Geological Storage of CO_2)

NATIONAL ENERGY TECHNOLOGY LABORATORY

CO_2-PENS is a modular, probabilistic, system model (& framework) for assessing potential risks at a CO_2 storage site.

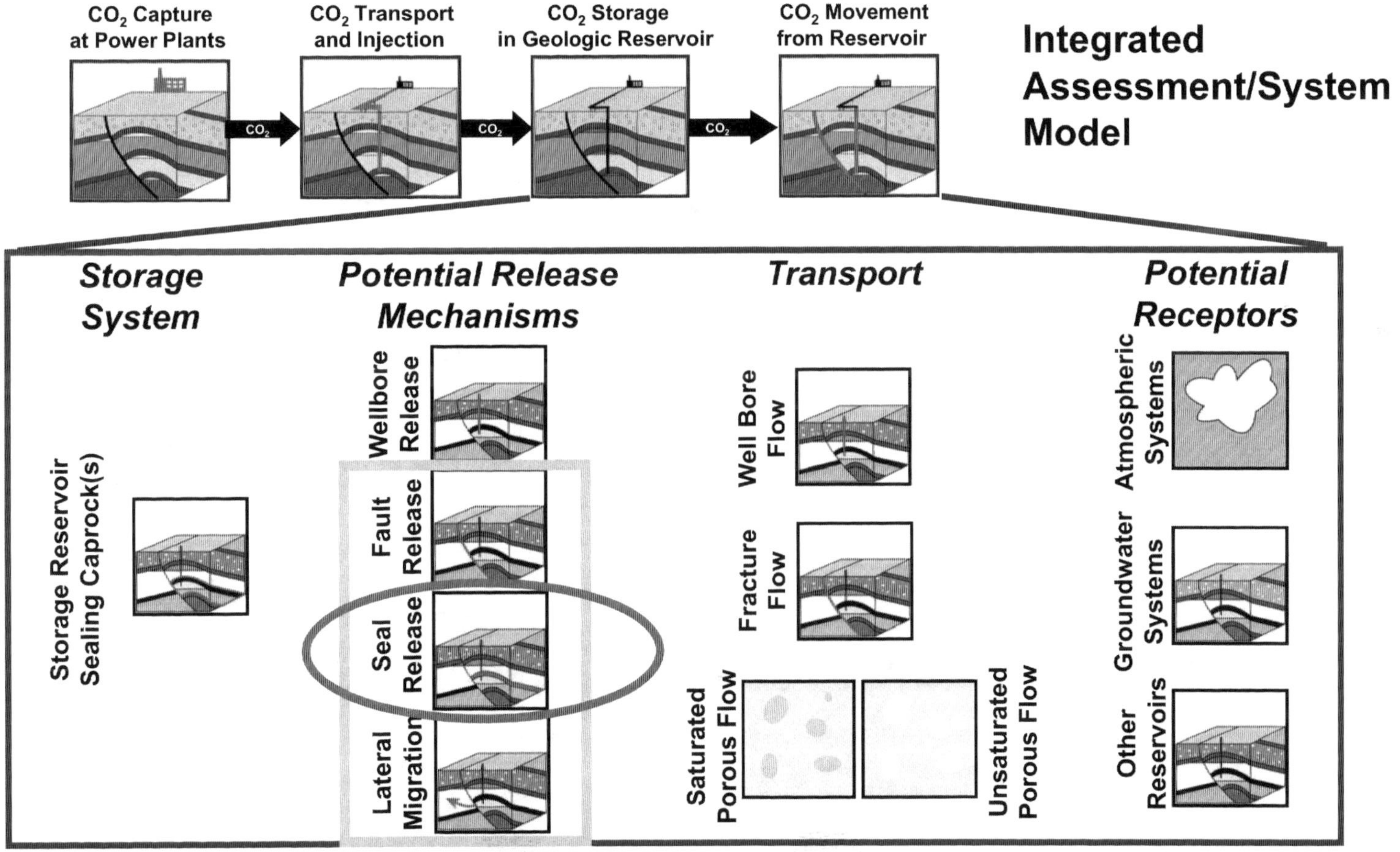

NATIONAL ENERGY TECHNOLOGY LABORATORY

How effective are storage formation seals?

- **What are potential fluid release pathways through formation seals?**
- **What physical and chemical characteristics affect release pathways under CO_2 sequestration conditions?**
- **What are the geochemical reactions that may affect seal rock integrity?**

NATIONAL ENERGY TECHNOLOGY LABORATORY

Study Objectives

- To survey demonstration sites considered for saline sequestration and identify physical and mineral characteristics relevant in assessing seal integrity and potential risk for leakage
- To develop conceptual models for CO_2 flow through idealized seal types that identify the most sensitive features and depict the kind of resolution necessary to sufficiently describe the extent of possible geochemical reactions and structural features on affecting seal integrity

Data Collection Methods

•Sites:

- Coconno and Apache Co., Northern Arizona
- Jackson and Adams Co. Mississippi
- Otsego Co., Michigan
- San Juan Co., Utah
- Belmont Co., Ohio
- Fayette, Macon, and Coles Co., Illinois
- Boone Co. Kentucky

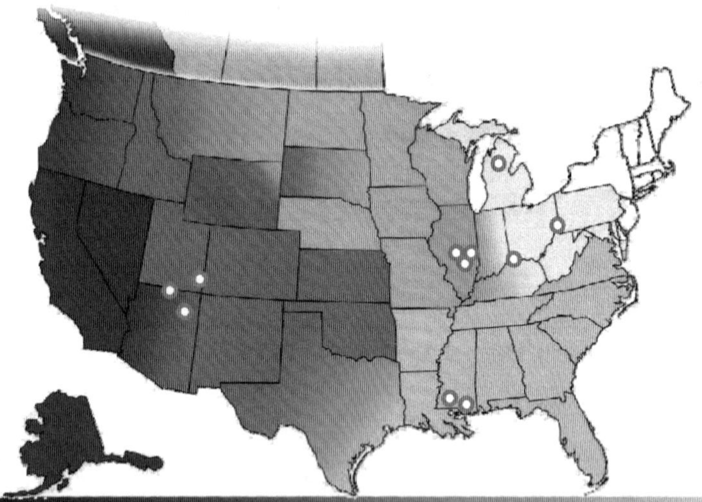

•Sources for regional mineral and physical characteristics:

- Open file reports
- Peer reviewed journals
- Thesis dissertations
- Regional partnership reports

•Limitations:

- Data collected on regional scale
- Few data available on sub-surface mineral and physical of seals
- Abundance of each mineral in seals not available
- Used equivalent outcrop units in surrounding region for mineralogy when subsurface data unavailable

NATIONAL ENERGY TECHNOLOGY LABORATORY

Data Collection Methods

- **Collection done in two parts:**
  - Mineralogical survey of 16 candidate seal formations within the surrounding regions of the 11 potential demonstration sites
  - Focus examination on 6 seal formations within the surrounding regions of 6 potential sites to study the structural and transport features of the candidate geological seals

Survey of mineral occurrence in seals across eleven potential sites

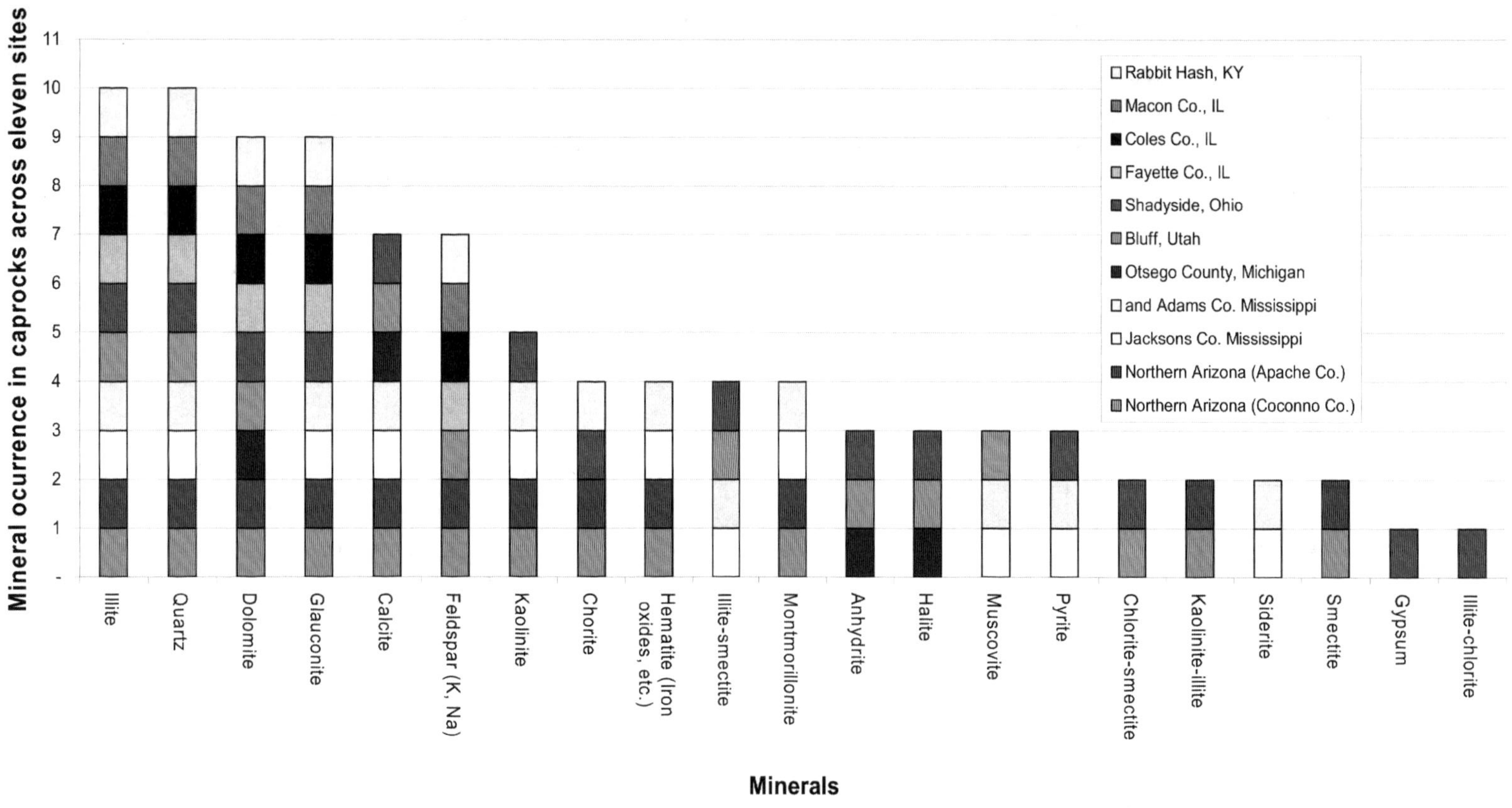

NATIONAL ENERGY TECHNOLOGY LABORATORY

Focus - Paradox Basin: San Juan Co., Utah

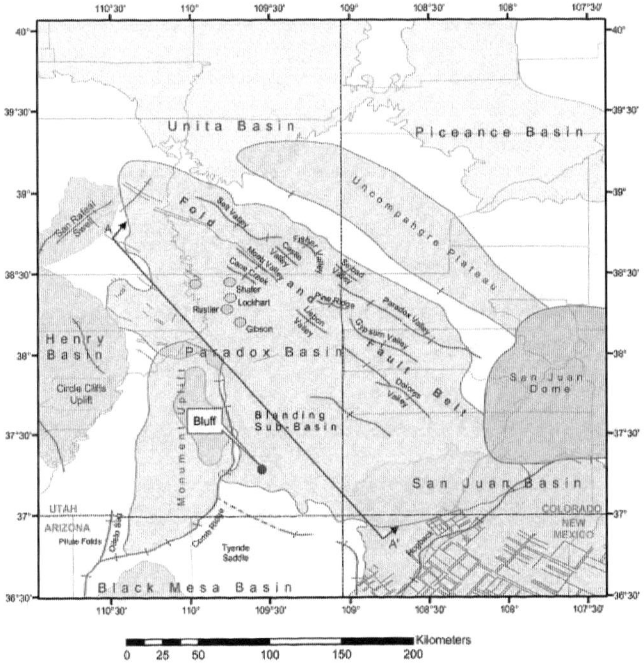

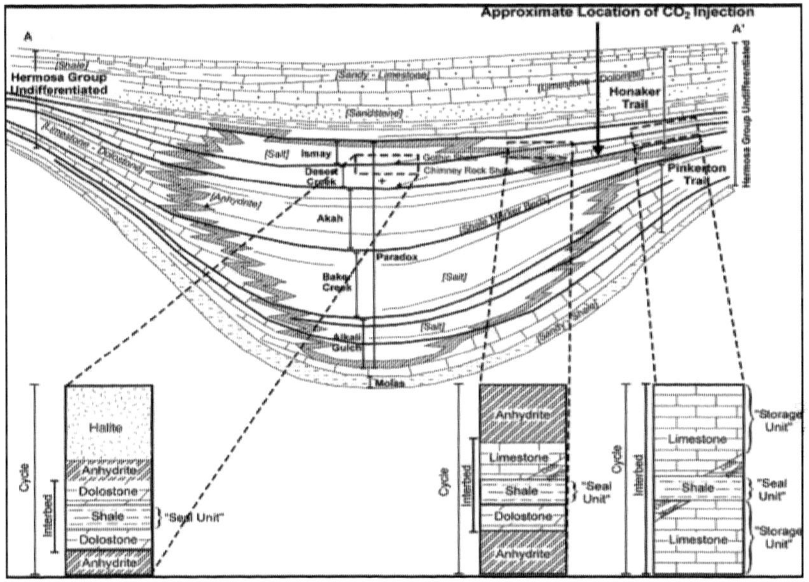

Characteristic	Gothic & Chimney Rock Shale
Physical Properties	
Depth (m)	1,707 – 1,768[1, 2]
Thickness (m)	61[9] 9 – 12[9]
Porosity (%)	10[15] 3 – 4[16]
Vertical Hydraulic Conductivity (10^{-3} cm/s)	0.1[15]‡
Horizontal Hydraulic Conductivity (10^{-3} cm/s)	0.1[15]‡ 5.7 – 7.0 x 10^{-5} [(16)]
Basin areal extent (km^2)	28,490[25]

NATIONAL

Focus - Gulf Coast: Jackson and Adams Co., Mississippi

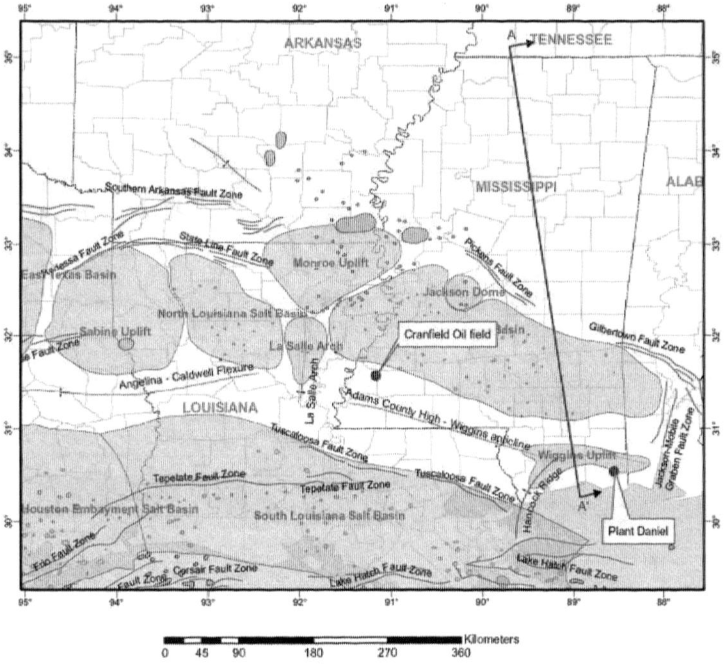

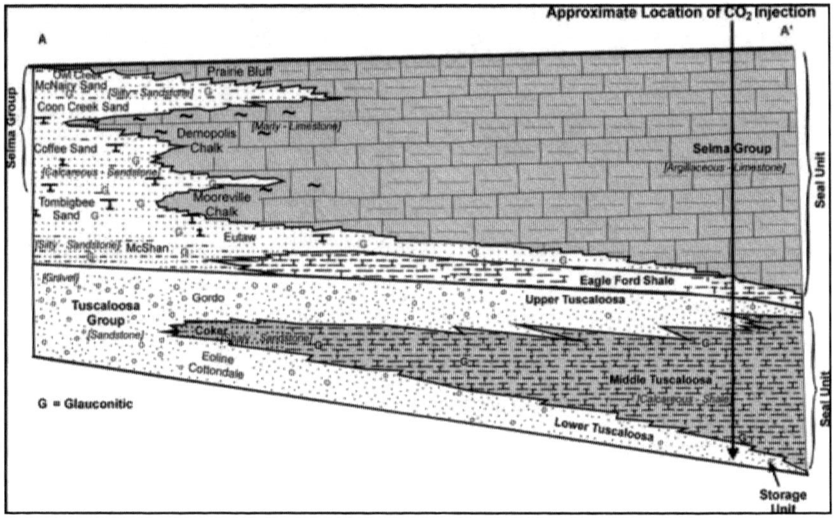

Characteristic	Marine Tuscaloosa Shale	Selma Group
Physical Properties		
Depth (m)	2,339 – 2,484[3a] 3,005 – 3124[4b] 3,078 – 3,120[5b]	1,676 – 2,377[6]
Thickness (m)	137 – 145[3] 114[4] 41[5]	305[6] 107 – 251[10, 11]
Porosity (%)	2.3 – 8.0[5]	9 – 45[17] 31 – 35[18]
Vertical Hydraulic Conductivity (10^{-3} cm/s)	4×10^{-6} – 4×10^{-7} [(20)]	4×10^{-6} – 4×10^{-7} [(20)]
Horizontal Hydraulic Conductivity (10^{-3} cm/s)	0.01 – 0.06[5] 3.5 – 27[21] 17 – 35.3[22] 4.9 - 29[23]	<0.05 - 10[17] 0.9 – 0.13[18] 3.5 – 27[21] 0.4 – 13[23]
Basin areal extent (km²)	119,139[6, 26]	

NATIO

Focus – Illinois Basin: Fayette, Macon, and Coles Co., Illinois

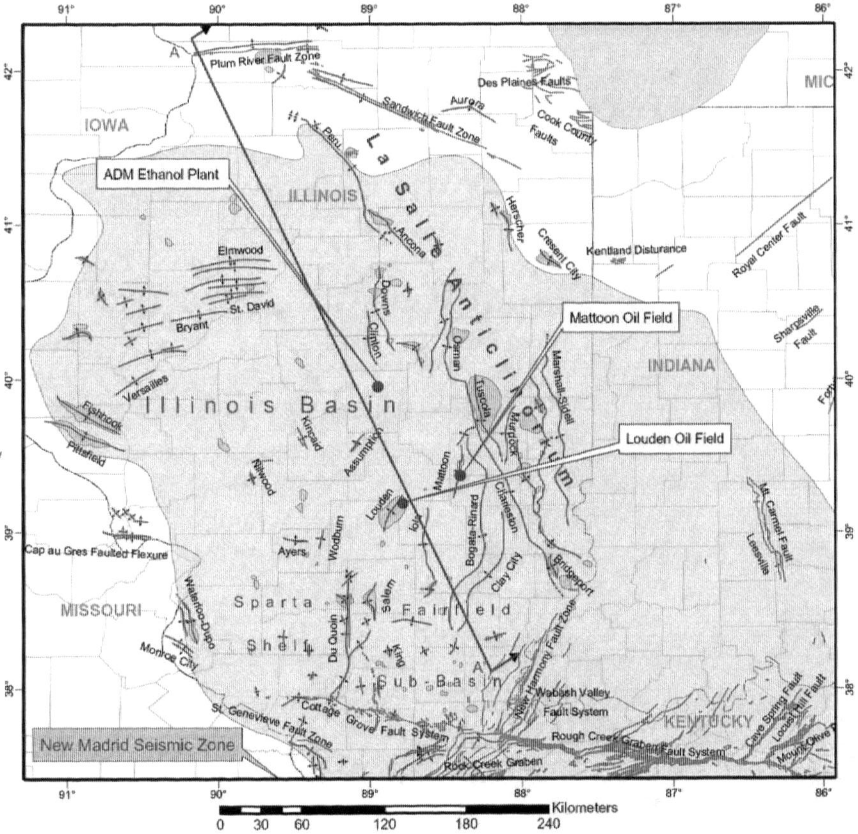

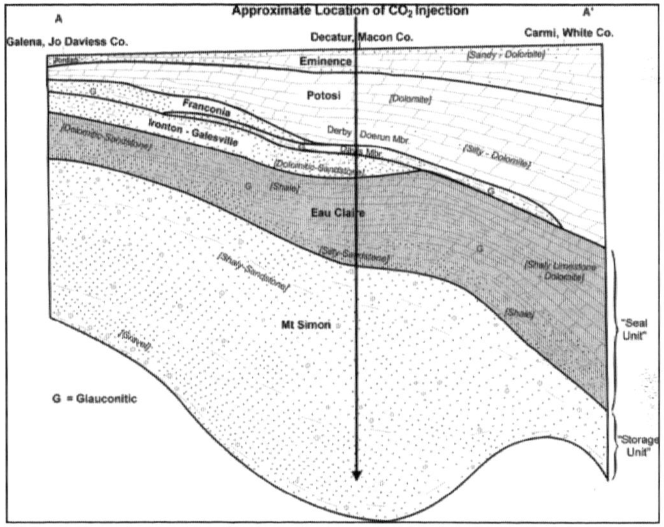

Characteristic	Mt. Simon Shale Interbeds	Eau Claire
Physical Properties		
Depth (m)	2,134 – 2,286[7, 8]	1,982 –1,753[7, 8]**
Thickness (m)	< 152[12]	91 – 152[13] ; 152 – 213[13, 14]
Porosity (%)	-	0.4 – 15.4 7.7[19]†
Vertical Hydraulic Conductivity (10^{-3} cm/s)	-	$1 \times 10^{-6} – 0.485$ 1.14×10^{-3} (19)†
Horizontal Hydraulic Conductivity (10^{-3} cm/s)	0.46 – 0.53[12] 0.53 – 1.3[24]	0.46 – 0.53[12] 0.53 – 1.3[24]
Basin areal extent (km²)	155,000[27]	

NATION Y

Seal rock lithology and physical features

- **Common lithological types of seal:**
  - Shale
  - Sandstone
  - Siltstone
  - Carbonate rocks
    - Limestone
    - Dolomite
- **Common lithologic features:**
  - Lateral facies changes
  - Variable thickness (30 - 1,155 ft)
- **Faults**
  - Near test sites that penetrate seals
- **Fractures**
  - Vein filled
    - Calcite
    - Dolomite
    - Anhydrite
    - Pyrite
  - Possibly open
  - Interconnected
  - Mircrofractures
- **Structural intrusions**
  - Domes and piercements

NATIONAL ENERGY TECHNOLOGY LABORATORY

Summary illustration of lithologic and physical characteristics of seals within site regions

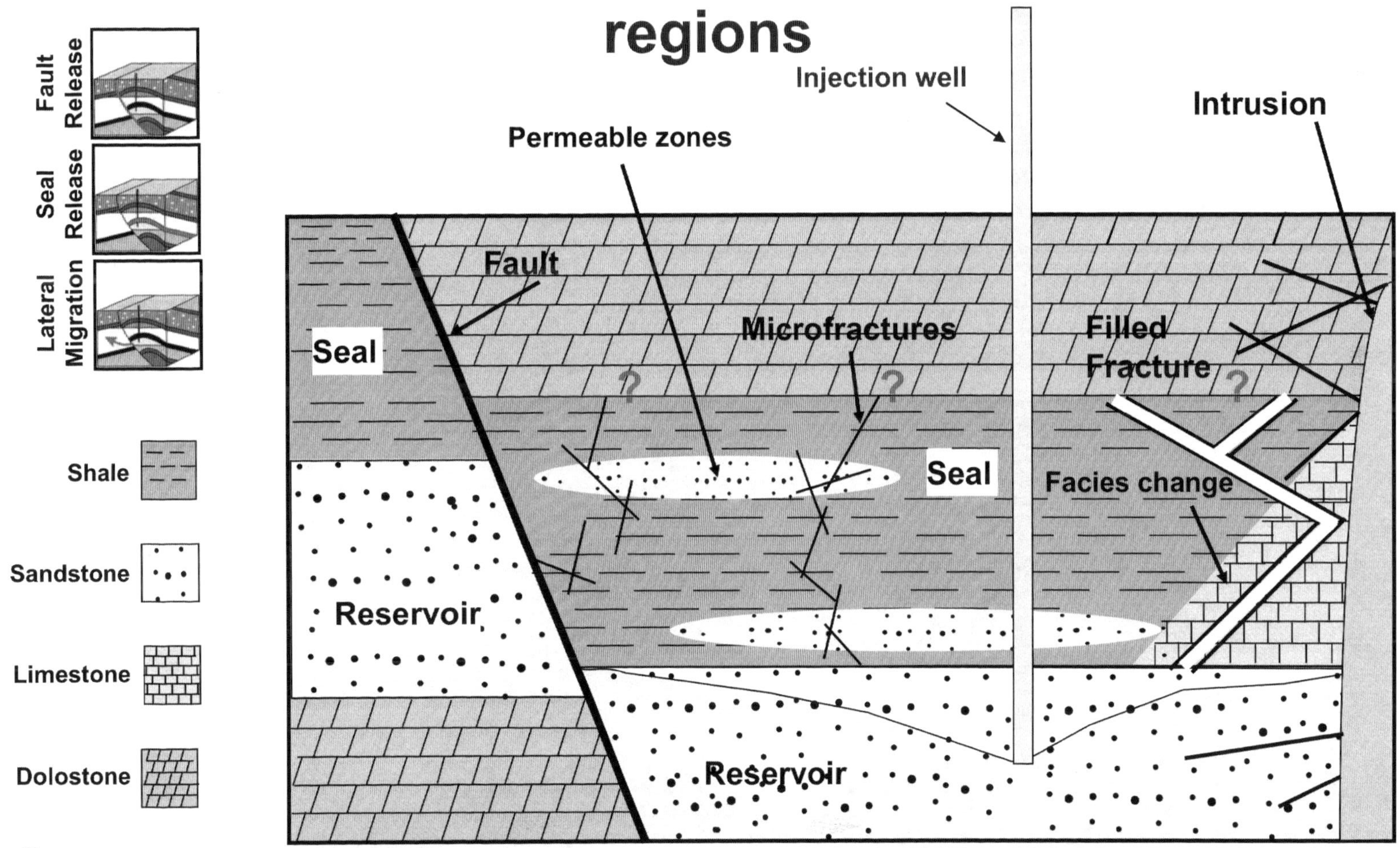

NATIONAL ENERGY TECHNOLOGY LABORATORY

Conceptual modeling of 1-D Vertical Darcy flow: Rock matrix

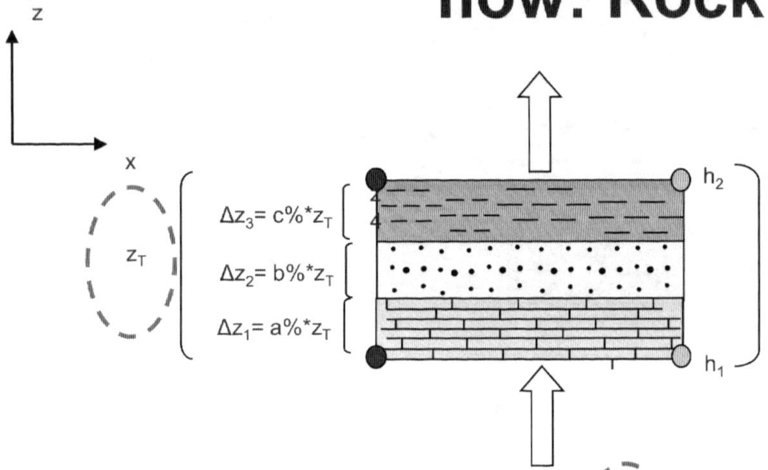

$$q = \frac{Q}{A} = K_z \frac{\Delta h_T}{\Delta z_T}$$

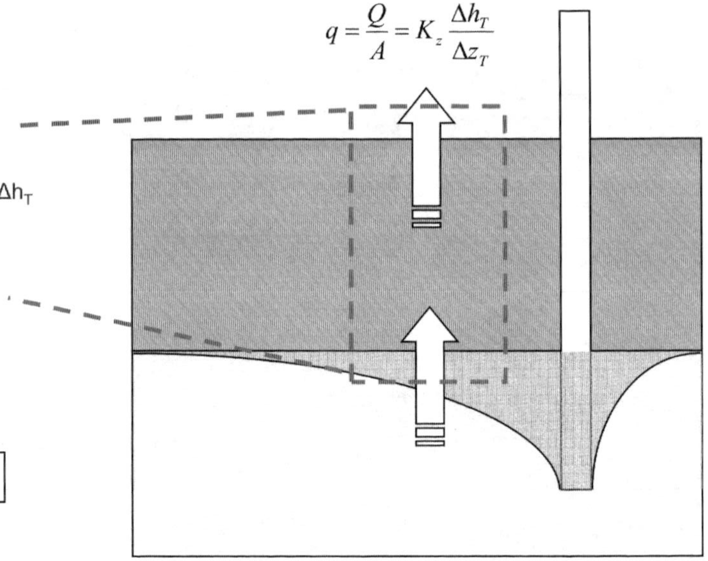

$$\Delta h_T = h_2 - h_1$$

$$h_i = z_i + \frac{p_i - p_o}{\rho_l g}$$

h_1 = 90%*Fracture pressure
h_2 = Hydrostatic pressure

$$K_{z,T} = \frac{v_z L}{\Delta h_T} = \frac{v_z z_T}{v_z \Delta z_1 / K_1 + v_z \Delta z_2 / K_2 + v_z \Delta z_3 / K_3}$$

$$K_{z,T} = \frac{z_T}{\sum_i^n \Delta z_i / K_i}$$

$$K_i = k\left(\frac{\rho_l g}{\mu}\right)$$

NATIONAL ENERGY TECHNOLOGY LABORATORY

Conceptual modeling of 1-D Vertical Darcy flow: Rock fracture

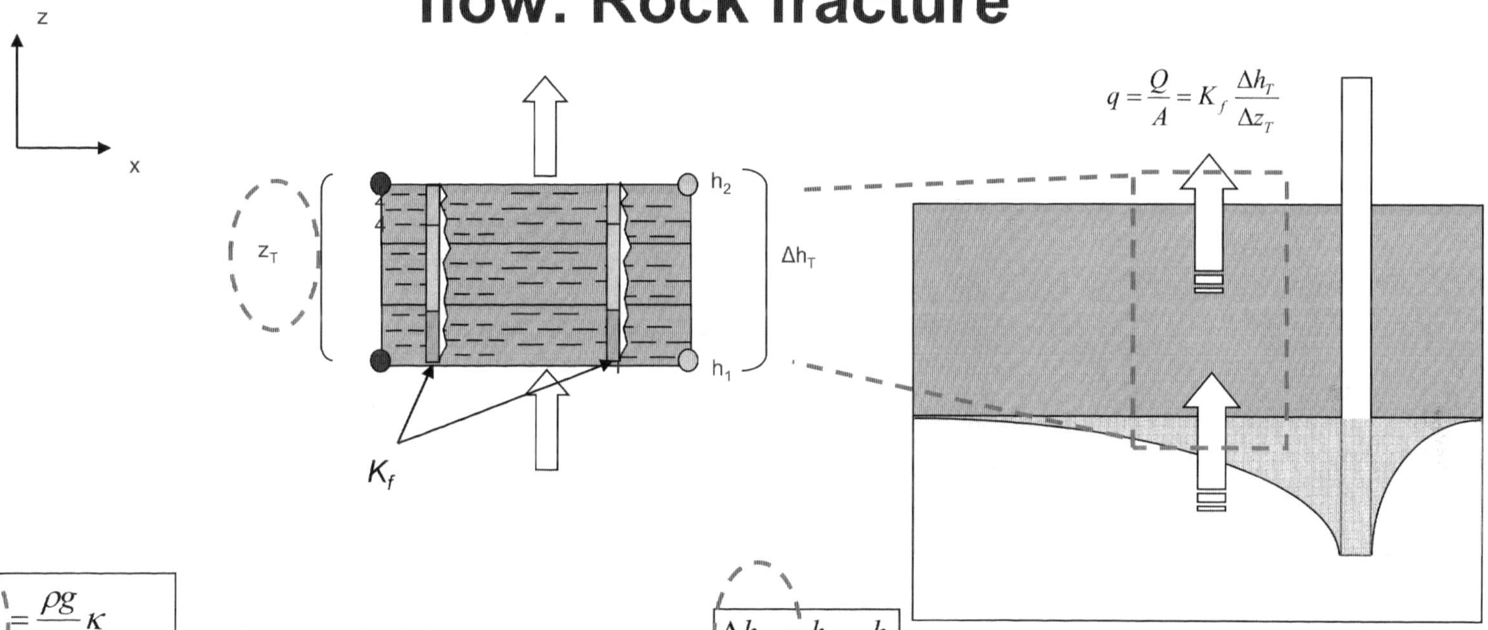

$$q = \frac{Q}{A} = K_f \frac{\Delta h_T}{\Delta z_T}$$

$$K_f = \frac{\rho g}{\mu} \kappa$$
$$\kappa = \frac{\phi_f b^2}{12} = \frac{n b^3}{12}$$
$$\phi_f = nb$$

If we split initial aperture into a sum of small ones

$$b = \frac{b_o}{n}$$
$$\kappa = \frac{b_o^3}{n^2 12}$$
$$\phi_f \propto b_0$$

$$\Delta h_T = h_2 - h_1$$

$$h_i = z_i + \frac{p_i - p_o}{\rho_l g}$$

h_1 = 90%*Fracture pressure
h_2 = Hydrostatic pressure

NATIONAL ENERGY TECHNOLOGY LABORATORY

Conceptual modeling of 1-D Vertical Darcy flow: Rock fracture

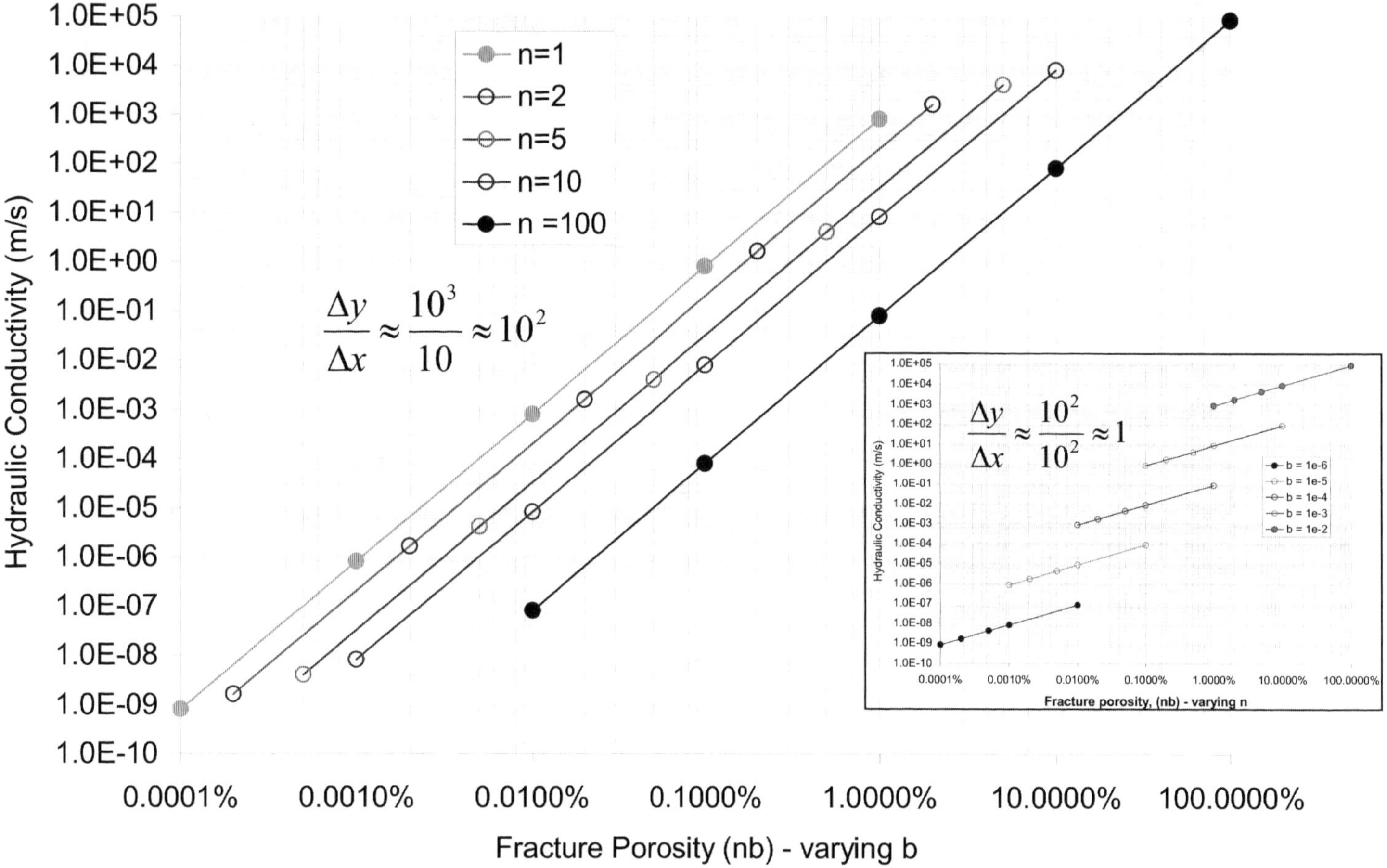

NATIONAL ENERGY TECHNOLOGY LABORATORY

Conceptual modeling of 1-D Vertical Darcy flow: Rock fracture

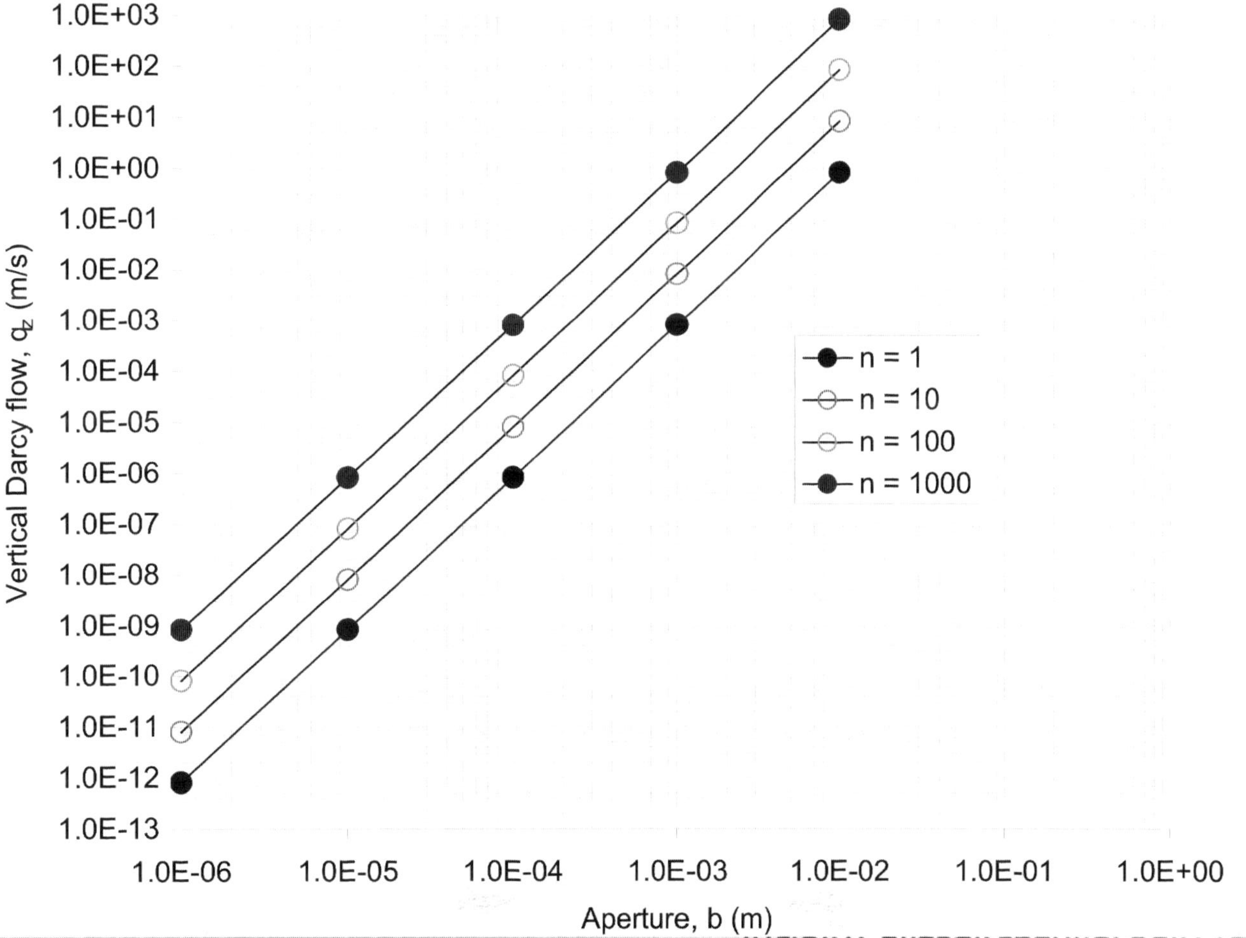

NATIONAL ENERGY TECHNOLOGY LABORATORY

Conceptual modeling of 1-D Vertical Darcy flow: Rock fracture

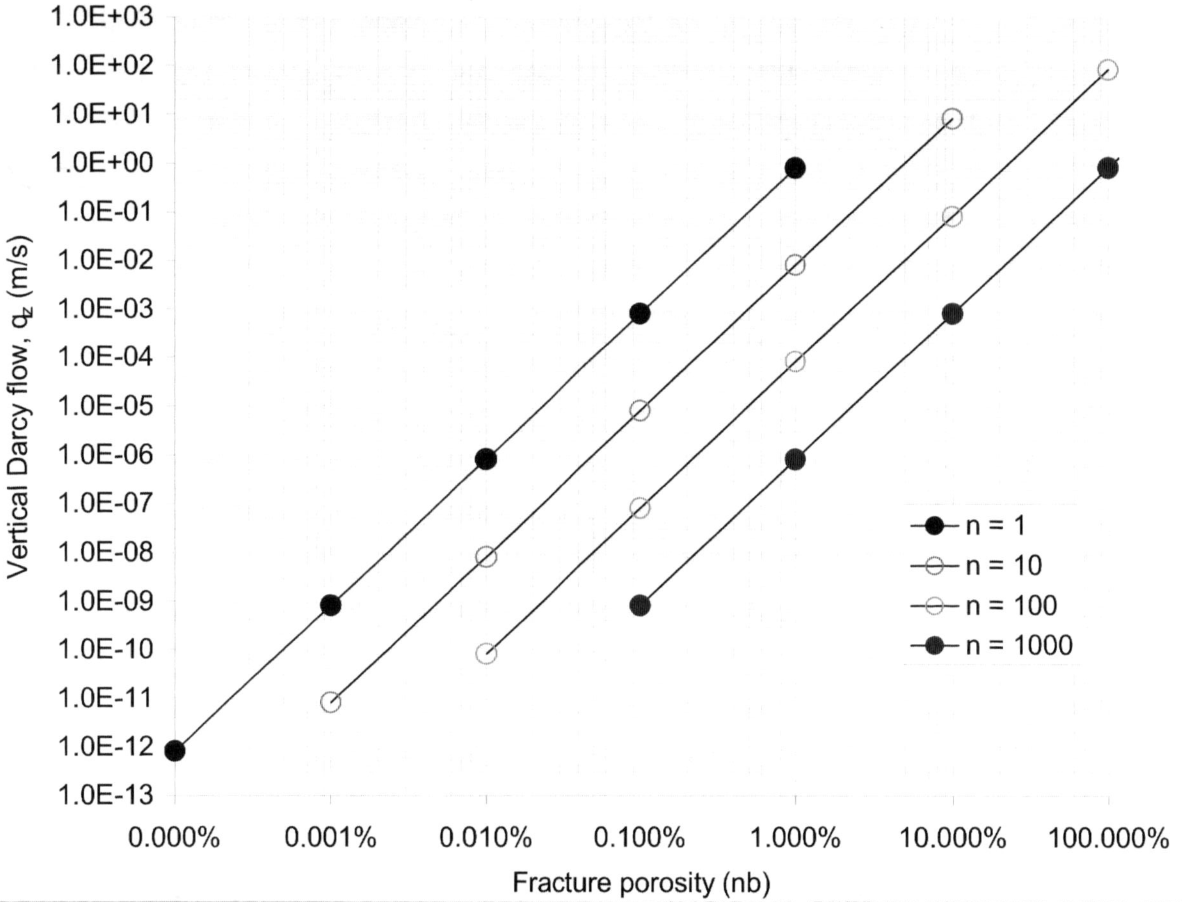

NATIONAL ENERGY TECHNOLOGY LABORATORY

Summary

- **Available data indicates the potential for seal leakage**
  - Common soluble minerals (in structure and fractures)
  - Abundant, multiple fractures
  - Regional faults
  - Regional intrusions through reservoirs and seals
  - Variable thickness
- **Very little data of type needed to asses leakage rates from seals is currently available**
  - e.g. permeability changes
- **Fracture permeability and porosity plays an important role in storage**
  - less than 1% change in porosity, due to geochemical reactions or other forces, can result in large changes in flow

Acknowledgements

Funding

Department of Energy, National Energy Technology Laboratory (NETL)

Collaboration

Carnegie Mellon University

Advanced Resources International, Inc.

DOE sponsored Regional Partnerships

NATIONAL ENERGY TECHNOLOGY LABORATORY

Supplemental slide: Conceptual modeling of 1-D Vertical Darcy flow: Rough rock fracture

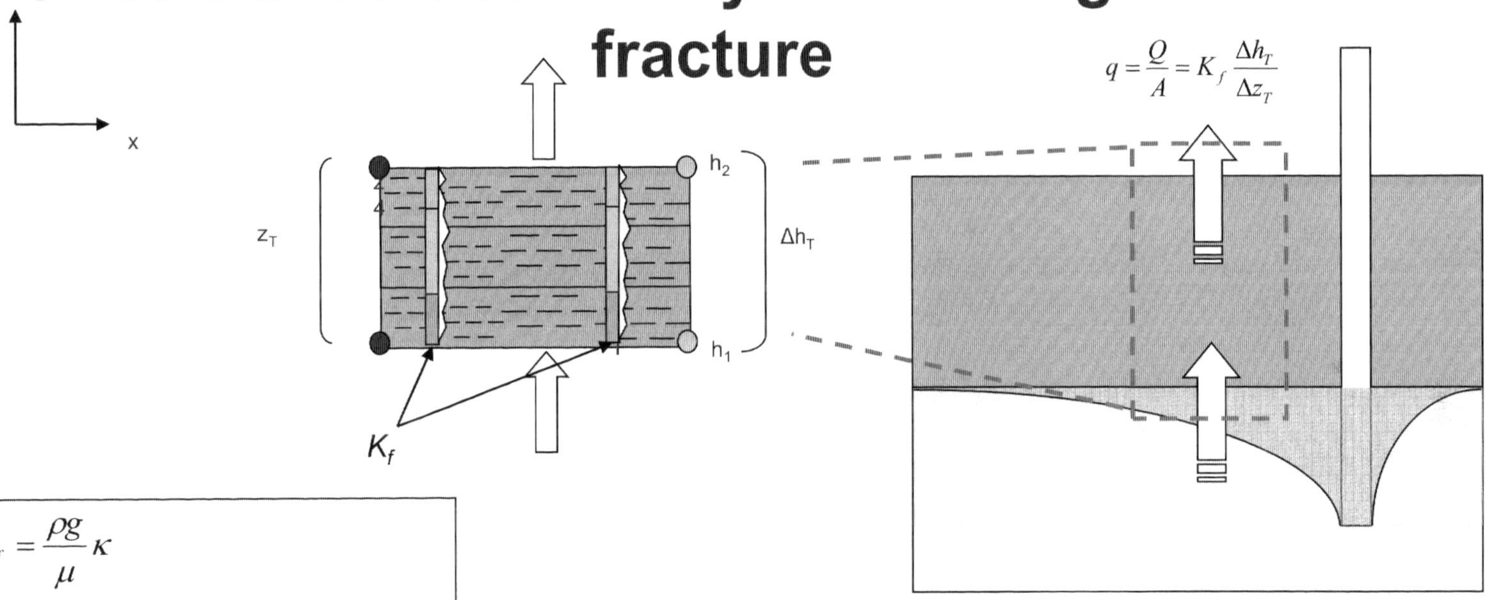

$$K_f = \frac{\rho g}{\mu}\kappa$$

$$\kappa = \frac{\phi_f b^2}{12} = \frac{n(b=e)^3}{12} = \frac{n\left[\frac{JRC^{2.5}}{E/e}\right]^3}{12}$$

$$\phi_f = nb$$

$$e = \frac{JRC^{2.5}}{E/e}$$

E = Physical aperture
e = Hydraulic aperture

Reduction of Wax & Hydrates in Deepwater Oil Pipelines Via Cold Flow

Jules J. Magda, Chemical Engineering, University of Utah, Salt Lake City, UT

Milind D. Deo, Chemical Engineering, University of Utah, Salt Lake City, UT

Kyeongseok Oh, University of Utah, Salt Lake city, UT

Mark Jemmett, Chemical Engineering, University of Utah, Salt Lake City, UT

Rich Roehner, Chemical Engineering, Roehner Consulting Services, Salt Lake City, UT

Husam El Gendy, Chemical Engineering, University of Utah, Salt Lake City, UT

When a waxy oil/water emulsion exits a wellhead and enters a deep undersea pipeline in the Gulf of Mexico, formation of methane hydrates and precipitated wax is likely due to the high pressure and cold temperature of the surrounding sea water. If this should occur, then wax will deposit on the pipeline walls and increase corrosion, whereas hydrates have the potential to plug the pipeline and completely block the flow. In the past, these problems have been avoided by insulating the pipelines, but this approach is becoming increasingly uneconomical as the need increases for longer pipelines. In this talk, we review a new strategy called "cold flow" in which the oil exiting the wellhead is transformed into a cold slurry with controlled particle morphology before entering the pipeline.

See more of this Session: Oil Characterization and Thermodynamics as Related to Flow Assurance
See more of this Group/Topical: Energy and Transport Processes

Effects of Precipitant Concentration On Asphaltene Deposition

Michael P. Hoepfner, Chemical Engineering, University of Michigan, Ann Arbor, MI

Tabish Maqbool, Chemical Engineering, The University of Michigan, Ann Arbor, MI

H. Scott Fogler, The University of Michigan, Ann Arbor, MI

Asphaltene deposition is a costly concern in flow assurance. Precipitation and deposition can occur in the porous rock formations, well-bore, production lines and in the refinery. Capillary deposition has been considered a useful tool to measure the conditions under which deposition occurs and the extent of deposition. Previous research has studied the effect of different precipitants, deposition near onset and past onset conditions, and the effect of shear rate [1,2]. Recently, it has been shown that a critical onset point is not sufficient and there are kinetic effects associated with the precipitation and aggregation of asphaltenes [3]. Both the time required to detect precipitated asphaltenes and the quantity of asphaltenes precipitated is a strong function of the precipitant concentration. Previous deposition work has not considered the effects of precipitation kinetics.

This investigation examines the effect of precipitant concentration, as a parameter to control the precipitation kinetics, on the deposition characteristics of asphaltenes in crude oil. Key measured deposition characteristics include rate of deposition, thickness of deposit and deposit location. In addition to varying the volume fraction of precipitant, temperature effects and types of precipitant are also considered. Previous work has considered the formed deposit to be uniform distributed over the entire length of the capillary; the validity of this approach is also under investigation. The primary goal is to develop a fundamental understanding of the deposition process and how aggregation effects can be incorporated into a model of the capillary deposition process.

[1] J. Wang, J.S. Buckley, and J.L. Creek, "Asphaltene Deposition on Metallic Surfaces," Journal of Dispersion Science and Technology, vol. 25, 2004, pp. 1-12.

[2] L. Nabzar and M.E. Aguilera, "The Colloidal Approach. A Promising Route for Asphaltene Deposition Modelling," Oil and Gas Science and Technology Revue De Linstitut Francias Du Petrole, vol. 63, 2008, pp. 21-35.

[3] T. Maqbool, et al, "Kinetics of Asphaltene Precipitation from Crude Oils," The 9th International Conference on Petroleum Phase Behavior and Fouling, Victoria, Canada: 2008.

See more of this Session: Oil Characterization and Thermodynamics as Related to Flow Assurance
See more of this Group/Topical: Energy and Transport Processes

Revisiting Asphaltene Precipitation From Crude Oils: A Case of Neglected Kinetic Effects

Tabish Maqbool, Chemical Engineering, The University of Michigan, Ann Arbor, MI

Arjames T. Balgoa, Chemical Engineering, The University of Michigan, Ann Arbor, MI

H. Scott Fogler, The University of Michigan, Ann Arbor, MI

The precipitation of asphaltenes from crude oils can lead to serious challenges during oil production and processing. This research investigates the kinetics of asphaltene precipitation from crude oils using n-alkane precipitants. For several decades, it has been understood that the precipitation of asphaltenes is a solubility driven phenomenon and the previous studies on the effect of time are usually limited to short time scales. By using optical microscopy and centrifugation based separation, we have demonstrated that the time required to precipitate asphaltenes can actually vary from a few minutes to several months, depending on the precipitant concentration used. Our results demonstrate that no single concentration can be identified as the critical precipitant concentration for asphaltene precipitation. Based on long term experiments, we have also been able to establish the solubility of asphaltenes as a function of the precipitant concentration and it is shown that the short-term experiments over-predict the solubility. Similarities between the current work and other research areas are also discussed briefly. This research opens up a new paradigm for understanding asphaltene precipitation.

See more of this Session: Oil Characterization and Thermodynamics as Related to Flow Assurance
See more of this Group/Topical: Energy and Transport Processes

Theory and Practice of Tie-Simplex Parameterization for Compositional Reservoir Simulation

Alireza Iranshahr
Denis Voskov
Hamdi Tchelepi

Introduction

Phase equilibrium calculations can be the most expensive kernel in a compositional flow simulation. Given the overall composition, pressure and temperature, the number of phases and their compositions must be determined for every control volume (gridblock) at each time step. The conventional approach is based on (1) performing a phase-stability test for single-phase gridblocks, and (2) solving the fugacity constraints together with the coupled nonlinear flow equations for gridblocks that have more than one phase [1].

The compositional space can be parameterized in terms of tie-simplexes [2, 3]. For example, a tie-triangle can be used, whereby its interior encloses the three-phase region and each of the three edges represents the various two-phase regions. The tie-simplex parameterization can be performed for different pressures, temperatures, and overall composition. For a given composition, parameterization in terms of pressure and temperature can be achieved using a multi-phase negative flash procedure. In order to correctly identify the phase-state for a composition that lies outside a higher dimensional tie-simplex, the parameterization can be performed for all lower dimensional phase states (e.g., tie-lines around a tie-triangle) at a fixed pressure and temperature.

Continuity of the tie-simplexes as a function of pressure, temperature, and overall composition is a necessary condition for the validity of the parameterization. The challenge is that all of these parameters can vary considerably during the course of a flow simulation. Here, we prove that the tie-simplexes change continuously with respect to pressure, temperature, and overall composition. We also show that by changing any of these three parameters, the computed tie-simplex through a given composition can degenerate to a lower dimensional tie-simplex, which also varies continuously in terms of the three parameters. The continuity of the tie-simplex parameterization allows for interpolation using discrete representations of the tie-simplex space as a function of pressure, temperature and composition. Here, we interpolate in tie-simplex space as a function of pressure and temperature using tables. For variations of composition, we use projection to the nearest tie-simplex, and if the tie-simplex is within a predefined tolerance, it is used directly to identify the phase-state of the composition.

Theory of dispersion-free compositional displacements, as well as computational experience of general-purpose (finite-volume, multi-dimensional) compositional flow simulation indicates that the solution path in compositional space is determined by a limited number of tie-simplexes [4]. Therefore, only a small number of tie-simplex tables is required to parameterize the entire displacement. The small number of tie-simplexes needed in the course of a simulation run motivates an adaptive tabulation procedure. Since a single tie-simplex identifies the phase-state of a large number of compositions in its vicinity, the efficiency gains of adaptive construction are considerable [2, 3]. This Compositional Space Adaptive Tabulation (CSAT) methodology has been extended recently to thermal-compositional [5] and multi-phase [6] displacements.

We compare our adaptive tie-simplex parameterization method with conventional EoS (Equation-of-State) procedures for two- and three-phase displacements in homogeneous and heterogeneous reservoirs. The results indicate clearly that the new method is significantly more efficient than conventional EoS methods for flow simulation.

Continuity of Compositional Space Parameterization

Here, we show that tie-simplex compositions change continuously with pressure, temperature, or along a continuous trajectory in the compositional space. Let us first consider a fixed composition in two-phase compositional space and study how the tie-lines through it vary with pressure (temperature is fixed).

We start by differentiating the equality of chemical potentials for each component, which defines the thermodynamic equilibrium for the tie-line:

$$\mu_{i,1} = \mu_{i,2}, \qquad 1 \le i \le N_c, \tag{1}$$

where, $\mu_{i,j}$ shows the chemical potential of component i in phase j. We note that the chemical potential of a component in a phase is a function of pressure, temperature, and the molar amount of the components present in that phase. For a given overall composition, changing the pressure (at fixed temperature) results in a redistribution of the components between the two phases. Consequently, the chemical potential of a component changes as a function of pressure and component molar amounts. Partial differentiation of Eq. (1) with respect to pressure yields:

$$\begin{aligned}&\sum_j \left(\frac{\partial \mu_{i,1}}{\partial n_{j,1}}\right)_{p,T,n_{k\neq j,1}} \cdot \left(\frac{\partial n_{j,1}}{\partial p}\right)_{T,n_{k\neq j,1}} + \left(\frac{\partial \mu_{i,1}}{\partial p}\right)_{T,n_{k,1}} \\ -&\sum_j \left(\frac{\partial \mu_{i,2}}{\partial n_{j,2}}\right)_{p,T,n_{k\neq j,2}} \cdot \left(\frac{\partial n_{j,2}}{\partial p}\right)_{T,n_{k\neq j,2}} - \left(\frac{\partial \mu_{i,2}}{\partial p}\right)_{T,n_{k,2}} = 0, \qquad 1 \le i \le N_c (2)\end{aligned}$$

The total amount of a component is given by:

$$n_{i,tot} = n_{i,1} + n_{i,2}, \qquad 1 \le i \le N_c, \tag{3}$$

where, we have considered partitioning of all the components in both phases. Therefore, for an overall composition in a single-phase state, one of the two molar amounts is negative for each component. Considering that the overall composition is fixed, we differentiate Eq. (3) to obtain:

$$\left(\frac{\partial n_{i,1}}{\partial p}\right)_{T,n_{k\neq i,1}} + \left(\frac{\partial n_{i,2}}{\partial p}\right)_{T,n_{k\neq i,2}} = 0 \tag{4}$$

which, when combined with Eq. (2), yields:

$$\begin{aligned}\sum_j \left[\left(\frac{\partial \mu_{i,1}}{\partial n_{j,1}}\right)_{p,T,n_{k\neq j,1}} + \left(\frac{\partial \mu_{i,2}}{\partial n_{j,2}}\right)_{p,T,n_{k\neq j,2}}\right] \cdot \left(\frac{\partial n_{j,1}}{\partial p}\right)_{T,n_{k\neq j,1}} \\ = \left(\frac{\partial \mu_{i,2}}{\partial p}\right)_{T,n_{k,2}} - \left(\frac{\partial \mu_{i,1}}{\partial p}\right)_{T,n_{k,1}}, \qquad 1 \le i \le N_c.\end{aligned} \tag{5}$$

Using

$$J^m_{i,j} = \left(\frac{\partial \mu_{i,m}}{\partial n_{j,m}}\right)_{p,T,n_{k\neq j,m}}, \tag{6}$$

Eq. (5) simplifies to:

$$\sum_j \left[J^1_{i,j} + J^2_{i,j}\right] \cdot \left(\frac{\partial n_{j,1}}{\partial p}\right)_{T, n_{k\neq j,1}} = \bar{V}_{i,2} - \bar{V}_{i,1}, \qquad 1 \le i \le N_c, \tag{7}$$

where, $\bar{V}_{i,m}$ denotes the partial molar volume of component i in phase m. Similarly, the equilibrium phase amounts of a given overall composition at fixed pressure change with temperature as follows:

$$\sum_j \left[J^1_{i,j} + J^2_{i,j}\right] \cdot \left(\frac{\partial n_{j,1}}{\partial T}\right)_{p, n_{k\neq j,1}} = \bar{S}_{i,1} - \bar{S}_{i,2}, \qquad 1 \le i \le N_c, \tag{8}$$

where the right hand side is the difference between the partial molar entropies of the i-th component. In order to prove the continuity of the compositional space parameterization, one needs to show that: (1) $J^1_{i,j} + J^2_{i,j}$ is invertible, and (2) $\bar{M}_{i,1} - \bar{M}_{i,2}$ has a finite norm, where M denotes volume or entropy. For this purpose, we note that the solution of the nonlinear set of equations for the negative flash problem, whose jacobian is $J^1_{i,j} + J^2_{i,j}$, is unique. Moreover, from

$$M = \sum_i n_i \bar{M}_i, \tag{9}$$

the partial molar property of each component has a finite value. Hence, tie-line compositions change continuously with pressure or temperature.

Next, we consider changes in the equilibrium phase behavior along a given trajectory in the compositional space. Differentiating Eq. (1) at fixed pressure and temperature yields

$$\sum_j \left(\frac{\partial \mu_{i,1}}{\partial n_{j,1}} . \frac{\partial n_{j,1}}{\partial \xi}\right)_{p,T,n_{k\neq j,1}} - \sum_j \left(\frac{\partial \mu_{i,2}}{\partial n_{j,2}} . \frac{\partial n_{j,2}}{\partial \xi}\right)_{p,T,n_{k\neq j,2}} = 0, \qquad 1 \le i \le N_c, \tag{10}$$

where ξ parameterizes the path in the compositional space. Along this path, we differentiate Eq. (3) to obtain:

$$\frac{\partial n_{i,1}}{\partial \xi} + \frac{\partial n_{i,2}}{\partial \xi} = \frac{\partial n_{i,tot}}{\partial \xi}, \qquad 1 \le i \le N_c. \tag{11}$$

Combining Eqs. (10) and (11) yields:

$$\sum_j [J^1_{i,j} + J^2_{i,j}] \left(\frac{\partial n_{j,1}}{\partial \xi}\right)_{p,T,n_{k\neq j,1}} = \sum_j J^2_{i,j} \left(\frac{\partial n_{j,tot}}{\partial \xi}\right)_{p,T,n_{k\neq j,1}}, \qquad 1 \le i \le N_c. \tag{12}$$

From the Gibbs-Duhem equation, we have:

$$\sum_j n_j J^m_{i,j} = 0, \qquad 1 \le i \le N_c, \tag{13}$$

and we conclude the continuity of the phase equilibrium behavior along ξ, as long as $\left(\frac{\partial n_{j,tot}}{\partial \xi}\right)_{p,T,n_{k\neq j,1}}$ has a finite norm (i.e., the path is continuous).

From numerical experiments, we can show that for a given overall composition at fixed temperature, there is a critical pressure value such that $J^1_{i,j} = J^2_{i,j}$ holds. Noting that $J^m_{i,j}$ is a singular matrix, the tie-lines degenerate to a point. Similar arguments can be made for changes in temperature, or along a trajectory in the compositional space. All of these derivations can be readily extended to three-phase systems, where tie-triangles degenerate to tie-lines. Figure (1), (2) and (3) illustrate the continuity and degeneration of two- and three-phase compositional space parameterization in terms of pressure and temperature for three- and four-component mixtures.

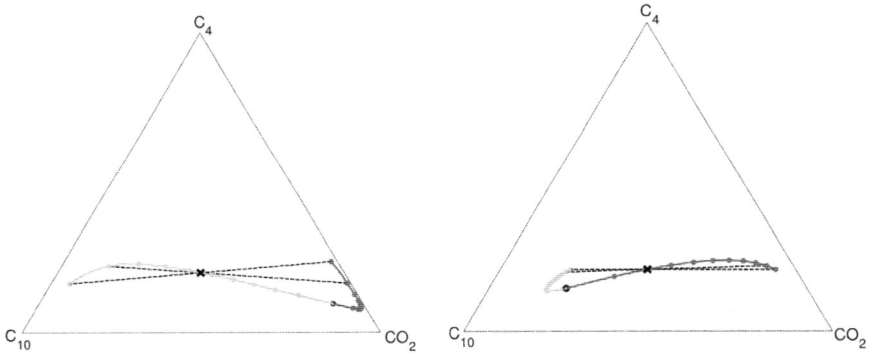

(a) $T = 375$ K, $p = (10, 145)$ *bar* (b) $p = 50$ *bar*, $T = (450, 597)$ K

Figure 1: Continuity of compositional space parameterization for three-component two-phase systems

CSAT for Three-Phase Systems

We now describe a generalization of the Compositional Space Adaptive Tabulation (CSAT) methodology to three-phase systems. We developed a bisection based negative-flash procedure to parameterize a composition in three-phase compositional space using a tie-triangle. We briefly illustrate different steps for the parameterization of three-phase systems.

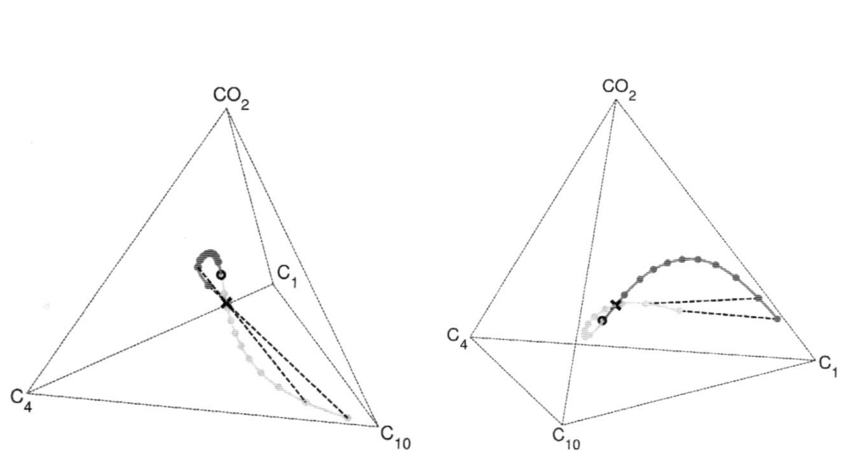

(a) $T = 450\,\mathrm{K}$, $p = (10, 175)\ bar$

(b) $p = 100\ bar$, $T = (250, 550)\,\mathrm{K}$

Figure 2: Continuity of compositional space parameterization for four-component two-phase systems

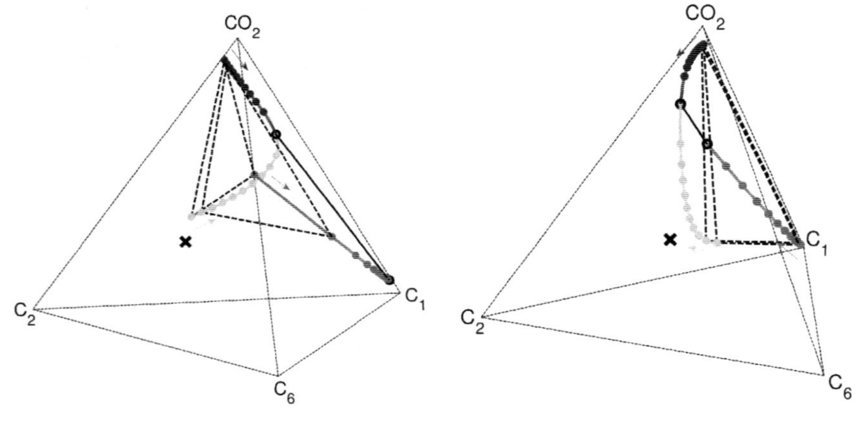

(a) $T = 180\,\mathrm{K}$, $p = (1, 18.5)\ bar$

(b) $p = 5\ bar$, $T = (150, 202.9)\,\mathrm{K}$

Figure 3: Continuity of compositional space parameterization for four-component three-phase systems

Plane Parameterization in Three-Phase Compositional Space

Given an overall composition in the multi-phase compositional space, we parameterize the space covered by the extension of the tie-simplex of the highest dimension. For a three-phase system, for example, we first compute the tie-triangle through a given overall composition and then parameterize the two-phase regions of the three sides of the tie-triangle using tie-lines. For this purpose, we start from the center of a tie-triangle and take offsets perpendicular to and outside the, tie-triangle to obtain a two-phase composition. We compute and store the parameterizing tie-line for that composition. Starting from the center of the tie-line, we proceed similarly in the two-phase region until the boundary of the space, or a critical tie-line, is encountered. Therefore, for a given overall composition, we store the tie-triangle and three sets of tie-lines which parameterize the sides of the tie-triangle. We perform this plane parameterization procedure at discrete temperature and pressure levels. Figure (4) shows an example of the parameterization of a three-phase space for a given composition.

State Identification in the Parameterized Three-Phase Space

Given a composition, we select the closest plane of a tie-triangle by searching in the table of parameterized planes. Interpolation in pressure and temperature is required for this purpose. If the distance between the closest plane and the given composition is farther than a pre-defined threshold, we parameterize a new plane for the composition. Otherwise, the state of the composition is determined by projection onto the plane of the tie-triangle (See Figure 5-a). We note that this is equivalent to performing a three-phase negative flash. If the projection of the composition is outside the tie-triangle region (single- or two-phase), we check the tie-line (tie-triangle sides) tables. This is achieved by finding the closest tie-line on a tie-triangle side (See Figure 5-b). The actual phase-state of the composition (single- or two-phase) is determined from projection of the composition to the closest tie-line. Projection to the closest tie-line is justified by the continuity of space parameterization along a continuous path in the compositional space (See Figure 4). Although the tie-lines on a tie-triangle side are not generally on the same plane as the tie-triangle, this procedure has proved to be quite successful for the wide range of problems studied so far.

Similar to the two-phase CSAT [3] approach, additional three- or two-phase negative flashes are required to accurately identify the state of compositions that

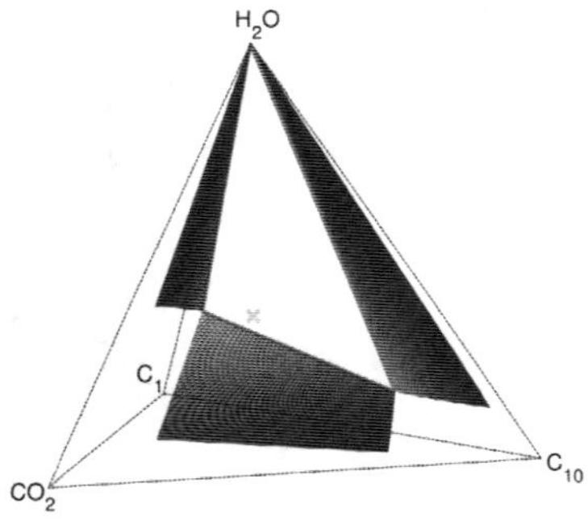

$\{500\,\mathrm{K}, 100\,bar\}$

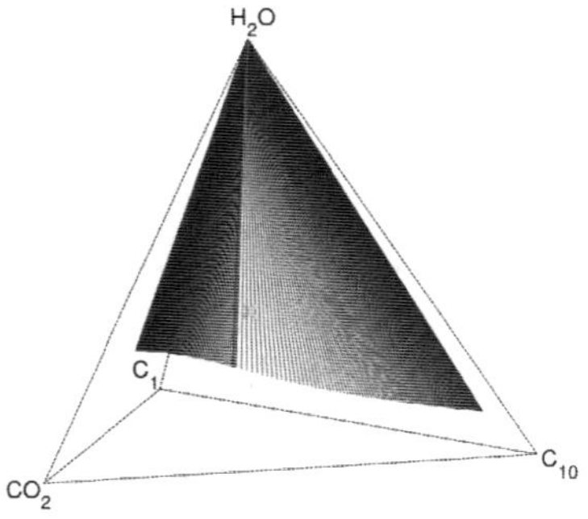

$\{500\,\mathrm{K}, 195\,bar\}$

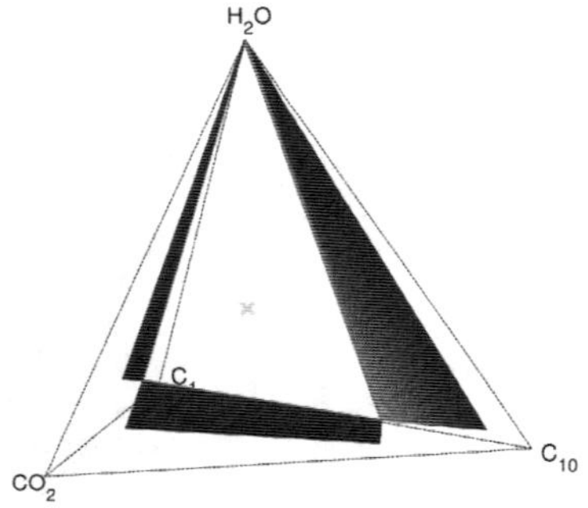

$\{450\,\mathrm{K}, 100\,bar\}$

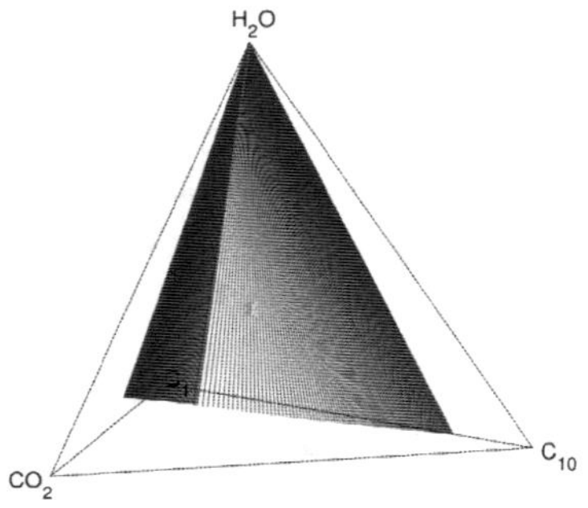

$\{450\,\mathrm{K}, 247.5\,bar\}$

Figure 4: Parameterization and degenerations in a three-phase compositional space

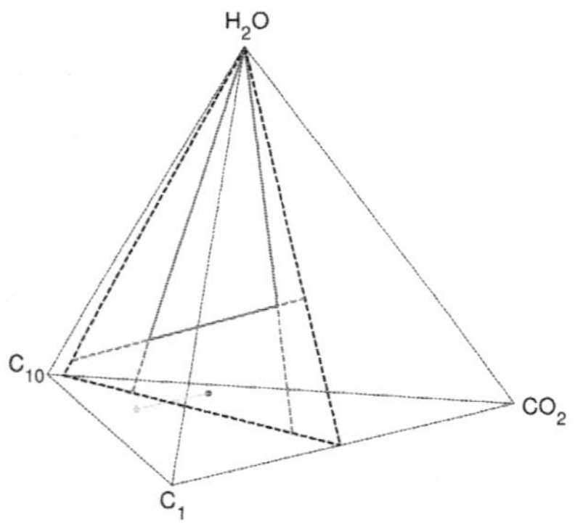

(a)

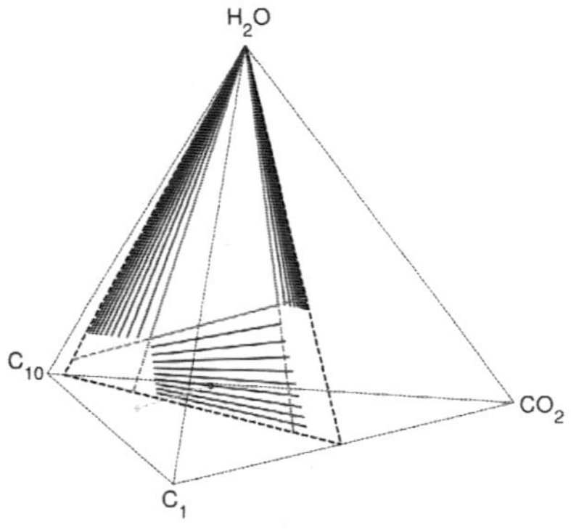

(b)

Figure 5: Projection-based phase identification, (a) first stage: project to the closest plane of tie-triangle; (b) second stage: project to the closest tie-line.

are close to phase boundaries (See Figure 6).

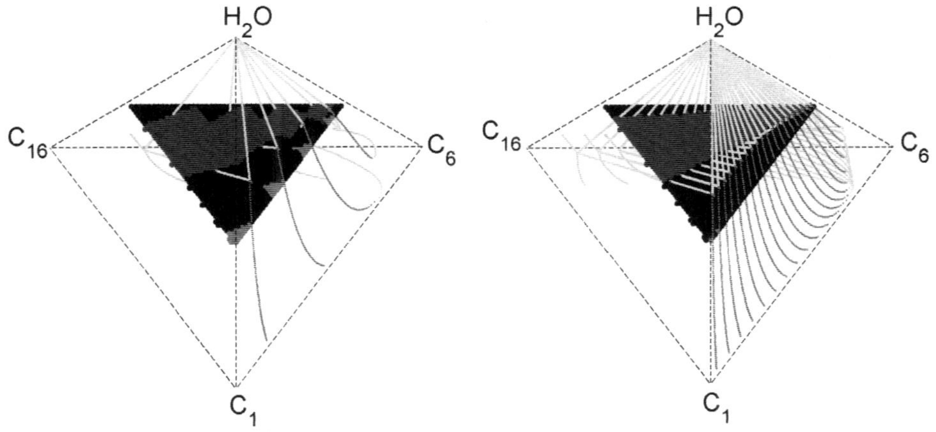

(a) (b)

Figure 6: Phase identification in compositional space with (a) low-density and (b) high-density parameterization. Test plane includes compositions in the single-phase gas (•), two-phase (•), and three-phase (•) states.

Numerical Examples

We illustrate the three-phase compositional space parameterization using various compositional displacements. As part of this finite-volume simulation framework, we developed a multi-stage negative-flash procedure. In addition to accurate determination of the number of the phases that are present, this approach also determines the phase compositions. We use CSAT for preconditioning of the standard negative-flash procedure. Considering the fact that the solution path in compositional space involves a few tie-simplexes, the resulting framework has proved to be quite efficient for multi-phase compositional flow simulation.

We consider two solution paths from a fully-compositional multi-phase flow simulator. Figures (7) and (8) show the displacement path in compositional space, as well as the tie-simplexes calculated in the course of the simulation. We note that although the path involves a three-phase region and different two-phase systems, only small numbers of tie-triangles and tie-lines are required to parameterize the entire solution path. For details on the improvement in computational performance of phase-equilibrium computations for large problems, see [2, 3] for isothermal and [5] for thermal systems.

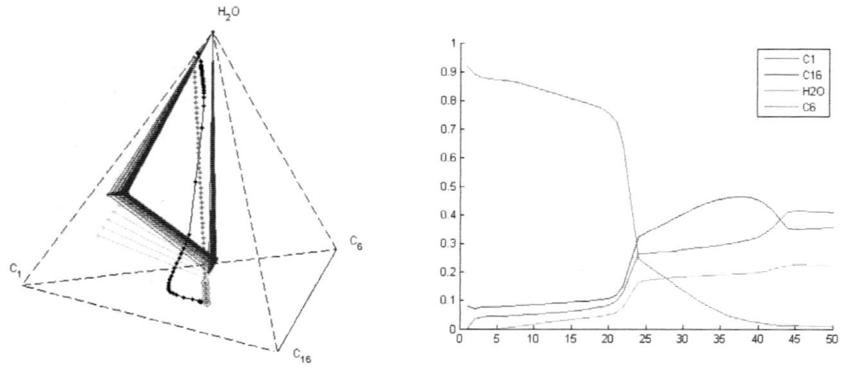

(a) (b)

Figure 7: (a) Solution path in the compositional space, and (b) overall composition profiles along the 1D domain. Injection of $\{C_1(0.1), H_2O(0.9)\}$ into $\{C_1(0.33), C_{16}(0.44), C_6(0.23)\}$ at $T = 523$ K, and $p = 140\,bar$.

References

[1] Cao H., Development of Techniques for General Purpose Simulators, Ph.D. Thesis, Stanford University, 2002

[2] Voskov D., Tchelepi H.A, Compositional Space Parameterization: Theoretical Background and Application for Immiscible Displacements, SPE Journal 14 (3), 2009, (paper SPE 106029 presented at 2007 SPE Reservoir Simulation Symposium, Woodlands, TX, February 26-28).

[3] Voskov D., Tchelepi H.A, Compositional Space Parameterization: Multi-Contact Miscible Displacements and Extension to Multiple Phases, SPE Journal 14 (3), 2009, (paper SPE 113492 presented at 2008 SPE/DOE Symposium on Improved Oil Recovery, Tulsa, Oklahoma, April 20-23).

[4] Orr, F.M., Jr., Theory of Gas Injection Processes, Tie-Line Publications, Copenhagen, Denmark (2007).

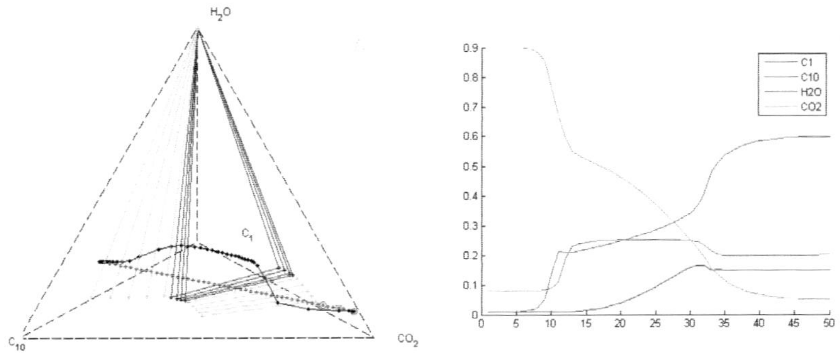

(a) (b)

Figure 8: (a) Solution path in the compositional space, and (b) overall composition profiles along the 1D domain. Injection of $\{H_2O(0.2), CO_2(0.8)\}$ into $\{C_1(0.2), C_{10}(0.6), H_2O(0.2)\}$ at $T = 473$ K, and $p = 140\, bar$.

[5] Iranshahr A., Voskov D., Tchelepi H.A, Phase Equilibrium Computations Are No Longer the Bottleneck in Thermal Compositional EoS Based Simulation, paper SPE 119166 presented at 2009 SPE Reservoir Simulation Symposium, Woodlands, TX, February 2-4

[6] Voskov D., Tchelepi H.A, Tie-Simplex Based Mathematical Framework for Thermodynamical Equilibrium Computation of Mixtures with an Arbitrary Number of Phases, Fluid Phase Equilibria, vol. 283 (1-2) pp. 1-11, 2009.

A Method to Determine the Solubility Curve of Wax in Crude Oil by Centrifugation and High-Temperature Gas Chromatography

Shanpeng Han, China University of Petroleum (Beijing), Beijing, China

H. Scott Fogler, The University of Michigan, Ann Arbor, MI

A method is proposed to determine the percent of precipitated wax in the crude oil at different temperatures using centrifugation and HTGC analysis on the crude oil and the centrifuged cake. The cake consists of precipitated wax and entrapped liquid oil. Two relationships are used to get the solid fraction in the cake and cooled crude oil. One is the ratio of solid fraction in the cake to the one in the cooled crude oil can be described by the ratio of mass of crude oil used for centrifugation to the mass of the cake. Another relationship is related to the fraction of light components in cake/crude oil. The light components of carbon number lower than a unique number will not present in solid phase according solid-liquid equilibrium. For the light component, its fraction in cake/crude oil, which could be obtained by HTGC analysis, can be described by its fraction in liquid part and the solid fraction in cake/cooled crude oil. Coupling these two relationships, the solid fractions in the cake and cooled crude oil can be obtained. It reveals that up to 94% liquid entrapped in the centrifuged cake. The newly developed solubility curve of wax in crude oil is used to simulate wax deposition experiments and the results are discussed.

See more of this Session: Oil Characterization and Thermodynamics as Related to Flow Assurance
See more of this Group/Topical: Energy and Transport Processes

Comprehensive Study On Gas / Crude Oil / Water Separation - Experimental and Numerical Analysis

Christine Noik, Applied chemistry and physical chemistry, IFP, Rueil-Malmaison 92852, France

Philippe Pagnier, Fluid Mechanics, IFP, Solaize 69360, France

Jean Trapy, Fluid mechanics, IFP, Rueil-Malmaison 92852, France

Malika Ouriemi, Fluid Mechanics, IFP, Solaize 69360, France

Alain Ricordeau, TOTAL, Pau, France

Benjamin Brocart, TOTAL, Pau, France

Jean-Philippe LeBrun, TOTAL, Pau, France

The increasing interest in the development of deep offshore separation units incited IFP and TOTAL to develop a large program of research to have a better understanding of three-phase (gas, oil, water) separation phenomena. In that program, a new test facility, named GOwSP Loop (Gas Oil water Separation Plat-Form) has been built in 2006 at Solaize (France), on the IFP-Lyon site. The topic consists in improving the design of separator faced to operating conditions, behaviour of fluids in deep offshore conditions and heavy crude oil.

The GOwSP plat-form is an industrial test facility designed for the separation research program, and also for testing multiphase equipments such as separators, pumps, monitoring or flow-meters for instance. The closed loop is under pressure and temperature control. Mixing devices (blade static mixer and controlled valves) may be used to manage emulsion flow shear rate. Used fluids are crude oil, water (pure or salted water) and natural gas.

Tests on GOwSP loop have been performed since the year 2007 on a basic three phase gas/oil/water horizontal separator with different crude oils. The separator has been designed and implemented with specific instrumentations : a perforated plate located at the inlet forms a tranquilliser area upstream the decantation zone ; two interface profiler gauges give the spatial distribution of water fraction and interfaces (water/emulsion, emulsion/oil, oil/gas) along a vertical axis at the entrance and near the weir of the separator. Finally, the global efficiency as well as the local description of flow within the separator, coupling with sampling, are analysed according to the change of operating conditions such as flow rates, water fraction, pressure or also fluids viscosity via temperature.

The aim of the research study is to couple physical models of coalescence and sedimentation with the hydrodynamic fluids behaviour. Those models are now implemented in 2D or 3D computational simulations performed by using a commercial CFD code based on three phases flow (gas/oil/water).

In the document, experimental results performed with one crude oil are described and compared to non-stationary CFD simulations. The predicted heights of the emulsion inside the tank due to several operating conditions are in good accordance with the measurements. Moreover, simulations with the CFD code adapted to separation study are able to predict the real efficiency of the separator with a good accuracy.

See more of this Session: Oil Characterization and Thermodynamics as Related to Flow Assurance
See more of this Group/Topical: Energy and Transport Processes

AUTHOR INDEX

AUTHOR INDEX